MODERN GEOGRAPHY of the Republic of CAMEROON

SECOND EDITION

Commissaire Ewu Ngeme LL.B. Hons.
First Asst. Prov. chief
Buea - S.W. Province
1987.

Aaron S. Neba

TEMPLE UNIVERSITY, PHILADELPHIA

Presented to Supt. Bill WEBER
of the DEVON and CORNWALL Constabulary.

NEBA PUBLISHERS, CAMDEN, N.J. 08101

Library of Congress Catalog Card Number: 87-70017

ISBN 0-941815-00-5 (Paperback)
ISBN 0-941815-01-3 (Hardcover)

For Information write:

In The United States of America
Neba Publishers
P.O. Box 3343
Camden, New Jersey 08101

In The Republic of Cameroon
Neba Publishers
P.O. Box 2094
Bafut, via Bamenda
N.W. Province, Cameroon

Dedicated to, and in memory of
my late father, Peter Neba Ngongalah.

PREFACE TO THE SECOND EDITION

Books grow in unique ways as a result of the experiences of their authors, and this is no exception. In late 1982, when I wrote this book, I did not know that it was going to receive the wide readership it later achieved.

At the time of publication, I was also not aware of the fact that many people would be eager to read about Cameroon. When they did, they became alarmed at the uniqueness of this country, and in particular, the many and varied aspects (of the society) which differentiate it from the rest of the other African countries. Despite that, some readers, including myself, were aware of the shortcomings of the book. Many topics were treated superficially, others were completely omitted, and the organization of the material was sometimes haphazard.

As I plunged into more serious studies, reading extensively on Africa, and Cameroon in particular, I acquired more knowledge of how the world operates. I realized that many issues on Cameroon can better be understood by going beyond the circumscribed field of geography to borrow from economics, history, sociology and politics.

In this edition, therefore, I have added three new chapters at the end of the book, which deal with politics (or the role of government) and development. Chapter 25 examines the principles for government policy, especially in economic activity; chapter 26 deals with *Government and Administrative units of Cameroon;* chapter 27 gives an outline of Cameroon's foreign relations, and the last chapter elaborates on what we mean by *Third World* and *Development*. The concept of development is given a broader and a more humanistic definition, which takes the reader beyond the narrow view of industrial (or economic) growth.

Elsewhere, a number of sections in the old edition are either new or completely rewritten, and a few topics of marginal importance omitted. In chapter 15, *Tourism in Cameroon,* I have taken a more critical approach by challenging the traditionally held view that the tourist industry is a profitable undertaking to less developed countries and stating the conditions under which the industry can be more beneficial to the society at large while at the same time remaining profitable.

Colour is now found throughout this edition; new photographs, coloured as well as black and white, are also included. Most significant, this edition has an elaborate glossary, a detailed bibliography and an expanded index. For the classroom teacher, there are a number of revision questions at the end of the text.

The second edition of this book, as would be expected, involves past and present contributions of numerous people. With the publication of this edition, I wish to acknowledge all those associated with the first edition. This book continues to reflect their ideas and expertise.

For this edition I would like to thank the following for permission to publish photographs: Cellucam for page 161; Sonel for page 109; Infocam for pages 7,8,11,12,13,14,18,21,33 bottom, 42 top, 56,86 bottom, 93 bottom, 94,110,111,113 bottom, 120,121 bottom, 123,125,136,137 bottom, 141 bottom, 145,147,155,160,167,171,178; United Nations Organization for pages 49 bottom, 86 top, 88 top, 91,103,122,177; Photo Prunet for pages 30,137 top; General Delegation for Tourism for pages 48,52 bottom, 55 and 88 bottom. My thanks also go to Carlisle W. Baskin of Randolph-Macon College, Ashland, Virginia, U.S.A., for permission to reproduce the illustration of "Central Places" on page 83.

Above all, it is a pleasure to acknowledge the help and encouragement I received from friends and family members. All other people who helped me to get the information I needed to write this text, are entitled to my heartfelt thanks. My undertaking would not have been successful without such assistance.

Aaron S. NEBA.

Philadelphia, U.S.A.
March 1987.

CONTENTS

Preface v
List of maps and diagrams ix
List of photographs ix
Important Facts on Cameroon x
Cameroon in the African Continent xi

PART ONE: PRESENTATION OF THE COUNTRY

Situation, Shape and Size 1

1. Brief History of Cameroon 2
The Birthday of Cameroon 2
Early Missionaries in Cameroon 2
More Europeans at the Coast 2
Anglo-German Agreements 3
Franco-German Agreements 3
Cameroon as a Mandate Territory 1922-1946 5
Cameroon as a Trust Territory of the U.N.O. 5
The Route to Independence 5
A) The French Cameroon 1946-1960 5
The 1955 Riots and the Ban on UPC 6
From ATCAM to ALCAM 1956 7
B) The British Cameroon up till 1961 9
The Lyttelton Constitution 10
The 1954 Election and its Consequences 10
The 1957 Election 10
The Plebiscites and Reunification 11
The Formation of One National Political Party 11
The Formation of the Unitary Government 12
From Ahmadou Ahidjo to Paul Biya 13
From CNU to CPDM 13

PART TWO: PHYSICAL STUDIES

2. Relief and Structure 14
1. The Coastal Lowlands 14
2. The Southern Plateau 16
3. The Adamawa Plateau 16
4. The Western Highlands 17
5. The Northern Lowlands 17
Some Major Physical Features of Cameroon 17
Mount Cameroon 17
Mount Kupe 19
The Mandara Mounts 20

3. Geology of Cameroon 22

4. Climate 24
The Main Climatic Elements: Latitudinal Extent, Temperatures, Precipitation, Air Masses 24
A. The Situation in the Dry Season 26
B. The Situation in the Rainy Season 26
Climatic Regions of Cameroon 26

5. Vegetation 30
1. The Dense Equatorial Forest 30
2. The Tropical Grassland 32

6. Soils 34
The Soil Profile 34
Factors Involved in Soil Formation 34
A. Zonal Soils 35
B. Azonal Soils 36

7. Hydrography 38
A. The Atlantic Basin 38
B. The Congo Basin 40
C. The Niger Basin 40
D. The Chad Basin 40
Lakes 41
Lake Chad 42

PART THREE: HUMAN STUDIES

8. The Peoples and Peopling of Cameroon 44
The Major Ethnic Groups of the South 45
The Pygmies 45
The Bantu 45
Major Bantu Tribal Groupings 47
The Douala 47
The Bakoko and Bassa 47
The Pahouins (Fang-Bulu-Beti) 50
The Semi-Bantu (Grassfield Bantu) 50
Major Semi-Bantu Groupings 51
The Bamileke 51
The Bamum 51
The Tikars: The Bafut, The Kom 53
The Nsaw, The Bali 54
The Widikum 55
Major Ethnic Groups of the North 55
The Sudanese 55
The Fulbe (Fulani) 56
The Shuwa Arabs 58

9. Population and Demography 59
Recent Population Trends 60
Birth Rate, Fertility Rate, Infertility, Death Rate (Mortality Rate) 60
Infant Mortality Rate, Growth Rate, Life Expectancy 61
Population Structure: Age Distribution Sex Ratio (Sex Distribution) 62
Nuptiality 63
Forms of Marriage; Marital Status 63
Migration 64
Population Density and Distribution 66
A. Areas of High Population Concentration 66
B. Sparsely Populated Zones 68
The Population Debate 69
Influence of Economic and Social Development on Population Growth 71
The Theory of Demographic Transition 72

10 Urbanization in Cameroon 73
What is Urbanization? 73
What then is an Urban Centre? 73
Rapidity of Urbanization in Cameroon 73

Evolution of Cameroon's Urban System 73
Distribution of Urban Centres 76
Urban Hierarchy, Urban Primacy 77
Urban Problems in Cameroon 77
Urban Structure and Development 79
The Concentric-zone Model 79
The Sector Model 81
The Multiple-Nuclei Model 81
The Central Place Theory 82

PART FOUR: ECONOMIC STUDIES

11. Agriculture 84
Traditional Agriculture 84
The Cultivation of Local Food Crops 84
a) Tubers b) Cereals c) Others 87
Cash Crop Farming 89
Plantation Agriculture in Cameroon 97
Plantation types in Cameroon:
a) The Large Capitalist Plantations 97
b)Peasant Plantations 98
Problems of Agriculture in Cameroon 98
The Role of Agriculture in Economic Development 99
Ways of Improving Agricultural Productivity . . . 100

12. Animal Raising in Cameroon 101
Factors Influencing Animal Raising in Cameroon:
a) Natural Conditions b) Human Factors 101
The Types of Animals 104
Government's Participation in Animal Raising 104
Problems of Livestock Raising in Cameroon 105

13. Minerals and Power 106
A. Minerals 106
B. Power: a) Hydro-electric Power (H.E.P.) 108
b) Thermal Electricity 109

14. Industries of Cameroon 110
A) Local Crafts Industries 110
B) Manufacturing Industries 111
Industrial Regions of Cameroon 114
a) The Douala Industrial Region 114
b) The Tiko-Limbe Industrial Region 115
c) The Edea Industrial Region 115
d) The Yaounde Industrial Region 115
e) The Garoua Industrial Region 116
Problems of Industrialization in Cameroon 116

15. Tourism in Cameroon 119
Factors Favouring the Development of Tourism in Cameroon 119
Touristic Sites 120
National Wildlife and Forest Reserves 122
a) The Waza Wildlife Reserve 122
b) The Benue Wildlife Reserve 123
c) The Bouba-Ndjida Wildlife Reserve 123
Regulations Governing Visits to National Wildlife Reserves 123
Forest Reserves 124
Critical Evaluation of the Role of Tourism 124

16. Fishing in Cameroon 129
A) Subsistence Fishing 129
B) Industrial Fishing 129
C) Inland Fishing 131

17. Forests and Timber Exploitation 133
Importance of the Forest 133
Forestry Legislation 133
Exploitation of the Forest 135
Transportation of Wood 136
Species of Trees 137
Measures Adopted for Forest Conservation 137
Problems of Timber Exploitation in Cameroon 138

18. Transport and Communications 139
A) Road Transport 139
B) Railway Transport 141
C) Air Transport 142
D) River Transport 145
Ports 145
River Ports 146
Telecommunications 146
Conclusion 147

19. Commerce 149
Internal Trade 149
Inter-Regional Trade, External Trade 149
Problems of Cameroon's External Trade 150
The UDEAC as a Trade Bloc 151

PART FIVE: REGIONAL STUDIES

20. The Southern Plateau Region 153
Agriculture, Animal Rearing 153
CAMSUCO, MIDEVIV 154
CTC, ZAPI 155
Towns: Yaounde, Bertoua 155

21. The Coastal Lowland Region 157
Physical Features, Climate, Vegetation 157
Population, The Economy: Agriculture 157
Livestock and Fishing, Forest Exploitation 159
Industries 159
Towns: Douala, Nkongsamba, Limbe 162
Kumba, Buea 163
Edea 164

22. The Western Highland Region 165
Physical Features, Climate 165
Population, The Economy: Agriculture 165
Animal Rearing, Industries, Infrastructure 169
Towns: Bafoussam, Bamenda 169

23. The Adamawa Region 172
Physical Features, Population, The Economy . . . 172
Transport Network, Industries 173
Towns: Ngaoundere 174

24. The Northern Lowland Region 175
Relief, Climate, Human and Social Aspects 175
The Economic Aspects 175

Agriculture and Pastoral Farming 177
Industries 177
Some Important Agro-Industrial Projects:
SEMRY, SODECOTON, CEREAL BOARD . . . 179
Towns: Garoua, Maroua 179

PART SIX: POLITICS AND GOVERNMENT

25. Principles for Government Policies in Cameroon 180
1) Planned Liberalism. 180
2) Balanced Development 181
3) Self-Reliant Development 181
Self-Reliance and International Trade 182

26. Government and Administrative Units of Cameroon 183
Administrative Divisions of Cameroon. 183
Local Government 183

27. Cameroon's International Relations 186
1. Relations with other African States 186
2. Relations with other World Regional Organizations 186
3. Relations with France 186
4. Relations with the United States of America 186
5. Relations with the Eastern Bloc Nations 187

28. The Notions of "Third World" and "Development" 188
What essentially is the Third World? 188
What really is development? 189
Conclusion 191

Questions 193
Glossary 195
Bibliography 198
Index 201

LIST OF MAPS AND DIAGRAMS

1. Cameroon in the African continent xi
2. Territorial Evolution of Cameroon 4
3. Mandate Territories (1922-1946) and Trust Territories (1946-Independence) 5
4. Relief of Cameroon 15
5. Sketch section of Mount Cameroon 19
6. Mt. Kupe 20
7. Geology 23
8. Precipitation and Temperatures 25
9. Prevailing winds in Cameroon 27
10. Climatic Regions 28
11. Vegetation 31
12. A profile of the Equatorial forest soils 35
13. A profile of the Tropical lateritic soils 36
14. Drainage 39
15. Capture of River Logone by Mayo-Kebi 41
16. Ethnic groups and some tribal groupings 46
17. Population Pyramid for Cameroon 1976 62
18. Internal migrations 62
19. Population densities in 1976 67
20. The Theory of Demographic Transition 72
21. Major Towns (1976) 74
22. Urban structure 80
23. Marketing Regions in a System of Central Places 83
24. Cash crops 90
25. Animal Raising 102
26. Minerals and Power 107
27. Industries 112
28. The Douala Industrial Region 115
29. The Tiko-Limbe Industrial Region 115
30. The Edea Industrial Region 115
31. The Yaounde Industrial Region 116
32. The Garoua Industrial Region 116
33. Wildlife and Forest Reserves 124
34. Fishing in Cameroon 130
35. Industrial Fishing 132
36. Forest Exploitation 134
37. Transport Network 140
38. Air Transport 143
39. The Southern Plateau Region 154
40. The coastal lowland REgion 158
41. The Western Highland Region 168
42. The Adamawa Region 173
43. The Northern Lowland Region 176
44. Administrative Divisions of Cameroon 184

LIST OF PHOTOGRAPHS

1. Andre-Marie Mbida 7
2. Proclamation of Independence for Cameroon .. 8
3. Dr. E.M.L. Endeley 11
4. John Ngu Foncha 11
5. S.T. Muna 12
6. President Paul Biya taking the oath of office 13
7. A Lagoon on the Douala Basin 14
8. Volcanic rocks at the rocky coast of Limbe ... 16
9. Mount Cameroon 18
10. Kapsiki Volcanic necks 21
11. Mangrove Forest on the coastal lowlands 30
12. Rain Forest 32
13. Guinea Savanna 33
14. Sudan Savanna 33
15. Sahel Savanna 33
16. Lake Awing: a crater lake 42
17. The Mefou Lake: 42
18. A Pygmy dwelling unit near Kribi 48
19. Rural dwelling units near Limbe 48
20. Rural dwelling unit near Yaounde 49
21. Residents of Sangmelima help a neighbour build his new house 49
22. Semi-Bantu dwelling unit in Bamenda 52
23. Semi-Bantu men in their traditional attire 52
24. The Palace of a Bamileke chief 52
25. A Musgum mud hut 55
26. Clustered settlements of the Mafa 56
27. A typical Bororo hut 57
28. Women in the grassland area prepare a field for planting 86
29. Terrace farming at Mora 86
30. Women husking maize 88
31. Millet farm in North Cameroon 88
32. Cocoa beans being sun-dried 91
33. Flowering coffee plant 92
34. Oil palm plantation in the Fako Division 93
35. Cotton farm at Guider in North Cameroon ... 93
36. Rubber Plantation at Tiko 94
37. Ndu Tea Estate 94
38. Herdsmen from Sangmelima bringing cattle from the north 103
39. Edea H.E.P. Station 109
40. Craftsmen at work in Bamenda 110
41. Brewery: Inside a Brasseries du Cameroun factory 111
42. Matches Factory in Douala 113
43. Fertilizer industry at Bonaberi 113
44. CIMENCAM: Cement factory at Bonaberi .. 115
45. The Wouri bridge 120
46. Partial view of a traditional ruler's Palace ... 121
47. Musicians of the Lamido at Banyo 121
48. A scene at a folk dance in Sangmelima 122
49. Giraffes in the Waza Park 123
50. Jubilant villagers with a dead elephant 125
51. Fishing at the coast of Limbe 129
52. Traditional method of Forest Exploitation ... 136
53. Felling of timber using a chain saw 136
54. Peeling of the logs 136
55. Transportation of timber on vehicles 137
56. Transportation of timber by water 137
57. Section of Bamenda-Bafoussam Road 141
58. Ferryboat on River Mbam 141
59. Kumba railway Station 142
60. The Abi falls in Mbengwi 145
61. Douala Port 146
62. Aerial view of the Zamengoe Station 147
63. Yaounde Central Market 150
64. Poultry farm at Mvog-Betsi 155
65. Partial view of Yaounde city 156
66. Petroleum Refinery at Cape Limbo 160
67. Reception and Storage tanks 160
68. The Paper pulp factory (CELLUCAM) 161
69. A partial view of Douala city centre 163
70. A vast and modern rice farm at Ndop 167
71. Bafoussam town 170
72. Bamenda town 171
73. Collection of hides and skins 174
74. Women walking to a community well 177
75. Cotton spinning in Garoua 178

IMPORTANT FACTS ON CAMEROON

Official name: Republic of Cameroon
Area: 475,000 km^2 **Capital:** Yaounde
Population: 10,446,400 (1986 estimate)
Official Languages: English and French
Largest Cities: Douala, Yaounde, Nkongsamba, Maroua, Bamenda, Bafoussam, Garoua.
Ethnic Composition: The main ethnic groups are the Bantu, Semi-Bantu and Pygmies in the south; the Fulbe, Sudanese and the Shuwa Arabs in the north. Tribal groupings are over 200 in number.
Religion: Christians — 33% (mainly in the south); Moslems — 15% (mainly in the north); and the rest are engaged in traditional religions. However, almost all Cameroonians, whether Christian or Moslem, adhere to some form of indigenous beliefs and practices.
Flag: Three vertical stripes from left to right — green, red, yellow — with a yellow star in the centre of the red stripe.
National Anthem: O Cameroon, thou cradle of our Fathers ...
National Motto: Peace, Work, Fatherland
Location: West-Central Africa, bordered by Lake Chad in the north, by Chad and the Central African Republic in the east, by the Congo, Gabon and Equatorial Guinea in the south, by the Republic of Nigeria in the west.
Physical Regions: Five main physical regions: 1) The coastal lowlands; 2) The southern plateau, covered by equatorial rain forest; 3) The western highlands; 4) The Adamawa Plateau; 5) The northern lowlands, all covered by tropical grassland.
Minerals and Power: Large deposits of bauxite, limestone, crude oil and natural gas. Main sources of power are hydro-electricity from Edea, Song-Loulou and Lagdo, and thermal electricity from many diesel engines.
Currency: African Financial Community Franc. (Communaute Financiere Africaine Franc; 50 CFA francs = 1 French franc)
Foreign Trade: Exports mainly cocoa, coffee, timber, cotton, rubber, tea, bananas and petroleum. Imports mainly vehicles, industrial equipment, electrical and pharmaceutical apparatus.
Principal Trade Partners: European Community countries (mainly France), the United States, Japan and neighbouring African countries.

The country derived its name from "Rio dos Camaroes" in the 15th century. Impressed by the swarming prawns in the mouth of the Wouri River, Portuguese explorers termed it *Rio dos Camaroes* ("River of Prawns") and this name was later extended to the whole territory.

Cameroon has long been one of the major racial crossroads of Africa as a result of Bantu migrations from the east and Fulani from the far west. Stretching for over 1200 km from the Equatorial region to the fringe of the desert, Cameroon is a cross-section of Africa from desert north through a central grassland elevation to a southern plateau covered by rain forest.

A former German protectorate since 1884, Cameroon was divided between Britain and France after World War 1, with the French getting about four-fifths of the area. In 1961, the country was reunited as a Federal Republic after a U.N. plebiscite, but at this time part of the British Cameroons was lost to Nigeria. The two parts, West Cameroon (formerly British) and a larger East Cameroon (formerly French), were governed as constituent states, each with its own legislature. After a referendum on 20th May 1972, a new constitution went into effect establishing a Unitary State.

Cameroon is the only African country in which English and French are official languages, and the *University of Yaounde,* established in 1962, is the first in Africa to teach courses in both English and French.

Cameroon has been one of Africa's *quiet* countries in which the single-party government has maintained a measure of political and economic stability. Prior to independence (1960), the former French part of the country was torn by an anti-colonialist revolt. Prolonged guerrilla activities, mainly in the forested central part of the country, were suppressed in 1970. Despite the diversity of its peoples, cultures and religions, Cameroon has escaped the civil wars and coups that have plagued neighbouring countries.

The country's great diversity of tropical products, and more importantly, the favourable economic and investment climate, have been the reasons for the steady economic growth. Though France is the main trading and investment partner, American, British, West German and Japanese interests are growing considerably. Petroleum production, which began in 1978, seems to hold most of the country's prospects for economic development. However, Cameroonians are aware of the dangers of depending solely on oil, and therefore share the common belief that agriculture, not oil, constitutes the backbone of their nation's economy.

Education and health facilities are also increasing tremendously and Cameroonians are proud of the fact that over 70% of the children attend school, a high rate for developing Africa.

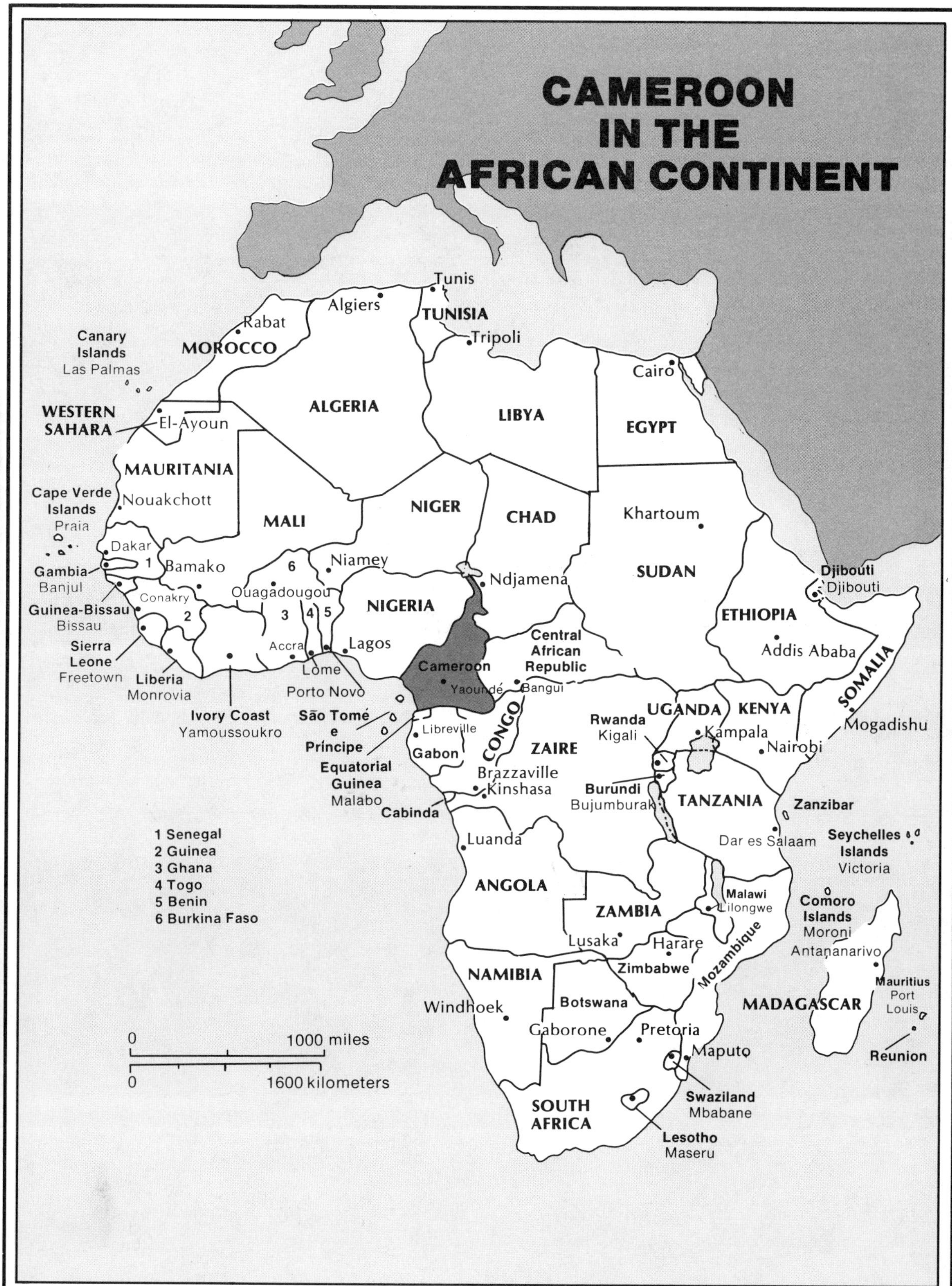
CAMEROON
IN THE
AFRICAN CONTINENT
Tunis
TUNISIA
Algiers
Rabat
MOROCCO
Canary
Islands
Las Palmas
Tripoli
WESTERN
SAHARA
El-Ayoun
ALGERIA
LIBYA
Cairo
EGYPT
MAURITANIA
Cape Verde
Islands
Praia
Nouakchott
MALI
NIGER
CHAD
Khartoum
SUDAN
Dakar
1
Gambia
Banjul
Bamako
6
Niamey
Ndjamena
Djibouti
Djibouti
Guinea-Bissau
Bissau
Conakry
2
Ouagadougou
3
4
5
NIGERIA
ETHIOPIA
Sierra
Leone
Freetown
Accra
Lagos
Central
African
Republic
Addis Ababa
Lome
Cameroon
Yaoundé
Bangui
SOMALIA
Liberia
Monrovia
Porto Novo
Ivory Coast
Yamoussoukro
São Tomé
e
Príncipe
Equatorial
Guinea
Malabo
Libreville
Gabon
CONGO
ZAIRE
UGANDA
KENYA
Rwanda
Kigali
Kampala
Nairobi
Mogadishu
Brazzaville
Kinshasa
Burundi
Bujumbura
TANZANIA
Zanzibar
Cabinda
1 Senegal
2 Guinea
3 Ghana
4 Togo
5 Benin
6 Burkina Faso
Luanda
Dar es Salaam
Seychelles
Islands
Victoria
ANGOLA
Malawi
Lilongwe
Comoro
Islands
Moroni
ZAMBIA
Lusaka
Harare
Antananarivo
Mozambique
Zimbabwe
NAMIBIA
Mauritius
Port
Louis
Botswana
MADAGASCAR
Windhoek
Gaborone
Pretoria
0
1000 miles
0
1600 kilometers
Maputo
Reunion
Swaziland
Mbabane
SOUTH
AFRICA
Lesotho
Maseru

PART ONE: PRESENTATION OF THE COUNTRY

SITUATION, SHAPE AND SIZE

The Republic of Cameroon is situated at the extreme northeastern end of the Gulf of Guinea. It lies between longitudes 8° and 16° east of the Greenwich Meridian and between latitudes 2° and 13° north of the Equator.

To the south the country is bordered by Equatorial Guinea, Gabon, and Congo; to the west by Nigeria; to the east by the Central African Republic and Chad; and finally to the north by a narrow portion of Lake Chad.

The territory has the shape of a carelessly drawn triangle with a base of about 700 km and a height of 1,200 km. The whole territory covers an area of 475,000 km^2, with a population of 10,446,400 inhabitants (following the 1986 estimate). This gives an average density of 22 inhabitants per km^2.

The country's boundaries were determined at the end of the 19th century by the colonial powers during the Partition of Africa. These boundaries in most cases are artificial, except in certain sections where natural boundaries are found, such as river courses, mountain chains, and the ocean to the southwestern part of the country.

Cameroon presents a lot of diversities in its physical, human and economic aspects. Physically, it has distinct regions which include mangrove swamps, coastal lowlands, plateaux, highlands, plains and several volcanic massifs. This diversity can also be noticed in its climate, vegetation and soils.

On the human aspect, the country has a number of ethnic groups who are involved in different economic activities especially at the local level. A more unique social aspect of Cameroon is its bilingualism, which resulted from a peaceful merger in 1972 of the former French-speaking East Cameroon State and the former English-speaking West Cameroon State.

CHAPTER I
BRIEF HISTORY OF CAMEROON

THE BIRTHDAY OF CAMEROON

The land of Cameroon is as old as the rest of the world, but when we talk of Cameroon in terms of a geopolitical entity with its present territorial boundaries and government, one may say that Cameroon did not exist before the arrival of the Germans.

In the ancient times, Hanno, the Carthaginian navigator, saw Mount Cameroon and called it the "Chariot of the Gods". It was the Cameroon mountain presumably under serious volcanic eruption as seen from what he wrote — "We saw at night," he said, "a land full of fire. In the middle was a lofty fire larger than all the rest touching seemingly the stars."

In 1472 a Portuguese navigator Fernando Po arrived on an island situated some thirty-five kilometres off the coast of Limbe and gave his name to it. More Portuguese citizens arrived and settled there with him. As time went by and especially during the second half of the 16th century, Portuguese traders who had come up the Wouri estuary found a great variety of prawns in the river and named the river "Rio dos Camaroes", meaning "River of Prawns". The Spanish version of Camaroes is Camerones which later gave rise to the name Cameroons. It is from this Spanish version that other spellings were derived following the colonial powers involved. The Germans spell it Kamerun, the French spell it Cameroun, and the English spell it Cameroon. Today there are two official forms: "Cameroon" and "Cameroun", used in English and French, respectively.

EARLY CHRISTIAN MISSIONARIES IN CAMEROON

The earliest evangelists to arrive and establish in Cameroon were Protestant missionaries. The very first one was a black Jamaican Baptist missionary called Joseph Merrick, who arrived at the end of 1843 and founded the Cameroons Mission in Bell-Town, Douala. In 1844, he founded another mission in Bimbia, translated the Bible into Isubu (the Bimbia dialect), set up a printing press, and developed a machine for making bricks out of local materials. Merrick was one of the early non-Africans to climb Mount Cameroon. On his way to England for leave in 1849, he died during the voyage.

The second Protestant missionary to Cameroon was Alfred Saker. He arrived and established a Christian mission at the Cameroon coast in 1845. His first baptisms were administered in 1849. Saker stayed in the country for thirty-one years and during this time he accomplished many tasks. He spread the word of God and opened many schools. After studying the Douala language, he translated and published the Bible on the press built by Joseph Merrick. Saker founded two other missions at Deido and Bonaberi. On June 5, 1858 he travelled and founded the town of Victoria, now Limbe. Saker finally retired back to England in 1876 where he died four years later.

In 1890 Catholic missionaries began arriving in the country. Pallotine fathers, Vieter and Walter, arrived and established the Marienberg Mission near Edea, situated at the mouth of the Sanaga River. At the outbreak of the First World War, the activities of the Catholic Mission had increased tremendously in the country. In 1922, the first Catholic Bishop, Monseignor Francois Xavier Vogt, was named, and in 1935 the first indigenous priests were ordained.

MORE EUROPEANS AT THE COAST

It was not until the nineteenth century, when Portuguese influence reduced considerably in West Africa, that other European powers began to come in. The Portuguese were followed by the Dutch, British and French. Many of them had established trading posts on the West African coast as early as the sixteenth century. The trade in human beings to serve as slaves in the plantations of the New World had become very lucrative to all European countries. Apart from slaves, West Africa also offered ivory, animal skins, gold and palm products. But the principal article of export from 1517 on along the coast remained the slave, and all European nations took their full part in the trade. In exchange the natives received gun powder, weapons, tobacco, spirits, beads, silk and woven goods. By the year 1850, many more European traders were already at the coast of Cameroon especially at Douala. British and German rivalry in particular had risen very significantly in this area.

Local chiefs — Douala and Bimbia — at the coast quarrelled among themselves and as a result were asking for protection of their territories from

European powers. Some of them applied for British protection, but their request was turned down by Britain because she was no longer interested in colonial expansion in West Africa.

British traders at the coast, however, also needed British protection against approaching French traders who were going to be a big threat to their trade. Through the persuasion of the British traders probably, the local cheifs, Akwa and Bell, addressed many letters to Queen Victoria of England. Conscious of the activities of the French and Germans at the coast, Hewett, the German Consul to Calabar, was commissioned by the British government in 1884 to travel and sign treaties of annexation with the local chiefs.

Hewett delayed and only reached Cameroon on July 19, 1884. Before his arrival, the German Consul, Nachtigal, had already arrived and hoisted the German flag on the territory on July 14, 1884. The hoisting of the flag was the formal acceptance and backing of the German government of the agreement signed earlier on July 12th, 1884 between local chiefs and German businessmen.

Until the 12th of July it was impossible to delimit Cameroon. By signing an agreement between he chiefs of Douala and German businessmen, Cameroon was delimited as follows as can be seen from the agreement.

"We, the undersigned, kings and chiefs of the territory named Cameroon, situated along the River Cameroon, between rivers Bimbia in the north and Kwakwa in the south, and up to longitude* 4° 10′ north, have in a meeting held today in the German factory on king Akwa's beach, voluntarily concluded as follows: We surrender this day our rights of souvereignty, legislation and administration of this our territory to messrs Edouard Schmidt, agent of the firm C. Woermann, and Johannes Voss, agent of the firm Jantzen and Thormahlen, both of Hamburg and merchants for many years among these rivers.

We have transferred our rights of souvereignty, of legislation and of administration of our territory to the above-mentioned firms, with the following reservations:
1. The territory may not be ceded to a third party.
2. All the treaties of friendship and commerce which have been concluded with other foreign governments shall remain fully valid.
3. The lands cultivated by us, and the sites on which villages are situated, shall remain the property of the present possessors and their descendants.
4. The payments shall be made annually, as in the past, to the kings and chiefs.
5. During the initial period of the establishment of an administration here, our local customs and usages shall be respected.

Cameroon, 12th July 1884.
Signed: King Akwa

Signed: Ed. Woermann
Signed: Ed. Woermann
Witnesses:
O. Busch, Endene Akwa, Ed. Schmidt, Coffee Angwa, John Angwa, Manga Akwa, Scott Jost, Lorten Akwa, Ned Akwa, David Meatom, Joh. Voss, King Bell, Joe Garner Akwa, Big Jim Akwa, William Akwa, Jim Joss, Matt Joss, David Joss, Jacco Esqre, London Bell, Barrow Peter, Elame Joss, Lookingglass Bell."

*This should read, 'latitude'.

The above-mentioned text as was drawn could not involve the rest of the country. It should be understood that the territory referred to was that portion of land occupied by the Douala people. It was not enough to sign these treaties without actually setting up territorial boundaries of lands acquired by each European colonising power. A conference at Berlin, called at the request of Von Bismarck and held from November 15th, 1884, to February 25th, 1885, specified conditions under which possessions could be made in Africa by the nations scrambling for territories on the continent.

Some of the important resolutions at the conference were as follows:
1. Nations were to own coastal lands before proceeding into the interior.
2. Treaties were to be signed with other powers to fix the territorial boundaries of land already acquired.
3. From the coast, explorers had to move into the interior. They would stop further acquisition of lands as they met other explorers, and there, boundaries could be established.

It is important to mention here that at this time the French were already in the Congo Basin, and the English were in Nigeria, so that the Germans had to acknowledge these two powers of their presence in Cameroon and consequently fix the boundaries with them.

Cameroon boundaries were fixed betwen 1885 and 1908 with adjustments coming later. Within this period the Germans signed agreements with the British and French delimiting each nation's territorial possessions.

ANGLO-GERMAN AGREEMENTS

1. The arrangements between Germany and Britain, which led to the first agreement on Cameroon-Nigerian boundaries, lasted from the 29th of April to the 16th of June 1885. There was an exchange of correspondence between the two governments fixing the southern frontier at Rio-del-Rey and the northern frontier up to the Cross River rapids in Mamfe. This frontier gave to the Germans the Cameroon mountain.
2. The second convention lasted from the 27th of July to the 2nd of August 1886. It extended the northern frontier up to the town of Yola.
3. In another Anglo-German agreement of July 1, 1890, the southern boundary line was modified.
4. The next agreement was on November 15, 1893. It extended the frontier from Yola to Lake Chad.

FRANCO-GERMAN AGREEMENTS

1. The first agreement was signed on December 24, 1885 at Berlin, Germany, fixing the southern and eastern frontiers between French and German possessions.

TERRITORIAL EVOLUTION OF CAMEROON

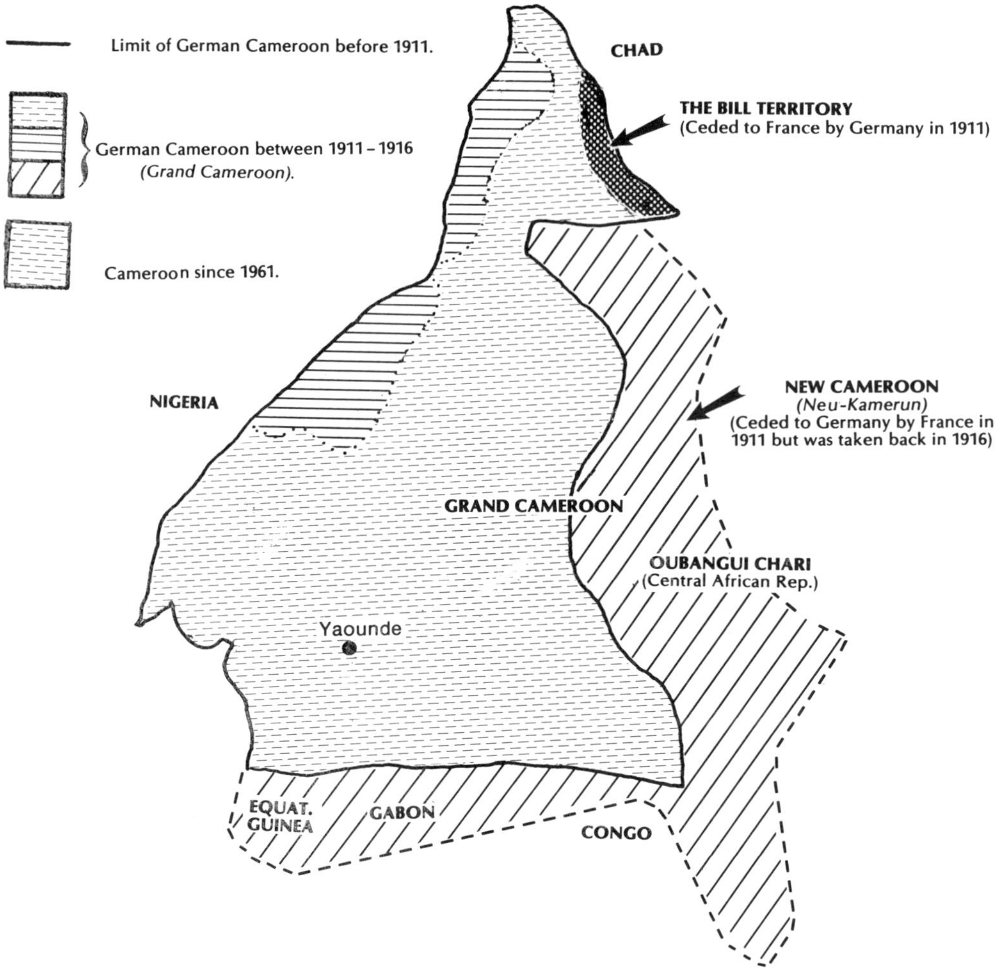

2. There was a draft agreement (protocol) on February 4, 1894 fixing the eastern line of demarcation in the region of Lake Chad, and settling questions pending between Germany and France in the region comprised between the colonies of German Kamerun and French Congo. The two powers then met at Berlin on March 15, 1894, to sign the agreement extending the boundary up to Lake Chad and giving formal recognition of each other's sphere of influence.

3. The next agreement was signed on April 18, 1908 fixing with precision the eastern geographical and political frontiers as determined by the rivers. The 1908 line did alter some portions of the March 15, 1894 boundary to establish the present Cameroon-Central African Republic boundary, between the Sangha and Mbere rivers.

The southern and eastern frontiers of Cameroon as fixed by the agreements were to be drastically altered later in 1911. This was because of Franco-German rivalry for the possession of Morocco. The Germans were experiencing transportation problems in their newly acquired lands of Central Africa. Their only outlet was through the Congo River, so the French proposed an exchange of territory for German recognition of French claim to Morocco. The two powers agreed and France ceded some of its territory to Germany. This newly acquired territory from the French was referred to by the Germans as *Neu-Kamerun.* The former German Cameroon together with Neu-Kamerun, but excluding the "Duck-Bill" territory were referred to as *Grand Cameroon.* The Germans gave up their claims to Morocco and the "Duck-Bill" territory situated between the Logone and Chari rivers in return for the French territory covering part of the Congo, Gabon, Ubangui Shari (Central African Republic).

Unfortunately, this *Grand Cameroon* was not

going to last long. In September 1914, at the outbreak of the First World War, combined French, British and Belgian armies attacked the Germans, who surrendered their claims in 1916.

By an accord signed between the French and British this Grand Cameroon was divided into three parts.
a) The territory *Neu-Kamerun,* which was ceded to Germany by France in 1911, was taken back and attached to French Equatorial Africa. The French retained the Duck-Bill territory by right of conquest.
b) The British took a narrow band of two discontinuous strips of land to the west along the border with Nigeria of about 90,000 km^2 in surface area.
c) The remaining territory of about 432,000 km^2 went to France.

CAMEROON AS A MANDATE TERRITORY: 1922-1946

As indicated above, after the expulsion of the Germans, Cameroon was shared between France and Britain on March 4, 1916. The division was later confirmed by the four allied powers — Britain, France, Italy and the U.S.A. — in 1919 through the Treaty of Versailles.

The League of Nations, created after the First World War, later mandated Britain to administer and develop West Cameroon while France had to do the same with East Cameroon. West Cameroon was administered from Nigeria by the British for the simple fact that they were already well established in Nigeria, whereas East Cameroon was administered directly from France. Cameroon remained under the French and British as Mandate Territories of the League of Nations until the end of the Second World War.

CAMEROON AS A TRUST TERRITORY OF THE U.N.O.: 1946 UNTIL INDEPENDENCE

After the Second World War the League of Nations was replaced by the United Nations Organization with headquarters in New York. The two territories of Cameroon merely continued to be ruled by their "masters", this time not as Mandate territories but as Trust Territories of the United Nations Organization. This status lasted until they achieved independence.

The political developments which finally ended in independence were pursued by the inhabitants of these two Cameroons separately.

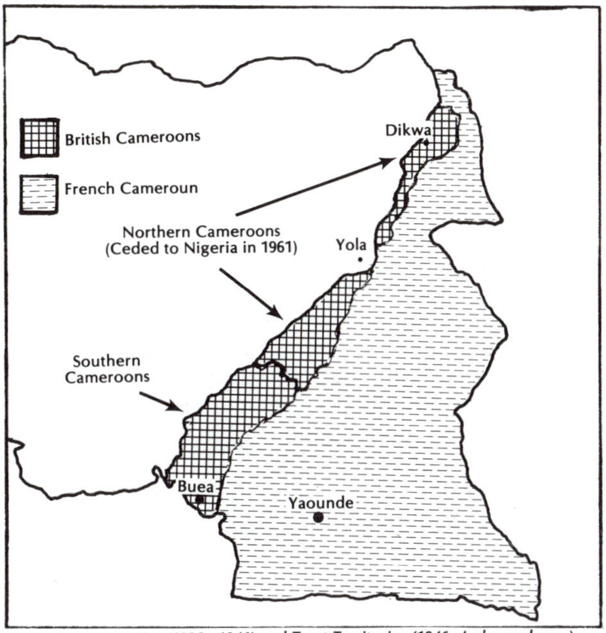

Mandate Territories (1922–1946) and Trust Territories (1946 – Independence)

THE ROUTE TO INDEPENDENCE:

A. THE FRENCH CAMEROON: 1946-1960

The French Constitution of 1946 outlined the position of every territory in relation to France. It established what was known as the French Union, composed of Metropolitan France (i.e. French Republic), the Associated States, and the Associated Territories. Under the 1946 constitution Cameroon became an Associated Territory of the French Union.

One of the Constitution's main objectives was to draw the inhabitants of France's overseas possessions closer to France and enforce the French doctrine of *assimilation.* In other words, it aimed at ruling out the possibilities of self-government and independence for Cameroon. In the French National Assembly, the deputies from the overseas territories represented their own territories. Cameroon was represented in the French National Assembly by Douala Manga Bell and Andre-Marie Mbida. Soppo Priso and Ahmadou Ahidjo were advisers in the French Union, and Arouna Njoya and Charles Okala were senators.

More significant was the creation of local assemblies in the overseas territories by the 1946 French Constitution. In Cameroon therefore, the decree of 25th October 1946 created the ARCAM (Representative Assembly of Cameroon) which comprised both Cameroonian and French citizens. In 1952, for example, out of 40 delegates, twenty-four were Cameroonians and sixteen were French. Among those twenty-four Cameroonians, significant ones were Abega Martin, Ahmadou Ahidjo, Charles Okala and Soppo Priso.

Superficially these reforms appear to be important but in reality they meant little. No reforms were actually implemented by the administrators. Charles De Gaulle wanted to reserve all political authority to France as could be evidenced by the actions of French administrators in the country. In 1958 he (Charles De Gaulle) declared that it was unimaginable for any overseas territory to be thinking of independence out of the French Union. Guinea defied his speech and declared herself independent in 1958. She broke all diplomatic links with France until 1976, when relations were re-established.

Cameroon experienced very little political change at this time except for slight increases in the number of representatives to the French Assembly (i.e. from 3 to 4 in 1952) and in the local Representative Assembly, ARCAM (from 40 to 50). Of these fifty representatives there were eighteen French and thirty-two Cameroonians.

From 1946 to 1952 the local assembly was called ARCAM but from 1952 it was called ATCAM (i.e. Territorial Assembly of Cameroon). This assembly concerned itself with public works, concessions, scholarships, infrastructure, the management of goods, urbanization problems, taxation, etc. In short, it had prerogatives in the administrative and financial domains. Unfortunately, it had neither legislative powers nor rights to embark on any political decisions.

It was the French High Commissioner who was responsible for giving instructions on matters which had to be presented to the assembly. The High Commissioner could nullify decisions of the Assembly. He had to judge whether decisions were legal, in which case it was he who had the political powers of the country.

In the local assembly two groups of Cameroonians could be distinguished. The *Evolutionaries* on one hand, included mainly chiefs, notables, syndicalists who though not entirely satisfied with the framework for political advancement by the French policy of assimilation were willing to work within it for the time being and to seek reforms later as the territory advanced. They favoured gradual development within the French Union towards independence.

The *Radical Nationalists* on the other hand, were those Cameroonians who were entirely dissatisfied with the framework to the extent that they were determined to have French policy altered in order to permit a political evolution towards self-government and independence. The Radical Nationalists were led by such people as Um Nyobe, Felix Moumie and Ernest Ouandie. They wanted immediate independence for Cameroon and reunification of the two Cameroons.

To the French, the Radical Nationalists were subversive because they wanted immediate and genuine self-government. These nationalists, under the leadership of Um Nyobe met and formed the *Union des Populations du Cameroun* (UPC) on 10th April 1948, with three stated goals, namely: (1) the suppression of the "artificial boundaries" created in 1916 between the two Cameroons; (2) abandonment by France of its policies of assimilation; (3) fixing of a time limit for ending the trusteeship, after which Cameroon would achieve independence. Such a platform was enough to set the UPC party at odds with the French government and surely it did.

THE 1955 RIOTS AND THE BAN ON UPC

The UPC , except for the mass following it was gaining, had hardly made any political successes when French administrators embarked on sabotaging its activities. The administration aided anti-UPC groups, such as the *Rassemblement du Peuple Camerounais* and on many occasions used the police and these anti-UPC groups to disrupt UPC meetings. For its part, the UPC stepped up its resistance against increased intimidation and repression especially at the arrival of Roland Pre, the new French Administrator who was perceived to be very ruthless and unsympathetic to the UPC.*

Between 22 and 30 May 1955, Cameroon was plagued with a series of violent riots which brought the UPC and the administration into direct confrontation. Many deaths and injuries resulted from these riots.

---

* *It is important to note that in recent history colonising powers of the 'west' have always attacked and destroyed anti-colonialist movements advocating self-rule (i.e. freedom fighters) by terming them either communists or terrorists. This has always led to a gross misunderstanding of liberation movements, and so has been the case with the UPC. Liberation movements have received support from both capitalist as well as communist nations and this does not necessarily make them capitalist or communist. These are ideological terms often used to confuse the quest for self-rule.*

In July 1955, at the request of Roland Pre, the UPC was banned by the French government, and it was no longer able to participate in political processes in the country.

FROM ATCAM TO ALCAM, 1956

In 1956, France made another giant step which would eventually lead to independence for its west African Colonies. The French National Parliament passed a bill granting self-government to her territories in West Africa. Each territorial assembly was transformed into a Legislative Assembly and was given the right to elect an Executive Council from its members and it was to be presided over by a Governor. The African member of this Council whose party had the majority vote during the election was to become Prime Minister.

The Legislative Assemblies had the powers to make laws on all subjects of local interest but leaving the economic policy, currency, defence, and foreign affairs to the Metropolitan Government in Paris.

The year 1956, therefore was a remarkable one for French Cameroon because on the 23rd of December, the Territorial Assembly was transformed into the Legislative Assembly of Cameroon, ALCAM (Assemblee Legislative du Cameroun).

Members of the new assembly had already arranged themselves into parliamentary groups according to their geographical areas of origin. The main results of the 1957 elections in the ALCAM were as follows:

a) *Union Camerounaise (UC),* had thirty deputies from the Central and Northern regions. The leaders of this party were Ahmadou Ahidjo, Arouna Njoya, and Jules Ninine.

b) *Democrats Camerounais (DC),* had twenty deputies from around Yaounde and its leader was Andre-Marie Mbida.

c) *Paysans Independants,* had nine deputies from the Bamileke lands. It was led by Mathias Njoumessi and Michael Njine.

d) *Action Nationale,* had eight deputies and was headed by Soppo Priso and Charles Assale. Its base was the coastal region.

It is worth recalling that at this time the UPC had already been eliminated from the political scene by the French administration, and that is why it could not be represented in the ALCAM elections of 1957.

Because none of the political parties had a clear majority in the above elections, a Coalition Government had to be formed, which was composed of members of the first three parties; Andre-Marie Mbida was chosen as Prime Minister on the 15th of May, 1957, and Ahmadou Ahidjo was his vice.

André-Marie Mbida: First Prime Minister of East Cameroon.

Mbida's government faced serious problems. First, Um Nyobe, on the 13th July 1957, petitioned the Prime Minister and High Commissioner in Cameroon to grant amnesty to the UPC party but Mbida gave a deaf ear to the request. This refusal led to a fresh outbreak of violent activities in the Sanaga-Maritime region. Instead, Mbida went to Paris and requested more French troops which he used against the UPC. Even with such assistance he failed to quell the violence.

Mbida's second problem came from his harsh and undemocratic attitude, which led to the resignation of many of the members of his cabinet. His third problem resulted from his agenda of priorities. He maintained that Cameroon was not ripe for independence, and worse still, he and his collaborators did not include reunification on their program. These were two themes that had become very popular to all political groups and nationalist leaders of the time.

In the midst of these problems, Mbida resigned and escaped to Conakry in Guinea on February 17, 1958. Ahmadou Ahidjo then took over on the

18th of February as Prime Minister after the departure of Mbida and was formally confirmed on the 19th by a vote of forty-five in favour and sixteen abstentions.

The government in the midst of the violence received reinforcement from France to fight the guerillas who, at this time, had intensified their activities. Many of them were killed, others fled into the forest, and some simply surrendered. Their leader, Um Nyobe, was killed near Boumnyebel, his birthplace, in September 1958.

As the political upheavals proved, no government in Cameroon at this time could have survived without the combined use of force and peaceful negotiations. Ahidjo did exactly this, using force and issuing a series of appeals to the rebels to return to legal ways. Many returned, including Mayi-Matip, a close associate of Um Nyobe.

When Ahidjo, the new Prime Minister, took over from Andre-Marie Mbida, he formulated his program, outlining his immediate goals, many of which were the demands of the U P C party. The main ones were: to seek internal autonomy of Cameroon, to establish a timetable for the independence of Cameroon, to seek national reconciliation, and to seek cooperation with France in a reciprocal manner.

In collaboration with the French government, Ahidjo discussed the possibilities of independence for Cameroon. On the 28th October 1958, France proposed the termination of the trusteeship and the granting of independence to the Fourth Committee of the United Nations General Assembly.

Dr. Felix Moumie of the UPC party, who had arrived in New York, opposed the resolution to grant independence on grounds that the government asking for independence was not the legitimate government for the people of Cameroon; that reunification, which was needed by the Cameroonian people, had not been guaranteed, that Ahidjo was a stooge for France; and that Cameroonians were still greatly deprived of their individual liberties.

A UN visiting mission was sent to Cameroon to assess the validity of Moumie's claims and after their investigations they made a contrary report.

On 21st February 1959, Ahidjo then led his government delegation to New York to discuss the question of Cameroon's independence in the United Nations Organization. Many other petitioners followed to protest against Ahidjo but all the petitions had no effect because the final decision was that the French Cameroon was to be granted independence on January 1, 1960. The few days before and immediately after the day of independence were sad and very unpleasant for many

Proclamation of Independence for Cameroon on 1st January 1960 by His Excellency Ahmadou Ahidjo.

Cameroonians.

It is worth recalling here that after the death of Um Nyobe on the 3rd of September, 1958, the UPC was then being organized and directed by three other leaders, namely Felix Moumie, Abel Kingue, and Ernest Ouandie, all from the western part of the country. On the 29th December 1959, these three men in exile at Conakry wrote condemning in the hardest words the independence that had to be granted to Cameroon, and on the 30th of the same month the UPC members invaded quarters in Douala, Yaounde and other major towns of the western and southern parts of the country. In the night on the eve of independence the rebels hit hard several times on the *Gendarmerie* brigade at Kekem. Many other incidents were recorded in other areas but they did not alter the situation. January 1, 1960, the date of independence, had both very happy and sad memories. While many people rejoiced for the independence achieved, others lamented the fire incidents, murders and deaths caused by the UPC members.

Sporadic incidents continued after independence up to the early 1970s, when another tense moment was experienced. That was after the arrest of Ernest Ouandie, leader of the radical wing of the UPC. He had returned from exile and was directing guerilla activities in the country. At the same time of arrest, the link between him and Monseignor Albert Ndongmo, Bishop of Nkongsamba, was discovered. Ouandie was arrested near the town of Mbanga in August 1970. He and his associates had two trials.

In the first trial which began on December 26, 1970, on charges of rebellion and mass killings, Ouandie and his two associates, Mathieu Njassep (Ouandie's private secretary) and Raphael Fotsing, were sentenced to death, and the bishop was given a life sentence. Thirteen other defendants were imprisoned and many others were acquitted.

In the second trial on charges of plotting to overthrow the then president, Ahmadou Ahidjo, a death sentence was imposed on the bishop and on two other associates, Gabriel Tabeu and Celestin Takala.

These sentences were immediately followed by appeals for clemency for the six people due to die, particularly from the Vatican, the Archbishop of Paris, Cardinal Marty, the International Commission of Jurists, Amnesty International, the Secretary General of the UNO, U Thant and many other sympathetic circles.

Ahidjo used the powers conferred on him by the Constitution to commute the death sentences of Monseignor Albert Ndongmo, and the two others to life in prison, while Ernest Ouandie, Njassep and Fotsing were transported in the early morning on June 15, 1971, from Yaounde to Bafoussam, their home town, where they were executed at 10:00 a.m.

B. THE BRITISH CAMEROON: 1946-1961

Things were not at rest in this part of the country. The British Cameroon was split into the northern and southern parts and attached to the Northern and Eastern Administrative regions of Nigeria respectively.

The beginnings of the independence movement in the southern British Cameroons were closely linked to those of the Nigerian independence movement. Earliest moves in this part of the country were headed by Paul Kale and others who founded the Cameroons Youth League (C.Y.L.) on 27th March 1940, in Lagos, Nigeria. Soon to become very significant in this part of the country were Dr. E.M.L. Endeley, with whom Paul Kale had formed the Cameroons Youth League, L. N. Namme, Namaso N. Mbile and many others.

Dr. Endeley began organising and reactivating tribal associations such as the Bakweri Land Committee. The main purpose was to articulate demands for the return of former German-owned lands ceded to the Cameroons Development Corporation (CDC). In 1949, Endeley formed a political organisation, the Cameroons National Federation (CNF) to support demands for a separate Southern Cameroons region within the Nigerian Federation.

In 1943, Sir Authur Richards became the Governor of Nigeria and remained in office until 1947. Within this time there had arisen a strong agitation from the nationalist elements of the country for constitutional reform. When Sir John MacPherson took over from Sir Authur Richards in 1948, he undertook to revise the constitution with a view to hastening up moves towards political independence for that country. Whatever changes the proposed new Constitution had to carry out were going to affect the status of Southern Cameroons.

In the Southern British Cameroons two meetings were held in Mamfe and Limbe by mid-1948 and a third one in Lagos to raise political awareness among Cameroonian citizens. The result of these meetings was a unanimous resolution to ask that the British Trust territory of Cameroon be given separate regional status with a House of Assembly and be directly responsible to the UNO Trusteeship Council.

This request was rejected at one of the Constitutional Conferences in Enugu. However, it was recommended that the Southern British Cameroons be represented both in the Regional House of Assembly and the Executive Council and in the Central Executive and Legislative Assemblies — a decision which was later confirmed at another Constitutional conference in 1950 at Ibadan. As a result of this decision the number and significance of south Cameroonian representatives in the various Houses in Nigeria increased. The MacPherson Constitution failed in 1953 and was replaced by the Lyttelton Constitution in the same year.

THE LYTTELTON CONSTITUTION

In August 1953, at a conference held in the Lancaster House, London, the Southern Cameroons was represented by E. M. L. Endeley, who wanted a separate autonomous region for Southern British Cameroons, while the Northern British Cameroons was represented by Mallam Abba Habib, who wanted that part of the territory to remain with Nigeria. Namaso N. Mbile, who represented the National Council of Nigeria and the Cameroons (NCNC), was in favour of the Southern British Cameroons remaining as part of the Eastern Region of Nigeria.

Among the many decisions arrived at in the conference, was the decision that should Endeley's party win the elections that were to be conducted in the Southern British Cameroons after the dissolution of the Nigerian Eastern Regional House of Assembly the Southern British Cameroons would become an autonomous region within the Central Nigerian Legislature and Council of Ministers. The Northern British Cameroons was to develop as part of the Northern Region of Nigeria.

THE 1954 ELECTION AND ITS CONSEQUENCES

The 1954 election in the Southern Cameroons saw Endeley and his party, the Kamerun National Congress (KNC) triumph over the Kamerun People's Party (KPP) led by Mbile. Endeley then became the leader of Government Business in the Southern Cameroons.

At the Lagos Constitutional Conference of 1954 it was consequently announced that the Southern Cameroons was separated from the Eastern Region of Nigeria and had to be made a quasi-Federal territory with its own House of Assembly. On the 26th October, therefore, the Southern British Cameroons House of Assembly met for the first time.

Not long after the granting of this quasi-Federal status to the Southern British Cameroons, Endeley changed his mind on the question of reunification with French Cameroon which was one of the pressing issues for many Cameroonian nationalists of the time. This abrupt change of mind, or more precisely, his inconsistency was later to be regretted because it led to a split in his party, the KNC. In 1955 John Foncha split with Endeley and organised the Kamerun National Democratic Party (KNDP) on a program of complete secession from Nigeria and reunification with the French Cameroon. Foncha saw these two issues as important ones and used them to secure popular support from South Cameroonians.

From May to June 1957, another Nigerian Constitutional Conference was held at the Lancaster House in London and the three Southern Cameroonian parties were represented: the Kamerun National Congress (KNC) was represented by Endeley, the Kamerun National Democratic Party (KNDP) was represented by Foncha, and the Kamerun Peoples' Party (KPP) was represented by P. M. Kale. Following the discussions at the conference it was decided that the term "quasi-Federal territory" be dropped and the territory be known as the Southern Cameroons. The Governor-General, the representative of Her Majesty in the Federation of Nigeria, was then to be responsible for all trusteeship matters in the region according to the British Government obligations as contained in the trusteeship agreement.

THE 1957 ELECTION

In the elections of 24th December, 1957, Foncha defeated Endeley and became the Premier of the Southern Cameroons Region within the Federation of Nigeria. Foncha and his party, the KNDP, from that time on started campaigning for complete secession from Nigeria and reunification with the French Cameroon. Endeley merged his KNC with the KPP, led by Mbile and Kale to form the Cameroon Peoples National Convention (CPNC). The CPNC was formed to counter the growing dominance of the KNDP. It campaigned for integration with Nigeria.

It is essential, however, to note from the attitude of most Southern Cameroons politicians at the time that they were nationalistic but at sometime had vacillated between one issue and the other, each trying to seize the most popular issue and work on it.

The Northern British Cameroons also had people of differing political views as regards the territory's future but, on the whole, the inhabitants of that region were less politically conscious because they had been intimately linked to the rule of the Emir who himself owed allegiance to the Sultan of Sokoto in Nigeria.

THE PLEBISCITES AND REUNIFICATION

Due to the disagreement between Foncha and Endeley on their political goals for the territory, a plebiscite was held to decide the future of the Southern Cameroons on February 11, 1961. Another one was held in the Northern British Cameroons on the same date. These plebiscites were organised and supervised by the United Nations Organisation. In the Southern Cameroons:
97,741 votes were in favour of joining Nigeria;
233,571 votes were in favour of joining the Republic of Cameroon.

After several talks between Ahidjo and Foncha following the results of the plebiscite the Southern Cameroons became independent on October 1, 1961, by joining the Republic of Cameroon.

The nation then came to be known as the Federal Republic of Cameroon with Ahmadou Ahidjo as President and John Ngu Foncha as Vice President and Prime Minister of West Cameroon. The former Republic of Cameroon became the East Cameroon state while the Southern Cameroons became the West Cameroon State.

Dr. E.M.L. Endeley:

Meanwhile, the Northern British Cameroons in the plebiscite voted as follows:
97,659 votes were in favour of reunification with Cameroon;
146,296 votes were in favour of joining Nigeria.

On June 1, 1961, the Northern British Cameroons formally joined Nigeria and that day was declared by Ahmadou Ahidjo as a national day of mourning in Cameroon for its citizens who were lost to Nigeria. From 1961 on, Cameroon prac-

John Ngu Foncha:

tised the Federal System of government until the birth of the Unitary State.

THE FORMATION OF ONE NATIONAL POLITICAL PARTY

The first move towards a unitary state was the formation of one national political party, the Cameroon National Union (CNU) in September 1966.

Before its formation we recall that in West Cameroon there was an increasing rate of party-formation especially due to conflicts within parties and frequent change of mind by leading politicians. To show the chaos and numerous defections in the political arena at this time, in 1955 for example, John Foncha split with Endeley and formed the Kamerun National Democratic Party (KNDP) on a program of complete secession from Nigeria and reunification with the French Cameroon because of the KNC's shift away from reunification. The KNC suffered another blow in September 1957, when Solomon Tandeng Muna resigned and joined the KNDP, reportedly accusing the KNC leadership of following a policy of "integration with Nigeria", even though at this time the party had not yet completely abandoned the idea of reunification. Muna again defected from the KNDP to form the Cameroon United Congress (CUC) in 1965 after Augustine Jua triumphed over him in the struggle for Premiership in West Cameroon. The CUC then stood in opposition to the governing KNDP/CPNC coalition in the West Cameroon Assembly after August 1965, and from that time on S. T. Muna maintained close ties with the Ahidjo government in Yaounde. Yet Muna was not the only one in these changes. The One Kamerun (OK), headed by Ndeh Ntumazah, was formed in 1957 as the southern Came-

S.T. Muna

roons rump of the U.P.C. after the latter was banned in East Cameroon. It was the only party in this part of the country that was totally opposed to union with Nigeria and never wavered in its commitment to unification with the French Cameroon.

Other short-lived parties included the Kamerun United Commoners' Party, the Cameroons Muslim Congress, The Cameroons Indigenes' Party, and the Kamerun United Party. The K.U.P. was formed in 1960 by P. M. Kale, one of the founders of the NCNC and former President of the KPP. Kale later dissolved his party to join the KNDP.

On the other hand, the *Union Camerounaise* (UC) under Ahmadou Ahidjo, was the dominant party in the East Cameroon State. The American political scientist, Victor T. Levine, maintains that there were some ninety registered political "parties" in this part of the country. Most of them were only small organisations run by local politicians. Others were merely groups of local or departmental organisations built solidly on ethnic bases with limited self and tribal interests.

In 1962 the four leaders of the other significant parties in East Cameroon who formed the opposition to Ahidjo's government were arrested for an alleged conspiracy against the government. In fact, they were bent on rejecting the concept of a unified party. They included, Charles Okala of the Cameroon Socialist Party (CSP); Theodore Mayi-Matip of the legalised wing of the UPC; Andre-Marie Mbida of the Cameroon Democrats (CD); and Marcel Eyidi of the Cameroon Labour Party (CLP). Although these opposition leaders were willing to work towards concrete solutions to national problems, they maintained that the unified party would undermine the other parties and lead to a dictatorship. These leaders were later arrested, tried and sentenced to various terms of imprisonment on charges of "inciting hatred against the government and public authority, inciting conflict between ethnic and religious communities, and disseminating news prejudicial to public authorities". They were later released in 1965.

In order to stop wasteful efforts in party struggle and to present a united force for nation-building the leaders of the UC, the KNDP, the CPNC, and the CUC unanimously agreed to merge their parties into a single national party, the Cameroon National Union (CNU) and on September 1, 1966, its constitution was drawn up by representatives of these parties. The exclusion of the other political party leaders was not intended to deprive them of power because after the formation of the CNU some of them were recruited in the government and some were given party positions. For example, Theodore Mayi-Matip was made a member of the CNU Political Bureau and was later elected Vice President of the National Assembly.

THE FORMATION OF THE UNITARY GOVERNMENT

Some six years after the formation of the CNU, the leaders of the nation thought it very uneconomical for a small country like Cameroon to be running three governments: the Federal Government, the East Cameroon State Government, and the West Cameroon State Government.

There were many other problems involved in the existence of the two federated states. Traders could not trade easily with their goods from one state to the other. To travel from West Cameroon to East Cameroon or from East Cameroon to West Cameroon individuals had to obtain a special pass, popularly known as the *Laissez Passer* as if one was travelling from one country to another. As a matter of fact, these two territories virtually existed as if they were two separate countries. Each one had a separate currency and language. They were kept apart by lack of roads, telegraph and telephone (a phone call from Limbe to Douala had to go through London or Paris). More than anything else, Ahidjo wanted to unite these two territories, even if it meant using unpopular methods.

On May 20, 1972, a referendum was conducted all over the territory to sound the people's opinions about having a unitary state and a unitary government. An overwhelming majority of the citizens supported the proposal and that gave birth to the United Republic of Cameroon.

In 1984, the word "United" was dropped. The reason offered was that from the time the name,

President Paul Biya taking the oath of office on November 6, 1982.

Kamerun, was applied to the territory in 1884, Cameroon has always been one nation and the division into East and West, in 1916, was an arbitrary undertaking by foreign imperialist nations at a meeting in which Cameroonians were never represented and their consent was not even sought.

FROM AHMADOU AHIDJO TO PAUL BIYA

Since independence (1960), Cameroon remained under the leadership of Ahmadou Ahidjo as President of the Republic for more than two decades. On November 4,1982, in an unexpected announcement, Ahidjo resigned from the Presidency, handing over powers to his constitutional heir, Paul Biya, who at the time was the Prime Minister. Despite the political upheavals that occurred in 1984, as a result of rivalry between Paul Biya and Ahmadou Ahidjo, Cameroon has continued to be one of Africa's most stable countries.

FROM CNU TO CPDM

The Cameroon National Union (CNU) formed in 1966 formally had its name changed during its Fourth Ordinary Congress held in March 1985 at Bamenda. The resolution changing the name of the CNU Party to the Cameroon Peoples Democratic Movement, CPDM, according to party officials, was in line with President Paul Biya's desire to render the party more responsive to the masses and more democratic in its decision-making. This change of name, in fact, did not entail any drastic shift in ideology or system. The various organs of the party remained the same. The change was essentially an attempt to break away from the previous regime. The CPDM, however, still maintained the essential ingredients of the one-party system as a uniting force for all Cameroonians and a melting-pot for people of all political ideals.

There is just no doubt that Cameroon is one of the few countries in recent times whose history has been marked by a lot of fragmentation and whose population has had to deal with three dominant European colonial powers — Germany, France and Britain — each of which has left behind differing administrative, economic, juridical, philosophical, political and social legacies. To compound matters, the peoples of this small country are divided into numerous tribal groupings, each with its own language or dialect and form of social organisation.

Yet Cameroonians seem to be determined to foster bonds, forge unity, and found a concept of Cameroonian nationality. Even in the midst of the most divisive challenges everyone seems to be saying, "We are all Cameroonians". As President Paul Biya once said, **"I am a Cameroonian; I was born a Cameroonian; I will live a Cameroonian; I will die a Cameroonian."**

PART TWO: PHYSICAL STUDIES

CHAPTER 2: RELIEF & STRUCTURE

A look at the relief map of Cameroon shows that it is a country of varied landscape with many plains, plateaux and highlands. The main physical units that can be distinguished are as follows:
a) The Coastal Lowlands
b) The Southern Plateau
c) The Adamawa Plateau
d) The Western Highlands
e) The Northern Lowlands.

1. THE COASTAL LOWLANDS

This vast sedimentary zone stretches from the mouth of River Akwa Yafe in the extreme west which forms the boundary with Nigeria to the mouth of River Lokounje. It is interrupted by the extension of the Cameroon Mountain into the ocean. The Coastal Lowland zone is bounded by the Atlantic Ocean to the west, the Western Highlands to the north and the Southern Plateau to the east. Between the ocean and the Southern Plateau it stretches out over a surface area of about 150 kilometres. It has an average altitude of 90 metres.

As this Coastal Lowland approaches the ocean some important characteristics are noticed. Close to the Atlantic Ocean there are mangrove swamps, creeks, sand bars and sand spits. The large rivers which drain into this zone account for the huge deposits of sand, silt and mud which in turn result in the marshiness of the coast. The marshes make it difficult for the creation of good harbours.

The Cameroon Coastal Region can be divided into two:
a) The Low-lying Coast
b) The Rocky Coast

A) THE LOW-LYING COAST: This zone can again be subdivided into the following sections, namely:

i) The Ndian Basin

It lies to the north-west of the Cameroon Mountain and is characterised by the presence of mangrove swamps, The rivers here, especially the Ndian River, split up into small *distributaries* which cut through sedimentary deposits before entering the Atlantic Ocean. Some significant distributaries here include Rio-del-Rey, Ngosso, and Andokat.

ii) The Douala Basin

This is a very low depression where a lot of sediments from the interior of the country are deposited. The high rate of deposition can be noticed at the Wouri estuary. The average altitude of this basin is just 30 metres. Like the Ndian Basin, it is characterised by the presence of creeks, sand bars and lagoons. Some of the creeks here include the Bimbia creeks, the Missellele creeks and the Mabeta creeks.

A lagoon on the Douala Basin

iii) The Southern Lowlands

The rest of the coastal lands extending from Douala down to Kribi is lowlying. Compared with the two preceding sections, the Southern Lowland zone has few creeks and the coastline has very few indentations.

iv) The Mamfe Depression

The Mamfe Depression forms a distinct zone located some 150 kilometres away from the coast. It is almost encircled by a range of mountains except to the west where its lowlands stretch out into Nigeria. The depression receives numerous streams which flow down from the surrounding mountains and join to form the Cross River which links Cameroon to Nigeria.

RELIEF OF CAMEROON

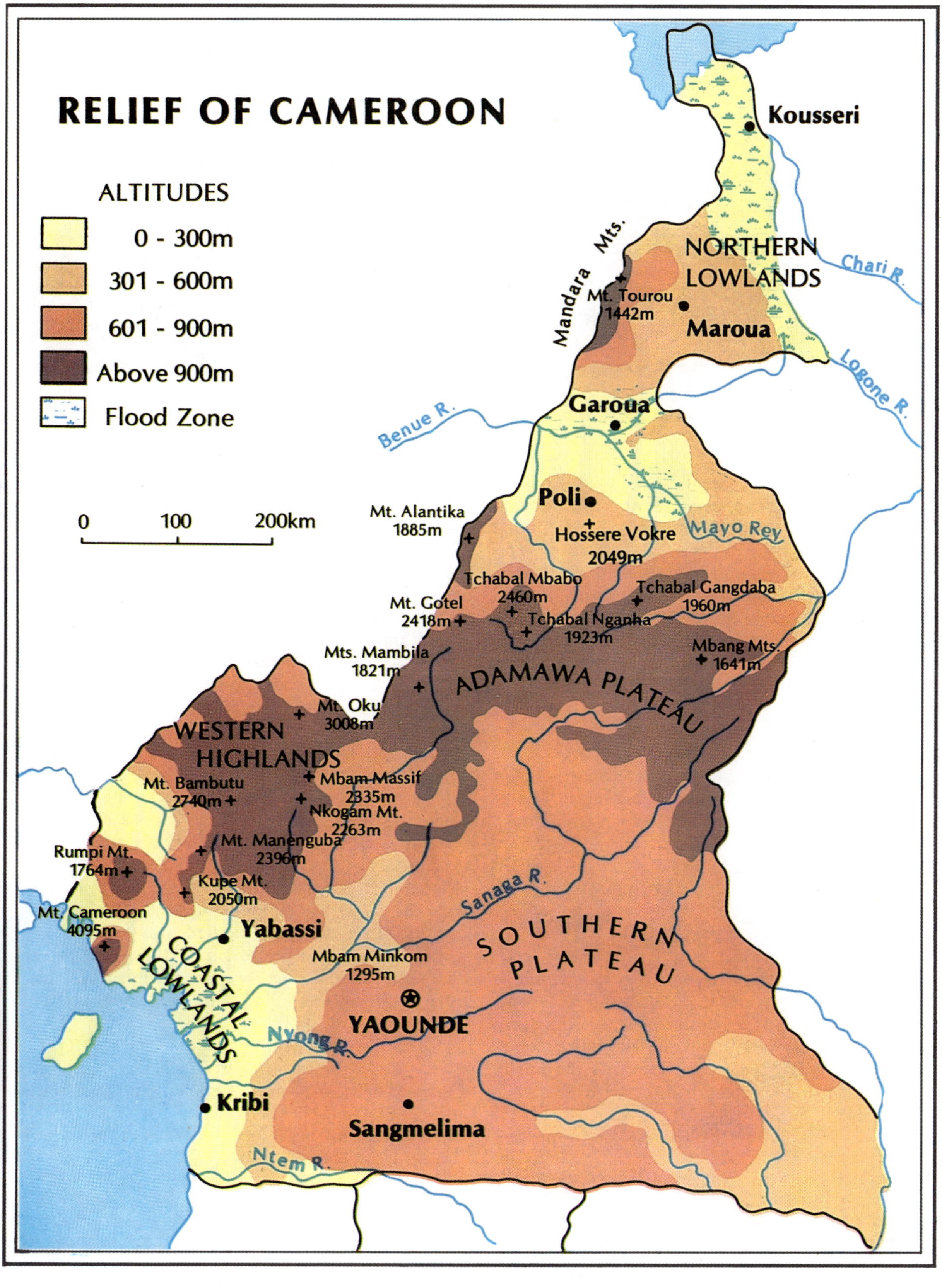

Volcanic rocks at the rocky coast of Limbe.

B. THE ROCKY COAST

The Rocky Coast presents itself in two sections. The first section lies between Bimbia and Idenau. The rocks here are volcanic rocks which spread down into the sea from Mount Cameroon. Volcanic activity in this area also resulted in the presence of volcanic stacks protruding above the sea some few kilometres off the coast of Limbe. This section forms a break in the low-lying nature of the coastal lands and it is characterised by the presence of bays, capes, cliffs and rocky islands.

The second portion of the Rocky Coast is found in the region of Kribi where the *continental platform* of the Southern Plateau comes close to the ocean. The region of Longi near Kribi is made of bare crystalline schists which sometimes stand as cliffs at the coast or split up into mighty rocky blocks. An important feature of this rocky portion is that there is a general absence of sand and silt deposits as is the case in the lowlying coast.

2. THE SOUTHERN PLATEAU

Situated east of the sedimentary coastal plain, it covers all of the south and south-east of Cameroon. It lies at the foot of the Western Highlands and the Adamawa Plateau and extends down to the frontiers of Cameroon. Towards the north-east the plateau rises gently into the Adamawa in the region of Bedzare and Meiganga. Towards the north-west it is interrupted by an important escarpment between Yoko and Linte. The altitudes on this plateau vary from 250m to 800m but averagely most of the hills are at an altitude of 650m.

The landscape of the Southern Plateau is monotonous, essentially of gently undulating hills of convex slopes. These dome-shaped hills are generally referred to as *half-orange* relief and such hills are well noticed in the Batouri and Belabo regions. On this landscape one occasionally comes across rocky domes which have concave slopes, some with no vegetation at all. In the environs of Yaounde these domes reach an altitude of 1,200m. Examples are the Mbam-Minkom and Mbankomo massifs. When these massifs stand out isolated on the plateau, they are known by different names, for example, *inselbergs* or *sugar-loaf* relief.

Three sections of this plateau can be identified:

a) The western part which has an accidented relief is composed essentially of gneiss. This section has deep valleys separating hills. Most of the hills have steep slopes and typical examples include the Mbam-Minkom (1,295m) and the Ngovayang chain. The Southern Plateau in this section ends at the Coastal Plain with an abrupt slope which can easily be seen at Kribi where the rivers tumble down in rapids and falls.

b) The eastern section on the contrary is a peneplain with a characteristic half-orange relief. The landscape here is more gentle than that of the western section. The lowest portions of this eastern section are found along the Sangha River valley.

c) The northern section has alternating depressions and granitic massifs which form the foreground of the Adamawa Plateau. It is actually the transition zone between the Southern Plateau and the Adamawa high plateau. The altitudes here range from 800m to 900m.

3. THE ADAMAWA PLATEAU.

The Adamawa Plateau is situated at the centre of the country and has an average altitude of 1,100 metres. It lies between latitudes 6° and 8° north of the Equator. It cuts across the country and penetrates far into the Central African Republic.

The Adamawa Plateau is a faulted and upraised block of the basement complex, and is composed mainly of granite. These granitic rocks are covered on the surface by thick basaltic flows which result from the volcanic eruptions that occur on this massif. The volcanic outpours have formed some sizeable peaks, the major ones are Tchabal Gangha (1,923m), Tchabal Mbabo (2,460m), and the Mambila mts. (2,418m). In the northern part of this high plateau the relief falls very abruptly into the Benue basin but in the south it descends gradually to merge with the Southern Plateau. The Adamawa Plateau is the main water-dispersal centre of the country.

4. THE WESTERN HIGHLANDS

Like the Adamawa Plateau, the underlying rocks here are old granite and gneiss of the pre-cambrian era which have a cover of basaltic rocks on the surface areas. The Western Highlands are, however, different from the Adamawa in that they are less extensive and possess higher mountain peaks and a more accidented relief than would be found on the Adamawa Plateau. Some of these high mountain peaks are of crystalline rocks, such as, the Nlonako, the Gotel mts., the Guingue mts., the Alantika and the Banglang mountains.

Apart from these crystalline massifs, we also notice that the main volcanic peaks of the country are located in this region. We must recall that this region falls on the great fracture line oriented in a SW-NE direction and stretching far into the country. On this fault line the following volcanic peaks can be identified — Annobon island, Sao Tome, Santa Isabel, located in the Atlantic Ocean. On land are the Cameroon Mountain, Mt. Kupe, the Manenguba mountains, the Bambutu, the Mbam massif, the Nkogam, the Mbapit Mountains and the Mandara Mountains. The Cameroon Mountain is still very active today.

The Western Highland region is rich in lakes which are of volcanic origin and the soils of this area are the richest in the country for agricultural purposes.

5. THE NORTHERN LOWLANDS

This region can broadly be divided into two parts, namely:

a) *The Benue Depression:* It is found on the northern piedmont zone of the Adamawa Plateau. It has an average altitude of 300 to 350 metres. The centre of this depression is formed by the large valley of River Kebi. The depression is separated from the Chad Plain and Logone Valley by a small elevation which has some significant peaks, such as, the Mindif, the Rumsiki (1,224m), and the Peske Bori (1,195). These peaks form part of the Mandara mountains.

b) *The Chad Plain:* Cameroon has only a small portion of this plain and the rest is owned by Chad, Niger, and Nigeria. The altitudes here are almost uniform and the land slopes gently towards the lake. A large portion of this plain suffers occasionally from floods.

SOME MAJOR PHYSICAL FEATURES OF CAMEROON

MOUNT CAMEROON

This is the most prominent of the West African volcanic mountains. The mountain is a heap of piled up lava reaching a height of 4,095m. It lies in a subsidence zone along the main Cameroon fracture line. Apart from the Cameroon Mountain this fracture line also carries most of the high peaks of the country such as, the Kupe, Manenguba, Bambutus, Kom and Oku mountains.

The Cameroon Mountain is very immense with a length of about 50km and a width of about 35km, covering therefore an area of some 1,750km^2. It starts from the sea and first rises into a small peak of 1,713m which bears the name *Etinde* or *Small Mount Cameroon.* From this small peak the mountain slopes down to a height of about 900m before rising again continuously until the summital plateau of 4,070m is reached. On this plateau, recent accumulations of solidified lava still raise the peak to a height of 4,095m.

The mountain is composed of different volcanic materials which, in many places, accumulate to form small cones numbering about 75 in all. Some of these cones have craters at their summits where volcanic ashes, lapilli and scoria can be noticed.

On some of the slopes of this mountain one comes across *fumeroles* which emit different gases, and *geysers* which send out water in jets. Through many vents lava found its way down to the foot of the mountain. An interesting example is the Ekona vent with lava of the *aa* type which flowed down during the 1959 eruption. This area covered with the *aa* lava is very difficult to walk on.

The Eruptions:

Mount Cameroon is an active volcano and this is evidenced by the permanent presence of fumeroles and geysers. It is one of the few peaks in West Africa that has experienced many eruptions within

Mount Cameroon

this century. Its last eruptions took place in 1909, 1922, 1954, 1959 and 1982, including earth tremors in 1977.

The eruptions are usually preceded by some type of underground rumblings and earth tremors. These are only the preliminary indications of the eruption. The explosion to blow off the vent and allow out magma then follows as a result of the great heat and pressure from the interior of the mountain. Some of the blown off materials are projected into the atmosphere forming thick clouds which, during the day, are dark but in the night appear luminous. These projected materials variously known as volcanic *bombs* or *scoria* fall back to form the numerous cones found around the mountain.

The 1909 eruption spread out lava from the northern part which flowed to the north-east covering a distance of some 6 kilometres. In 1922 it was more catastrophic. Lava extended down to the ocean burning down hectares of forests through which it flowed and destroying many farms. The 1954 eruption emitted no lava, only thick columns of gas rose vertically from the principal crater at the summit. Some strong shocks also rocked buildings in the Buea area. The 1959 eruption emitted lava which went for about four kilometres with a width of over 100m and a thickness of about 20m. The lava was not from the main vent but from a secondary vent located at Ekona in the north-eastern part of the mountain. It caused some destruction to certain portions of Ekona town. The 1982 eruption began on the 16th October and lasted a number of days. The main crater of the mountain emitted smoke and red-hot molten rocks (or lava) which spread in three directions, on the southern, south-western and south-eastern slopes of the mountain, destroying all vegetation and animals on its way.

The 1982 eruption took the same course as that of 1922. The nearly 300 inhabitants of Bakingili village, on the south-western slope of the mountain fled their village during the 1982 eruption.

Gradient, Temperatures, and Vegetation of Mt. Cameroon:

The principle that the higher one goes the colder one feels is easily noticed as one climbs up the Cameroon mountain. The types of vegetation at different altitudes of this mountain merely reflect the climatic and soil conditions which prevail at each altitude.

The Cameroon mountain presents itself in a series of terraces from its base to the summit. From the coast up to an altitude of about 50m is a sedimentary plain which extends to Tiko where it is limited by a small escarpment located between Tiko and Mutengene. Beyond Mutengene the land rises gently up to the altitude of Buea (800m).

At the foot of this mountain one experiences the typical hot equatorial climate with temperatures rising above 23°C in such places as

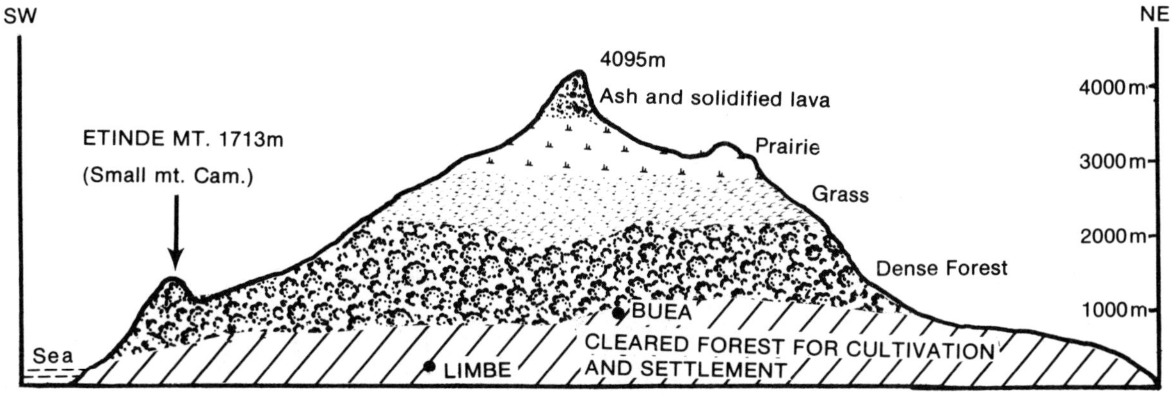

SKETCH SECTION OF MOUNT CAMEROON

Tiko and Limbe. Here the natural forest vegetation has been cleared down to create settlements, farms and plantation lands for rubber, palm trees and tea.

From about 915m above Buea town one climbs a very steep slope covered by thick and evergreen forest. The forest extends up to an altitude of 1,700m and gives way to typical savanna vegetation. At about 1,600m is the location of Hut One. From the end of this forest, between Huts One and Two, one is faced with climbing a much steeper section of the mountain than the one below. This very steep section ends at an altitude of 3,000m where Hut Two is found and is covered by savanna. Here, one remarks a considerable fall in temperatures.

From 3,000m to 3,500m the slope is gentle, measuring about 30° and is still covered by grass which is much shorter than the preceding savanna and can be termed *prairie.*

Between 3,600m and 4,000m the slope is very steep and the typical vegetation is composed of lichens and mosses. It can be termed the *summital plateau* and it has small crests of scoria of over 50m, large funnel-shaped craters, and fumeroles whose temperatures measure between 60° C to 80° C.

Temperatures on the summital plateau are extremely low. Average daily temperatures are about 4° C in the wet season and about 8° C in the dry season. Present in this zone are very strong winds blowing at a speed of 240km per hour. Getting to the culminating point which is formed of solidified lava one comes in contact with snow caps and temperatures are at freezing point.

The Cameroon Mountain has a crater of 100m in diameter and about 50m deep with vertical inside walls. The mountain is formed of piled up layers of lava. From base to summit are a succession of basalts, trachyte, phonolite and other volcanic materials. Rainfall on the mountain is very heavy, especially on the seafacing slopes.

Hydrographically, this mountain disperses water to two directions if one considers only those streams and springs which take their rise from above 3,000m. First are those streams of the south-east which flow parallel to each other before meeting to form the Ombe River. Second are those of the north-west which flow to form the Onge River. On the other hand, when one considers the mountain as a whole, one may say that it disperses water in a radial pattern to all directions.

MOUNT KUPE

The Kupe Mountain (2,064m) overlooks the towns of Loum and Tombel. It is built up essentially of crystalline materials and of layers of volcanic flows. The mountain has slopes which are almost vertical.

One remarks at certain places that the thick layers of hardened lava flows have been split up perpendicularly into large blocks which stand aloof from the main massif and some of them carry thick vegetation. At the foot of this mountain, especially around Mbanga, Nyombe and Nkongsamba and more specifically between Tombel and Penja, are over eighty small volcanic cones with an average altitude of 50m. They are made up essentially of volcanic materials such as

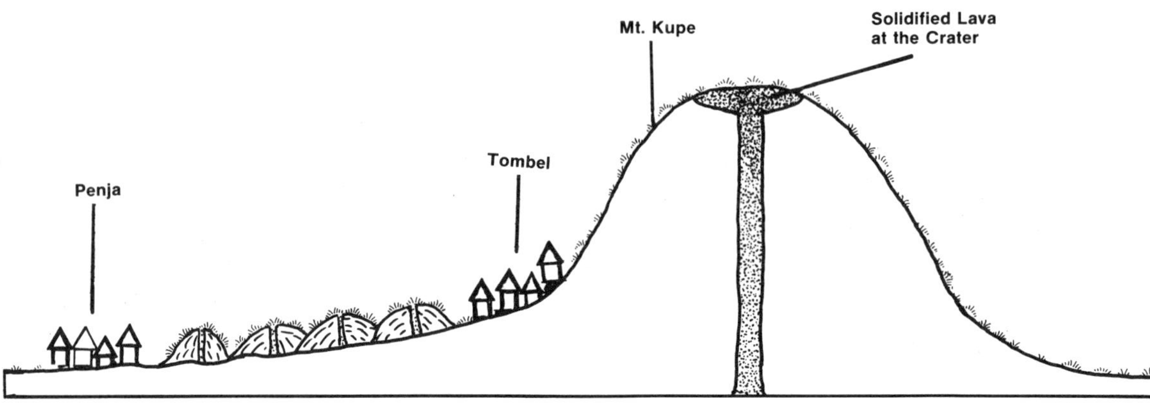

ashes, lapilli and scoria. Such cones resulted from a type of *adventitive eruption*. The volcanic materials are frequently used for road maintenance in this area.

The Adventitive Eruption:

This type of eruption occurs when the main volcanic vent on the mountain does not emit lava any longer, as a result of a blockade of the vent by solidified lava. The boiling heat and magma beneath find their way out by developing several small vents at the base of the main volcanic mountain. The materials emitted through these small vents accumulate to build up small adventitive cones around the main mountain. This type of eruption produces a feature which is different from the *composite cone* in that the cones are produced right at the base of the main volcanic mountain while the *subsidiary cones* of the *composite volcano* are produced on its flanks.

THE MANDARA MOUNTS

The Mandara Mounts in north Cameroon form an immense block on the Cameroon-Nigerian frontier. They are composed essentially of very old granitic rocks. Three sections of the mounts can be distinguished, namely:
a) The middle plateau
b) The mountainous northern and southern parts
c) The surrounding inselbergs

a) The middle plateau:

This zone lies between the latitudes of Mokolo and Tchevi. The southern part of this zone is a real plateau with a generally flat landscape ranging from 800m to 900m. There are landforms of *necks* and *dykes* formed by the volcanic intrusions through the crust. These volcanic features are composed of trachyte and rhyolite, and could be found in the north of Rumsiki.

Kapsiki Volcanic necks on the Mandara mounts.

b) The mountainous northern and southern parts:

These two zones resemble each other in that they are extremely hilly. The massifs here, composed principally of granite, have been greatly dissected. The northern section, which is much higher than the southern part, has the highest part of the Mandara mountains — the Tourou Mt. (1442m). The dissection of the Mandara Mounts is as a result of internal earth movements and river action on the surface.

c) The surrounding inselbergs:

Morphologically it is difficult to dissociate the surrounding inselbergs from the Mandara mounts. The most significant inselbergs are found in the south-western part of the massif, e.g. Popologozom (1165m), and Peske Bori. Some of these inselbergs are even located far into the plains and stand out isolated, e.g. the Mindif inselberg (769m), the Waza hills and many other smaller ones.

CHAPTER 3

GEOLOGY OF CAMEROON

Four-fifths of the geology of Cameroon is composed essentially of hard crystalline rocks which form the bed rock. This bed rock is covered on the surface in certain zones either by sedimentary materials or volcanic soils. Most of the volcanic soils were thrown out during volcanic eruptions of ancient times.

Metamorphic Rocks: The southern part of the country is made of hard crystalline rocks which have suffered a lot of transformation to form such metamorphic rocks as gneiss, quartzites and migmatites. These rocks are millions of years old and are noticeable in the environs of Yaounde where they have been quarried and used for building constructions.

Plutonic Rocks: Due to igneous activity, plutonic rocks originated from the interior of the earth and solidified within the crust. They have subsequently been exposed by erosion to form crystalline massifs. This is the case with the granitic massifs of north Cameroon such as the Peske Bori. The Adamawa is essentially formed of these plutonic rocks in which granite predominates. Here, the hard granitic rocks are covered on the surface by volcanic soils which result from lava flows.

FRACTURES

An important phenomenon provoked some volcanic activity in this part of the world. Towards the end of the Cretaceous, a period of disturbances and dislocation of the earth's crust (diastrophism) was manifested by faulting, uplifts, and depressions. In Cameroon, this great earth movement created a major fault line, which Passarge (1910) termed the "Cameroon line." This line extends in a north-east direction from the Gulf of Guinea towards Chad. This tectonic line, composed of several parallel fractures, carries the large volcanic massifs: Mts. Cameroon, Kupe, Manenguba, Bambutus etc. This axis extends south-westwards into the Atlantic towards St. Helena; it is marked out by a set of volcanic islands: Fernando Po, Sao Tome and Annobon. North-eastwards this tectonic trough passes into the Chad Basin in the direction of Tibesti. Elsewhere in the southern part of Cameroon, some faulting of the crust was also experienced, especially in the localities south of Yaounde and Mbalmayo.

The volcanic activity which occurred along this line created volcanic massifs composed of basalt and smaller cones built of ashes, lapilli, etc. Isolated volcanic *necks* are of trachyte as is the case with the Kapsiki. Apart from the slopes of the active volcanic mountain at Buea, Limbe, Tiko and other nearby localities, one still comes across ancient volcanic flows which have been weathered into fertile black soils. Volcanic rocks and soils are found in the localities of Tombel, Loum, Nkongsamba, Dschang, Foumbot, Wum, Njinikom, Ndop, etc.

Sedimentary Rocks: Sedimentary materials are found mainly in two regions, namely, the Coastal Lowlands and the Benue Depression. The sedimentary materials of these regions were deposited mainly within the Cretaceous and Tertiary eras; that is, during the separation of the African and South American continents. Marine transgression at the time caused the deposition of many materials in these zones. During the quaternary era, new sediments were once more deposited on the older layers. Other depressions, including the Chad basin and the Mamfe depression, are areas of intense deposition of alluvium, clays, limestones and sandstones.

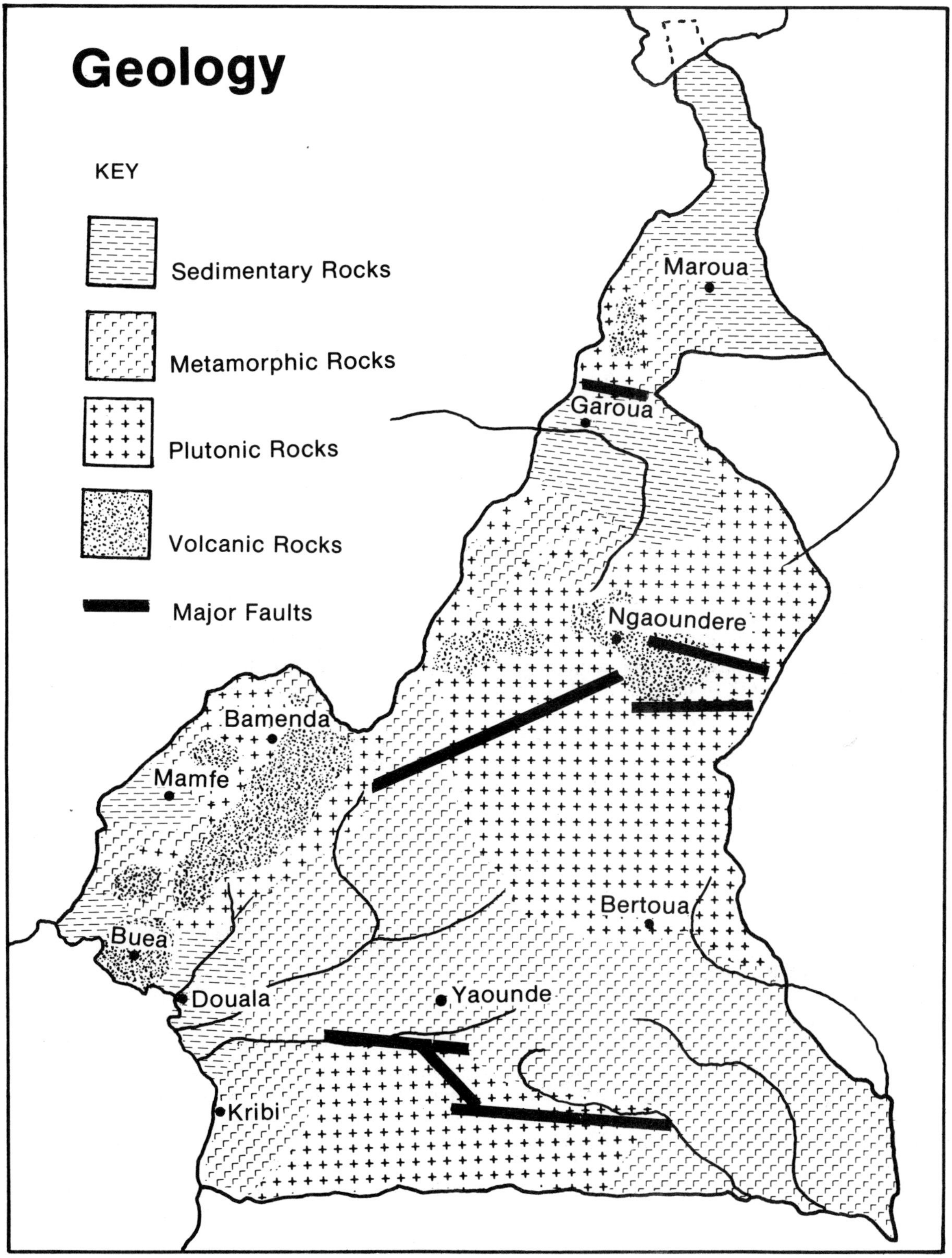
Geology
KEY
Sedimentary Rocks
Metamorphic Rocks
Plutonic Rocks
Volcanic Rocks
Major Faults
Maroua
Garoua
Ngaoundere
Bamenda
Mamfe
Buea
Douala
Yaounde
Bertoua
Kribi

CHAPTER 4:

CLIMATE

The Republic of Cameroon is found completely in the inter-tropical zone but the climate is, however, not uniform. Different climatic elements account for this lack of uniformity.

THE MAIN CLIMATIC ELEMENTS

1. LATITUDINAL EXTENT.

Cameroon extends from latitude 2° to 13° North of the Equator. This location gives the country almost all the characteristics of the inter-tropical climates which include generally hot, moist and dry conditions. The equatorial climate is found in the south up to latitude 6° N and the tropical climate between latitudes 6° and 13° N. One would have expected to find a zonal arrangement of the intertropical climatic domains on the territory from south to north if local factors such as proximity to the sea, continentality, and altitude did not interfere. These local factors have introduced climatic differences between neighbouring areas. A clear example can be noticed with the two close towns of Buea and Limbe. Buea, located at a higher altitude than Limbe, has a cool and wet climate while Limbe has a hot and wet climate.

2. TEMPERATURES:

Cameroon, like most countries in the intertropical zone, is a hot country with its average yearly temperatures varying from 20° to 28° C. On the whole, the temperatures and temperature ranges increase as one moves from the south towards the north and from the coast to the hinterland. However, altitude exercises much influence in lowering the temperatures of certain areas as can be noticed with these two towns — Yaounde (1,120m) with 23.5° C is situated on hills while Garoua on the Benue depression has an average temperature of 28° C. In the southern part of the country, temperatures remain relatively constant. The temperature ranges in the north are much greater than in the south.

As regards the rate of insolation, it is much greater in the north where the climate is hotter and drier than in the south: 1,023 hours of insolation per year at Douala, 1,841 hours at Yaounde and 2,969 hours at Garoua. The reduced insolation in the south is caused by more cloudy skies, greater precipitation and higher relative humidity.

3. PRECIPITATION:

Generally, rainfall in Cameroon reduces from the coast towards the north and interior of the country. Highland areas, however, receive more rainfall than areas of low altitude. Douala receives 4,016mm of rainfall annually, Yaounde 1,596mm, Lomie 1,654mm, Bamenda 2,596mm, Garoua 1,000mm, Kousseri 630mm. The influence of local factors such as continentality, relief, and air masses on these rainfall figures is worth noting.

Relative humidity also varies in the same order as rainfall, being greater at the coast than in the hinterland.

As regards the seasons, there are four distinct seasons in the southern and central regions — the long dry season at the beginning of the year, the long wet season from September to December, the short dry season in August and the short wet season between the months of March and June. The north, which begins from the Adamawa to Lake Chad, has two distinct seasons: a dry season from November to April and a wet season from May to October.

4. AIR MASSES:

The anticyclone of the Azores in the northern hemisphere and that of St. Helena in the southern hemisphere control the flow of air masses on the whole length of the Cameroon territory. Two air masses from these high pressure centres converge in a low pressure zone often termed the Intertropical Convergence Zone (ITCZ). The convergence zone has the nature of a front, which is often termed the Intertropical Front.

The shifting of this Intertropical Front following the movement of the pressure centres as a result of the position of the sun controls the cycle of the seasons. The type of season experienced in Cameroon depends essentially on the type of dom

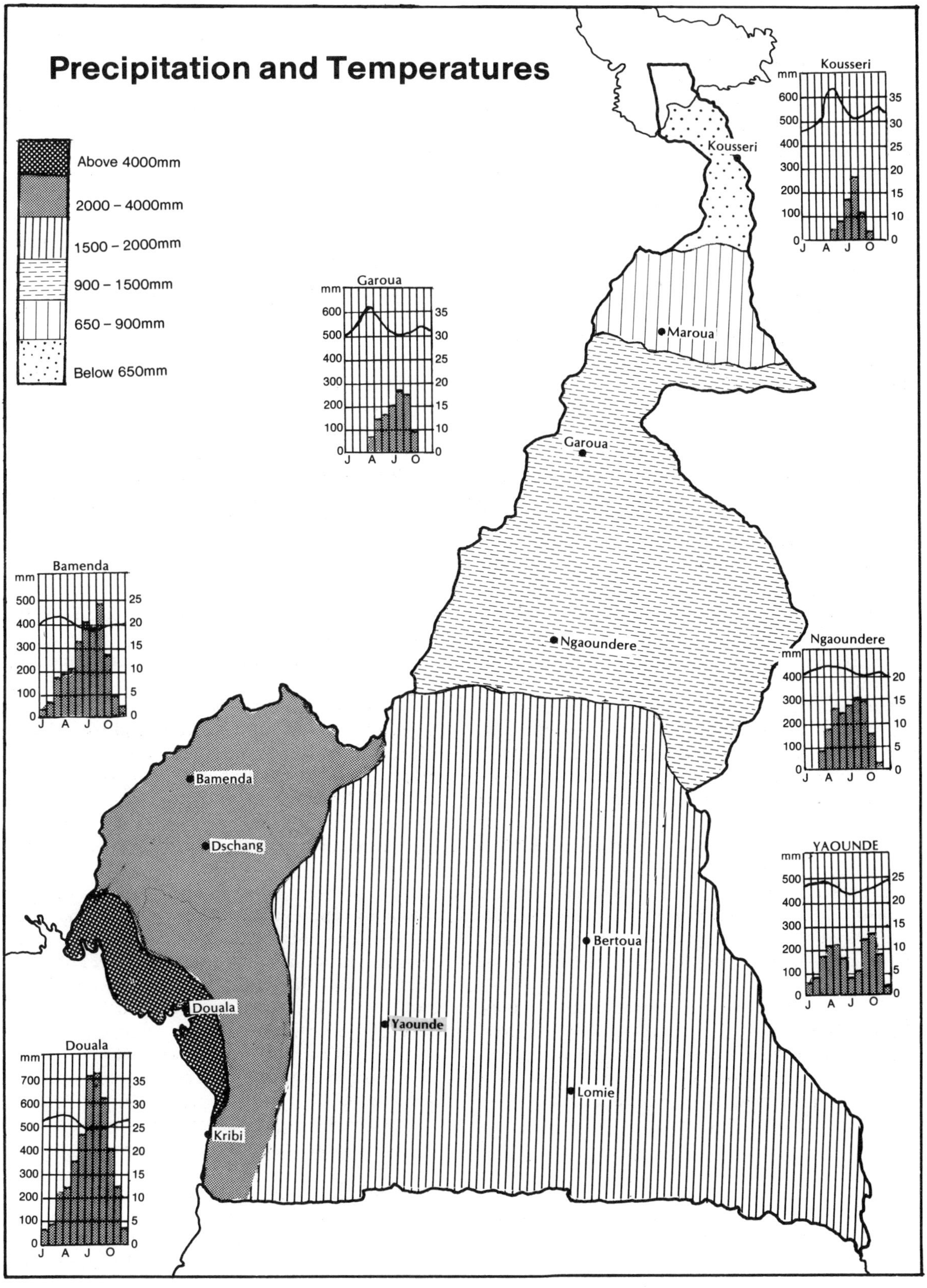

Precipitation and Temperatures
Above 4000mm
2000 – 4000mm
1500 – 2000mm
900 – 1500mm
650 – 900mm
Below 650mm
Kousseri
Maroua
Garoua
Ngaoundere
Bamenda
Dschang
Douala
Kribi
Yaounde
Bertoua
Lomie
YAOUNDE
mm
J A J O

inant Trade Winds blowing at the time.

Trade Winds are regular winds which flow from the two main pressure centres mentioned above. From the anticyclone of the Azores flows the *Harmattan,* and from the St. Helena anticyclone flows the *Monsoon.* The characteristics of these two winds are very contrasting and depend mainly on their source region, whether they are from the maritime south or desert north.

A. THE SITUATION IN THE DRY SEASON:

The Harmattan is the name given to the North-East Trade Winds which are hot and dry because of their journey over the hot and dry Sahara desert. These winds are also very stable. They begin to blow across the territory from October till the month of June.

In January, the St. Helena anticyclone is very far in the south. That of the Azores is enforced and the Harmattan becomes stronger than the Monsoon so that the Intertropical Front is pushed more southerly and situated around latitude 5° N. The whole of north Cameroon is invaded by the Harmattan, thus bringing about the dry season. The Monsoon winds at this time are very weak.

The effects of the Harmattan are very drastic in the north but become less severe as one moves southwards. Because this wind blows over the Sahara desert, it carries along fine particles of sand that fill the atmosphere thereby leading to poor visibility. Some small streams dry up completely while many big rivers reduce in volume. The vegetation turns brown and is scorched at certain places due to very low atmospheric humidity. Not only do rivers and vegetation suffer but human beings as well. People's lips, feet and skin crack; passengers on unpaved roads easily become bathed in dust especially in the north where few people at this time of the year look clean and fresh; the days are usually very hot while the nights are very cold.

In the south, especially at the coastal areas, these adverse conditions are greatly reduced because the Monsoon Winds, though weak at this time, occasionally provide some showers of rain. The atmospheric humidity in this area is also fairly higher than in the north.

B. THE SITUATION IN THE WET SEASON:

In July (August) the Intertropical Front is above Lake Chad and the St. Helena anticyclone is found practically on the Equator. The Monsoon Winds invade the south of Cameroon and this is the rainy season. The winds from the St. Helena anticyclone blow first in a south-easterly direction and as they cross the Equator they are deflected to the east (s.w. direction) by the effect of the *coriolis force* exercised on the right of the movement in the northern hemisphere. The air mass is thus deflected in a south-westerly direction and it then adopts the name *Monsoon.* The Monsoon is very warm and humid due to its movement over the Atlantic Ocean. It is unstable and susceptible to bringing heavy rains especially when the ascending movement is caused by relief. This explains the heavy rains on the sea-facing slopes of the Cameroon Mountain (Debundscha 10,000mm).

CLIMATIC REGIONS IN CAMEROON

The whole territory of Cameroon can broadly be divided into two climatic domains, namely:

a) The Equatorial Climate domain

b) The Tropical Climate domain

1. The Equatorial Climate Domain:

The southern part of the country extending from latitude 2° to 6° N experiences the equatorial climate. This climate can be sub-divided into two:

a) The *Guinea type:* It starts from the coast at Kribi and covers the whole of the Southern Plateau. It is characterised by its four seasons in a year (2 rainy and 2 dry) and abundant rains ranging from 1,500 to 2,000mm. It has double rainfall maxima e.g. Yaounde, whose first maximum is in September (long rainy season) and the other in March-April (short rainy season); the first minimum is around December-January (long dry season) and the second in July-August (short dry season). This type of climate is characterised by high and fairly constant temperatures, 25° C on the average.

b) The *Cameroon type:* This type of climate is found on the south-western coast at the neighbourhood of Mt. Cameroon and extends right down to the mouth of the Sanaga River and on the Western and Bamenda highlands. It is quite hot and humid. It is also characterised by the existence of only two seasons — one wet season of about eight months during which time rains are abundant throughout, and a comparatively short dry season. This type can further be divided into:

i) The *Maritime Cameroon type* which reigns at the coastal region and extends down to the mouth of the Nyong River. The sea-facing slopes of the Cameroon Mountain which receive the Monsoon Winds at right angle are even reckoned with as the second wettest place in the world with Debundscha (10,000mm of rainfall) coming only after Chirrapunji in India. It also has high and constant temperatures, and of course, high atmospheric humidity.

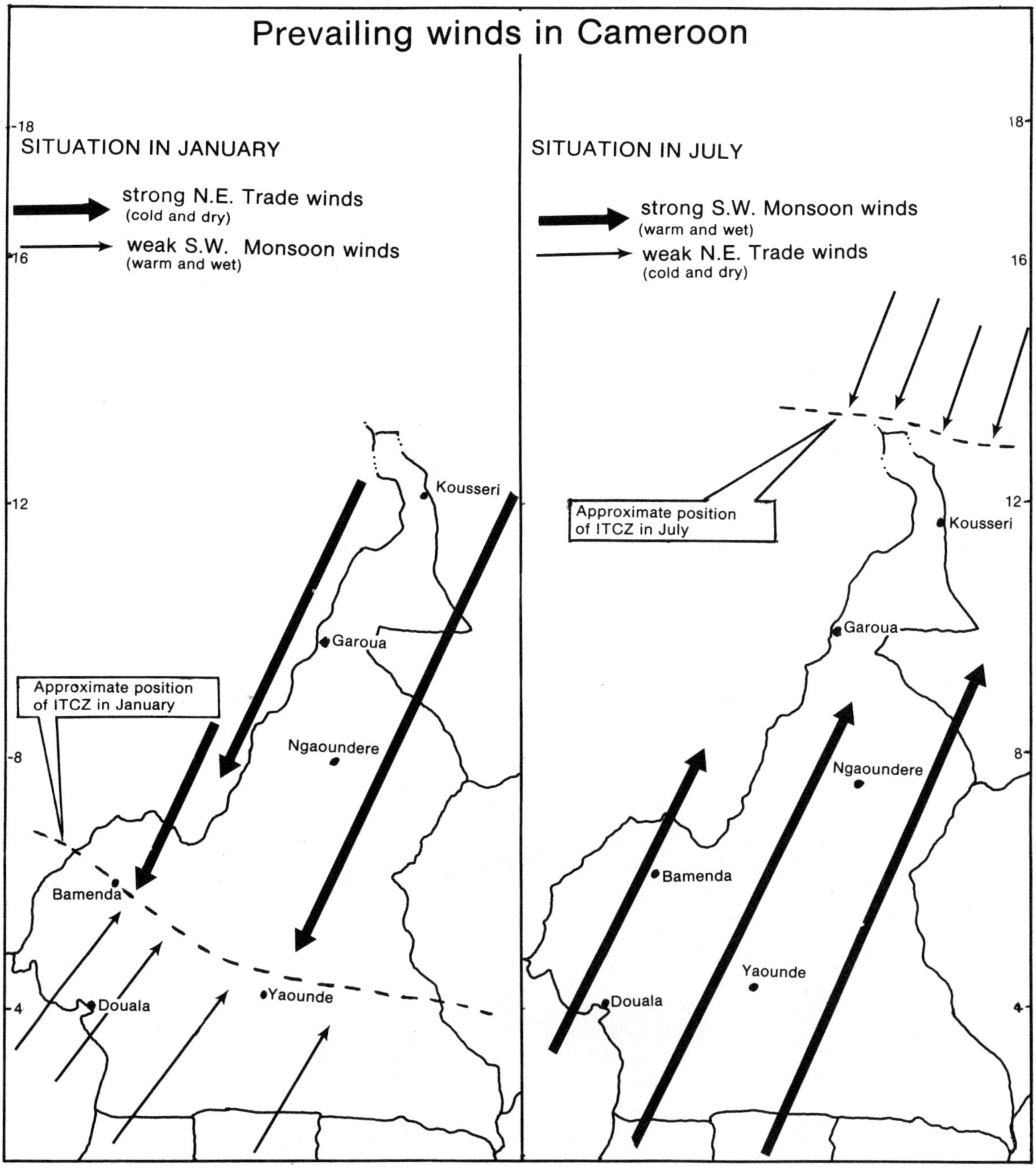
Prevailing winds in Cameroon
SITUATION IN JANUARY
strong N.E. Trade winds
(cold and dry)
weak S.W. Monsoon winds
(warm and wet)
Approximate position of ITCZ in January
Kousseri
Garoua
Ngaoundere
Bamenda
Douala
Yaounde
18
16
12
8
4
SITUATION IN JULY
strong S.W. Monsoon winds
(warm and wet)
weak N.E. Trade winds
(cold and dry)
Approximate position of ITCZ in July
Kousseri
Garoua
Ngaoundere
Bamenda
Yaounde
Douala

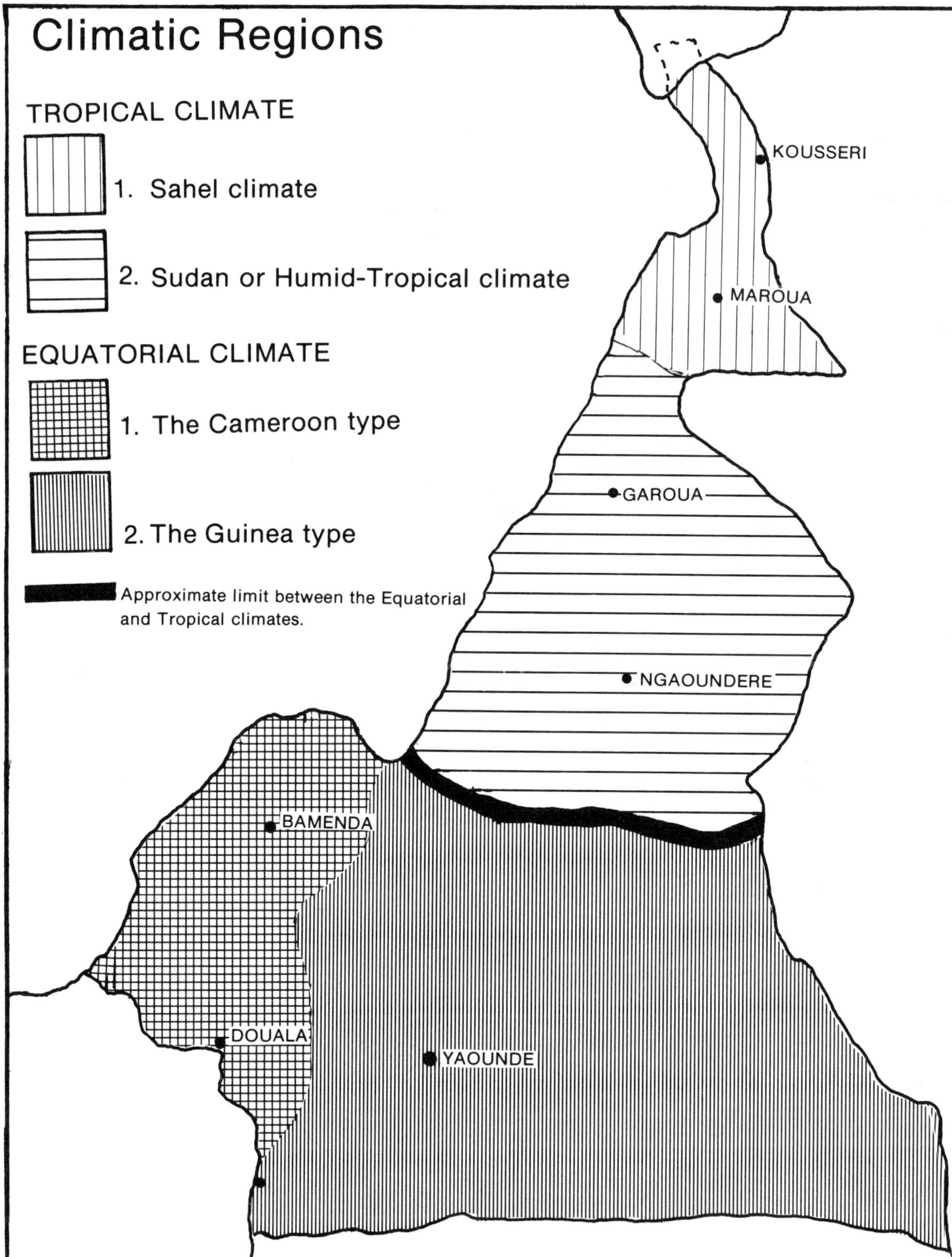
Climatic Regions
TROPICAL CLIMATE
1. Sahel climate
2. Sudan or Humid-Tropical climate
EQUATORIAL CLIMATE
1. The Cameroon type
2. The Guinea type
Approximate limit between the Equatorial and Tropical climates.
KOUSSERI
MAROUA
GAROUA
NGAOUNDERE
BAMENDA
DOUALA
YAOUNDE

ii) The *mountain Cameroon type* — Found on the Western Highlands; it is marked by fresh temperatures which, in the past, always attracted colonialists to establish their settlements in such places as Buea, Dschang and Bamenda.

The Cameroon type of climate degenerates gradually into the tropical Sudan climate as one moves northwards.

2. The Tropical Climate Domain:

The domain of the tropical climate can be divided into two:

i) The Sudan or Humid Tropical Climate

ii) The Sahel Climate.

a) *The Sudan or Humid Tropical Climate:* The area covered by this type of climate extends from latitude 6° to 10° N. It is characterised by two seasons — a rainy season which lasts seven months and a dry season of about five months. On the Adamawa range, however, rains are heavier than elsewhere in the north because of the influence of relief (Ngaoundere 1,500mm), and these rains are often accompanied by great storms. They last from March to November.

Temperatures are also fresh, 21°C on the average, but the average annual ranges (6°4') are often greater than in the case of the Cameroon type of climate.

Rainfall is generally low and the rainy season often lasts six months. At the start of the rainy season the rains are usually torrential and accompanied by tornadoes.

b) *The Sahel Climate:* The domain of this climate starts from the north of the Benue basin and covers the plains of Mayo-Danay, the Diamare and the Mandara Mountains. It is characterised by very low precipitation, usually below 900mm and an increased length of the dry season which lasts at least seven months. The low atmospheric humidity increases the average annual temperature ranges (7° 7'C) and the level of dryness.

The Mandara Mountains present a slight difference from the rest of the area due to its altitudinal situation. This results in fresher and more humid conditions (Mokolo 967mm) though the duration of the dry season still remains long.

In the Chad Plain, north of latitude 11° N, arid conditions are quite intense (Kousseri 630mm) and the rainy season lasts barely three months. The rate of evaporation in this region is equally very high.

CHAPTER 5:

VEGETATION

It would be realized that the major vegetation zones follow the climatic regions because humidity is a very important factor in the development of vegetation. However, this correspondence is distorted where man, soils and relief intervene. Two major vegetation zones can be distinguished in Cameroon though the change from one to the other is usually very gradual.

They are:

a) The Dense Equatorial Forest
b) The Tropical Grassland

1. THE DENSE EQUATORIAL FOREST:

This covers the Coastal Lowlands and the southern part of the Southern Plateau. It can be subdivided into two types:

a) **The Mangrove Forest:** This forest occupies the two swampy regions at the coast — the region around Rio-del-Rey and the Douala Basin.

Mangrove Forest on the coastal lowlands of Cameroon.

The region around Rio-del-Rey extends from the Akwa Yafe river in the west to the foot of the Cameroon Mountain, just east of Bamusso and inland as far as to Isangele, and then stretching down to the coast. This region is characterised by swamps, creeks and estuaries. In the very swampy zones raffia-palms can be found.

The second zone starts from Bimbia through Tiko to Modeka and down to the whole of the Douala Basin. The trees in this region develop very long roots which plunge from the stem down into the swamps. It is a region highly infested by mosquitoes.

The mangrove forest is sometimes known as mangrove swamps, tidal forest, or simply as mangrove. The most important constituents of the mangrove forest are the *Rhizophoria* and *Avicennia,* but other tree types are also to be found. Numerous breathing roots project from the mud, those of the Avicennia generally grow straight up. These breathing roots serve for gaseous exchange between the atmosphere and the plant parts beneath mud and water.

b) **The Rain Forest:** This is variously known as Equatorial Rain Forest, Tropical Rain Forest or Wet Evergreen Forest. This evergreen forest lies inland just after the mangrove forest. It is situated between altitudes 200m and 800m and covers a very large part of the Southern Plateau.

This forest has been subjected to serious human attack for lumbering, farms and urban expansion. Where these exploitations have taken place, a type of *secondary forest* may then replace the primary forest as seen in the surroundings of Yaounde.

The primary forest in its unhampered form can still be found in the south-eastern sector of the country, due to inaccessibility and very little human habitation. The forest is immense, luxuriant and has a continuous canopy of leaves. It is dark and damp inside and has a very thin undergrowth because little or no sunlight ever reaches the underneath.

Throughout the area, temperature, rainfall and humidity are high, not only in yearly total, but all and every day. Seasonal and diurnal fluctuations are much less than other zones. Trees in this forest develop buttresses. These buttresses are large

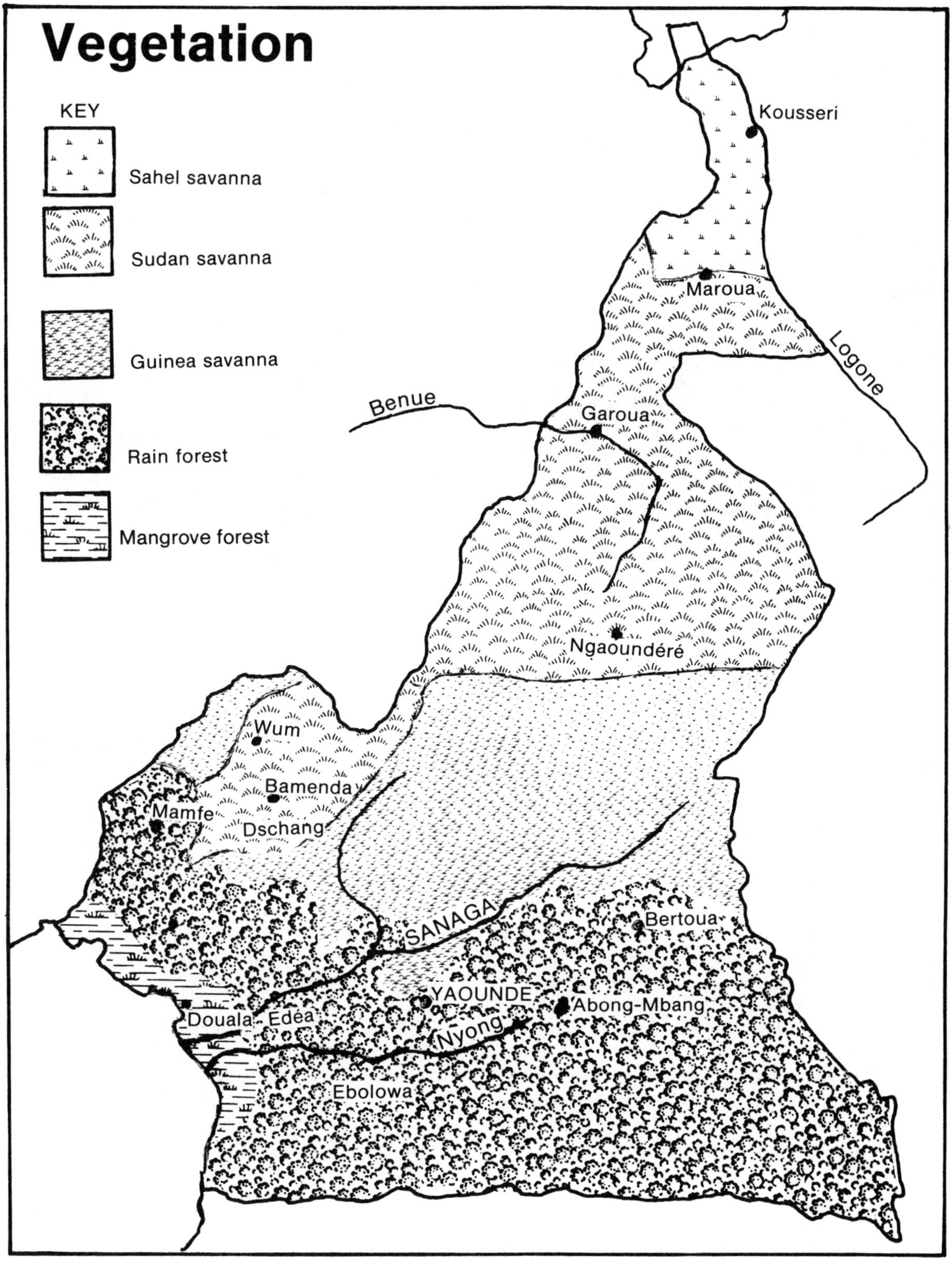
Vegetation
KEY
Sahel savanna
Sudan savanna
Guinea savanna
Rain forest
Mangrove forest
Kousseri
Maroua
Logone
Benue
Garoua
Ngaoundéré
Wum
Bamenda
Mamfe
Dschang
SANAGA
Bertoua
YAOUNDE
Abong-Mbang
Douala
Edea
Nyong
Ebolowa

Rain Forest in the southern part of Cameroon.

extensions of trunks and roots at and around the bases of the trees often reaching a height of ten metres up to the trunk. They are of varying shapes and sizes. Some are plank-like, closed or open at the base. While the forest as a whole is described as evergreen, there are some trees which shed their foliage at one or other period of the year.

It contains trees of economic importance such as iroko, mahogany, obeche, ebony and many others. Remnants of this primary forest are also found in such less accessible zones as the Djerem and Mbam depressions, the accidented slopes of Mounts Febe and Yeye in Yaounde. In some areas, for example, the Widikum and Bafang slopes where there are a lot of palm trees, the forest looks very much like the primary forest but the fact is that some of these palms were planted by human beings. Most of them are wild and grow on their own.

The rain forest is the home for many animals: elephants, monkeys, chimpanzees and many rodents. As one moves north from the south the forest gives way gradually to savanna.

2. THE TROPICAL GRASSLAND:

There is no precise distinction between the forest and grassland zones. The change is gradual. The areas covered by this grassland correspond to the two domains of the tropical climate. The general term of Savanna is given to this grassland vegetation which covers the northern part of Cameroon. The appearance of this savanna changes as the annual precipitation diminishes northwards and also with altitude, so that one has the problem of exactly describing what the natural vegetation is. Obviously, the smaller the amount of precipitation, the less capable are such areas to support a luxuriant vegetation cover.

Moreover, the vegetation of many areas has been modified by human activities and bush fires for years, and it is believed that some of these areas in the south supported much denser woodland in the past than exists today. The former vegetation has been replaced by some type of *derived savanna.*

Three main varieties of this savanna can be distinguished, namely: the Guinea Savanna, Sudan Savanna, and Sahel Savanna.

This division into three types is a useful conventional classification, but it tends to exaggerate the sharpness of the transition from one type to the other. As mentioned earlier, the transition is in fact gradual, sometimes complicated and often very difficult to describe.

a) **The Guinea Savanna:** This vegetation lies immediately north of the rain forest and it is a mixture of high savanna and trees. The trees form a thick woodland and the Savanna presents a grassy undergrowth. More to the north, the trees give way to grass. To the south one meets trees which belong to the rain forest. The southern slopes of the Adamawa in particular and the whole of the Adamawa in general are covered with

Guinea Savanna.

Sudan Savanna.

this vegetation. Bush fires in the dry season applied by cattle rearers, and other human activities help to savannise this area thereby extending the grassland southwards.

b) **The Sudan Savanna:** This type of grassland covers the Western Highlands, the Benue Depression, the Diamare plain and the Mandara mountains. It is a type of wooded savanna whose shrubs shed off their leaves in the dry season in order to withstand the bush fires and dry conditions. They also have thick barks for the same purpose.

It is this savanna that gave rise to the term *Grassfields* applied to what was formerly known as the Bamenda Province. This vegetation when young provides the beauty of the hills and cattle rearers like it. This is the case with the grazing lands at Nkambe, Wum and Fundong. Standing out very distinct above the grass are stunted trees, which are well adapted to the harsh climatic conditions of the dry season.

Another interesting type of vegetation that occurs in this Sudan Savanna zone is the raffia palm bush, which is limited to the valleys and depressions. Such palms are common in the North-West Province and the Bamileke lands.

c) **The Sahel Savanna:** With the drastic decrease in precipitation as one approaches the far north of the country, the vegetation suffers from drought. The savanna in this region becomes very low and scorched by the heat of the sun. The grass degenerates more and more until it ends up in small patches of grass around the fringes of the Sahara desert. These harsh conditions are met in the extreme north of the country towards Lake Chad. However, the land bordering Lake Chad is swampy and can support a taller and thicker growth of grass.

As human activities increase, the destruction of vegetation is likely to render the land completely bare in the north in future. It is for this reason that the government has in recent years, embarked on an afforestation programme in the north termed *Operation Green Sahel.*

Sahel Savanna.

CHAPTER 6

SOILS

Soil is the loose material which forms the upper layer of the earth's mantle. It is a combination of mineral and organic matter, air and water. It is that portion of the *regolith** that supports the growth of plants. About one-half of the total volume of good quality surface soil is a mixture of disintegrated and decomposed rock (mineral matter) and the remains of animal and plant life (organic matter). The remaining half consists of pore spaces, where air and water circulate.

The Soil Profile

If we were to dig a trench, or cut a section downward through the soil, we would see that its walls consisted of a series of horizontal layers. The individual layers are called *horizons,* and collectively make up the soil profile. The *soil profile,* therefore, is a vertical section through a soil showing its succession of horizons and the underlying parent material. Every well-developed, undisturbed soil has its own distinctive soil horizons. The boundaries between soil horizons may be very sharp, or the horizons may blend gradually from one to another. In every mature soil, three basic horizons are identified and from top to bottom are *A, B,* and *C,* respectively. The *A* and *B* horizons may be subdivided further.

The uppermost layer in a soil profile, the A horizon, is often called the *surface soil.* It is that part of the soil with the greatest biological activity and is therefore the horizon where organic matter is most plentiful. Since falling rain reaches this horizon first, soluble materials and tiny particles such as clay are *leached* (washed out) by percolating water.

Lying immediately below the surface soil is the B horizon, or *subsoil.* It is in this zone that much of the material removed from the *A* horizon is deposited. Because of this, the *B* horizon is referred to as the *zone of accumulation.* The *A* and *B* horizons together constitute the *solum,* or "true soil". It is in the solum that the soil-forming processes are active. Living roots and other plant and animal life are largely confined here.

Below the solum is the *C* horizon, a layer characterised by partially altered parent material. This is the layer that has not been sufficiently enriched in humus and has not had important illuviation of clay or humus.

Soils are formed as a result of the interplay of many factors which are interdependent in their functions.

FACTORS INVOLVED IN SOIL FORMATION

Parent Material

Soil, for the most part, is parent material that has been altered by various processes, and its ultimate nature depends upon the original composition of the parent material. The parent material may either be the underlying bedrock or a layer of unconsolidated deposits. Soils formed on bedrock are termed *residual soils,* while those developed on unconsolidated deposits are called *transported soils.*

The nature of the parent material influences soils in two ways. First, the mineral composition of the parent material determines the rate of weathering and thus the rate of soil formation. Also, unconsolidated deposits (alluvial deposits, wind-blown sand, etc.) will easily form soils because they are already weathered and do not need to go through the process of disintegration anymore. Second, the mineral composition of a parent rock will affect the fertility and texture of the soil. Some materials are rich in lime, others in iron, others in quartz, and others in feldspar. All these elements produce different kinds of soil.

Climate

Climate is believed to be the most important control of soil formation. It determines the rate at which weathering, leaching and organic decomposition take place in the different climatic regions of the world. It determines whether chemical or mechanical weathering will predominate. In the equatorial region with hot and wet conditions these processes proceed much faster than in the cool, dry temperate regions. Finally, climatic

---

* *Regolith: the layer of rock and mineral fragments produced by weathering.*

conditions are an important determinant of the type of vegetation cover and animal life which, in turn, affect soil type.

Plants and Animals

Plants and animals essentially provide organic materials to the soil. Some soils are composed almost entirely of organic matter, while others (e.g. desert soils) may contain only a small portion of organic matter.

Decaying plants and animals produce humus which increase soil fertility. Plants also contribute to soil acidity. Trees and grasses protect soil from sun's rays and erosion.

Small animals, micro-organisms including fungi, bacteria and single-celled protozoa, play an active role in the decay of plant and animal remains. The end product is humus. Living creatures within the soil such as earthworms and termites mix the various soil horizons. Earthworms, in particular, feed on organic matter which they pass out and enrich the lower layers of the soil. By making passage holes they also aid the passage of water and air through the soil.

Topography (Slope)

Slope and topography determine how much water percolates into the soil. In other words, they determine the amount of erosion and water content of soil. On steep slopes soils are not easily formed because weathered materials are quickly washed away. Leaching is also limited on steep slopes.

Time

Given a type of parent material under certain conditions, the time during which the parent material is exposed to those conditions matters for a type of soil to be formed. The longer a soil has been forming, the thicker it becomes and the more transformed the parent material from which the soil is formed.

It is important to note that none of these factors can be solely responsible for a soil being as it is. The combined influence of parent material, time, climate, plants and animals, and topography determine the type of soil formed.

We can attempt a broad classification of the soils of Cameroon as follows:

a) Zonal Soils — 1) The soils of the Equatorial forest domain.
2) The soils of the tropical domain.

b) Azonal soils: alluvial and volcanic soils.

A) ZONAL SOILS:

1) The Ferrallitic soils of the Equatorial forest:

The typical soil which covers more than four-fifths of the forest region in the south is the ferrallitic soil. The ferrallitic soils are red or red-brown in colour. They are essentially clayey, resulting from the decomposition of either crystalline rocks such as granite, gneiss, schists and michachists, or from the sedimentary rocks of the Coastal Lowland region.

In the areas of prolonged humid climate, they appear to have a brownish-yellow colour. From Bertoua through Nanga-Eboko to Yaounde the soils are reddish in colour. Generally, they are deep, often about ten metres. The silica content of these soils is usually washed out (leached) by percolating water. They usually do not support farming for a long time because they quickly get exhausted and thereby need very careful methods of farming with a lot of manure application. It should be stressed, however, that it is the quick exhaustion of the soil as a result of the hot and humid conditions facilitating quick decomposition of humus, that makes agricultural practices difficult, rather than the poor nature of the soil as is often implied from the writings of many geographers. This problem is partially solved by the practice of shifting-field cultivation, where the farmers of the southern part of Cameroon easily move from an area that has been exhausted to another uncultivated area.

Two simultaneous processes, eluviation and illuviation, are common in the soils of this southern region. *Eluviation* consists of the downward transport of fine particles from an upper soil horizon (usually the lower part of the A horizon. *Illuviation,* on the other hand, is the accumulation of materials in a lower horizon, brought down from the upper horizon.

A PROFILE OF THE EQUATORIAL FOREST SOILS

The A horizon can be subdivided into A°, A^1, and A^2.

A° Debris: This portion of the A horizon is very thin and composed of debris of leaves and animal materials. Though located in the area of the thickest forest in the world, the debris decomposes very rapidly because of constant heat and humidity. Over ninety percent of it is transformed within a year.

A¹ Humus: The debris of the above horizon decomposes to increase this layer.

It is usually very acidic with a pH of 5.

A² Eluvial Zone: It is exactly at this horizon that the problem of tropical soils is found. Although this horizon is deep, it is often deprived of its hydroxides by leaching. The colour is almost light yellow.

B. Illuvial Zone: The B horizon is illuvial and has a reddish colour with ferrogenous concretions. It contains a lot of iron and aluminium oxides leached from the above horizon. The horizon is also clayey and rich in Kaolinite. Quite often this layer is clear and variegated due to the presence of reddish and whitish spots of iron and Kaolin.

C. Parent Rock: This is the zone made up of the weathered parent rock.

This profile is only the ideal type because for one reason or the other there are variations in different localities.

Also found in isolated areas are what we call *cuirasses*. They are indurations which result from the enrichment of the superficial soil layers in iron and aluminium hydroxides. These cuirasses are what we at times term "hard-pan".

2) The soils of the tropical domain:

We must note here that unlike in the equatorial domain, where heavy rains lasting for a fairly long period favour rock decomposition, the tropical domain has a short rainy season whose rainfall is not as abundant as in the preceding case. Consequently, the period of rock decomposition is shortened and the soil is not as thick or deep as in the equatorial domain.

North of the country, there is a great variety of soils. In the transitional zone occupied by the Guinea Savanna we still find the ferrallitic soils, and in the extreme north with the driest conditions we find brown soils, but the typical soil of the tropical domain can be said to be the sandy-clay ferruginous soil otherwise known as *laterite*.

In this tropical domain where two almost equal seasons alternate (dry and wet seasons), the situation is different from that in the humid equatorial forest region.

During the rainy season, soluble compounds are leached by percolating water (especially silica, which is quickly leached in the presence of abundant moisture and high temperatures); but during the dry season this same water is drawn to the surface again by capillary action before being evaporated into the atmosphere by hot and dry air.

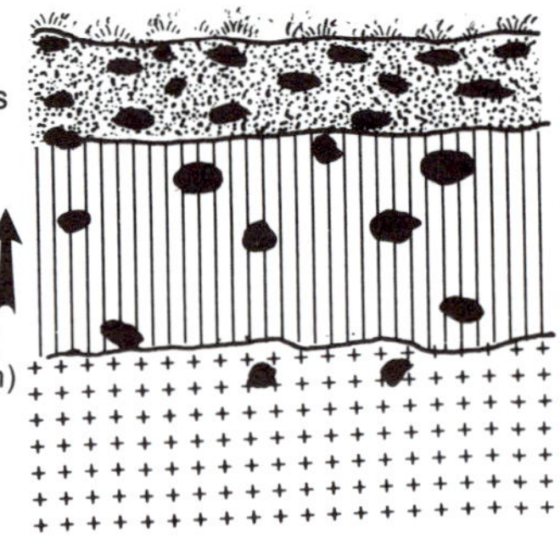

A PROFILE OF THE TROPICAL LATERITIC SOILS

As evaporation takes place, the already dissolved compounds are again deposited near the surface, but this time in a precipitated form. These precipitated compounds continue to accumulate so that with time small and strong concretions of iron and aluminium oxides are formed. Sometimes they appear quite near the surface of the ground and sometimes at a depth of several feet with a thin soil cover over them.

When an area contains enormous quantities of these concretions, they can, after some period and under certain conditions harden to form a type of very hard rock-like substance which are referred to as "hardpan" or "cuirasse." Such rock-like substances are usually quarried and used for several purposes because their iron, bauxite and manganese content are very high.

These *lateritic soils* are typical of the Sudan climate domain and are brown or brownish-red in colour. Their organic matter content is comparatively very low and cannot therefore support forest or thick vegetation. These soils are very important for the cultivation of two seasonal crops during the year. Areas with such soils are important for cereal cultivation, especially millet and groundnuts.

B) AZONAL SOILS:

Azonal soils are young soils which have not been sufficiently subjected to the effects of the soil-formation processes, agencies and influences to develop mature characteristics. These are immature soils and therefore do not have distinct horizontal layers, or horizons.

In Cameroon the two main types of azonal soils are volcanic and alluvial.

Volcanic and Alluvial Soils:

Besides the two domains discussed above, we can also mention that the volcanic regions of Cameroon, especially the one extending from the Cameroon mountain to the western highlands, are covered by black or brown soils which result from basaltic rocks. These soils are extremely fertile and are responsible for the numerous agricultural plantations found in these regions.

Again, alluvial soils are found in some isolated zones especially in the plains, (particularly the coastal plains) the bottoms of the major valleys, such as, the Wouri valley, the Benue and Logone valleys, the Mbo and Ndop plains, etc. The flood plains of the Chad Basin are covered by black clays.

Still in the forested equatorial domain in localised areas, we find some azonal soils. On the steep slopes of massifs or inselbergs, such as Mt. Yeye and Mbankomo around Yaounde, a type of soil exists, which is commonly known as *Rankers* because it looks like the Rankers of the temperate zones. The rankers result from the physical and chemical weathering of the outcropping rock, and usually support some little vegetation at the tops of the rocky domes.

CHAPTER 7
HYDROGRAPHY

Cameroon has a dense network of rivers and most of these river courses are interrupted by falls and rapids.

Another common characteristic of the rivers is that their regimes are never regular. They reduce in size in the dry season and greatly increase in the rainy season. This is one of the aspects that distinguishes tropical rivers from rivers in the temperate zones.

Cameroon has one major watershed which is the Adamawa Plateau. Other minor ones include the Southern Plateau, the Kupe and Manenguba mountains, the Bambutus and Bamenda highlands.

From the Adamawa highlands, rivers flow to the north into Lake Chad; to the north-west into the Benue; to the south-west into the Atlantic ocean; to the south-east into the Ngoko and Sangha; and westward into Nigeria.

These rivers drain into four drainage basins, namely:

a) The Atlantic Basin,
b) The Congo Basin,
c) The Niger Basin,
d) The Chad Basin.

A) The Atlantic Basin:

The rivers of this basin can be divided into three groups:

i) *The Sanaga and its tributaries:*

The flow of the Sanaga river has been well recorded because of its importance for power generation. It measures 918 km. and covers a drainage basin of about 140,000 km^2. The upper course of this river bears the name of *Djerem* and takes its rise from the Adamawa highlands. The Djerem receives on its right flank other tributaries descending from the Adamawa, such as, the Vina of the South, the Meng and the Mekie. More to the south the Djerem meets the Lom. It is from the confluence of the Djerem and Lom that the Sanaga takes its name. From the confluence the Sanaga flows in a NE-SW direction for many kilometres before receiving another major tributary, the Mbam, which rises from the west of Tibati.

Up to the confluence, the Djerem tumbles down the Adamawa at many places in rapids and falls, but from there the Sanaga flows on a gentle bed till the Nachtigal falls are reached.

The Mbam meets the Sanaga at Nkolkosse and it is very important because it supplies not less than 37% of the Sanaga's debit at Edea. At Edea the Sanaga has an average debit of 2,300m^3 per second. It is not very navigable except downstream for the last 65 km from Edea.

ii) *The Rivers of the Western part of the Coast:*

The *Cross River* (600 km), also called the *Manyu* in its upper course drains the extreme south-western part of Cameroon. It appears to have attained its equilibrium profile in recent years because at certain parts it flows in a deep gorge. Its main affluents flow from the Rumpi and Bambutus mountains. It flows through the Mamfe basin and enters the Calabar Sate in Nigeria.

The Akwa Yafe, the Ndian, the Lokele and *Meme Rivers:*

These are small and short rivers which drain the extreme south-western part of the country, taking their rise from the Rumpi hills and the neighbouring highlands to the west. Their upper courses are full of rapids and falls. In their lower courses they break up into small *distributaries* before entering the sea because of alluvial deposits brought down from the hills. The Akwa Yafe forms the natural boundary line for the country in the west. All these rivers enter the sea through very large estuaries, especially that of the Ndian River which is known as Rio-del-Rey.

The Mungo River, which has a length of about 150 km, flows from the Rumpi mountains and increases its waters by receiving the streams descending from Mount Kupe and the Bakossi lands. On the sedimentary lands found below the Mungo falls, the river becomes navigable for about 100 km before entering the mangrove swamps. The course is completely split into very small and numerous channels (or distributaries) which all find their way into the sea at the Bay of Modeka. This river, for its greater length in the south, forms the boundary between the Littoral and South-West Provinces.

The Wouri River, which is called the Nkam at its upper course, has a length of 250 km and takes its rise from the Bamileke plateau. The floods of the Nkam are usually a menace to the inhabitants of the Mbo plain and the Kekem village. In the heart of the rainy season, the floods exceed this minor valley and occupy the large major valley so that one hardly knows where the river bed is situated.

After tumbling down from the Bamileke plateau, the Mungo flows on a gentle bed in the Mbo plain and down at Ekom it produces a fall of 80

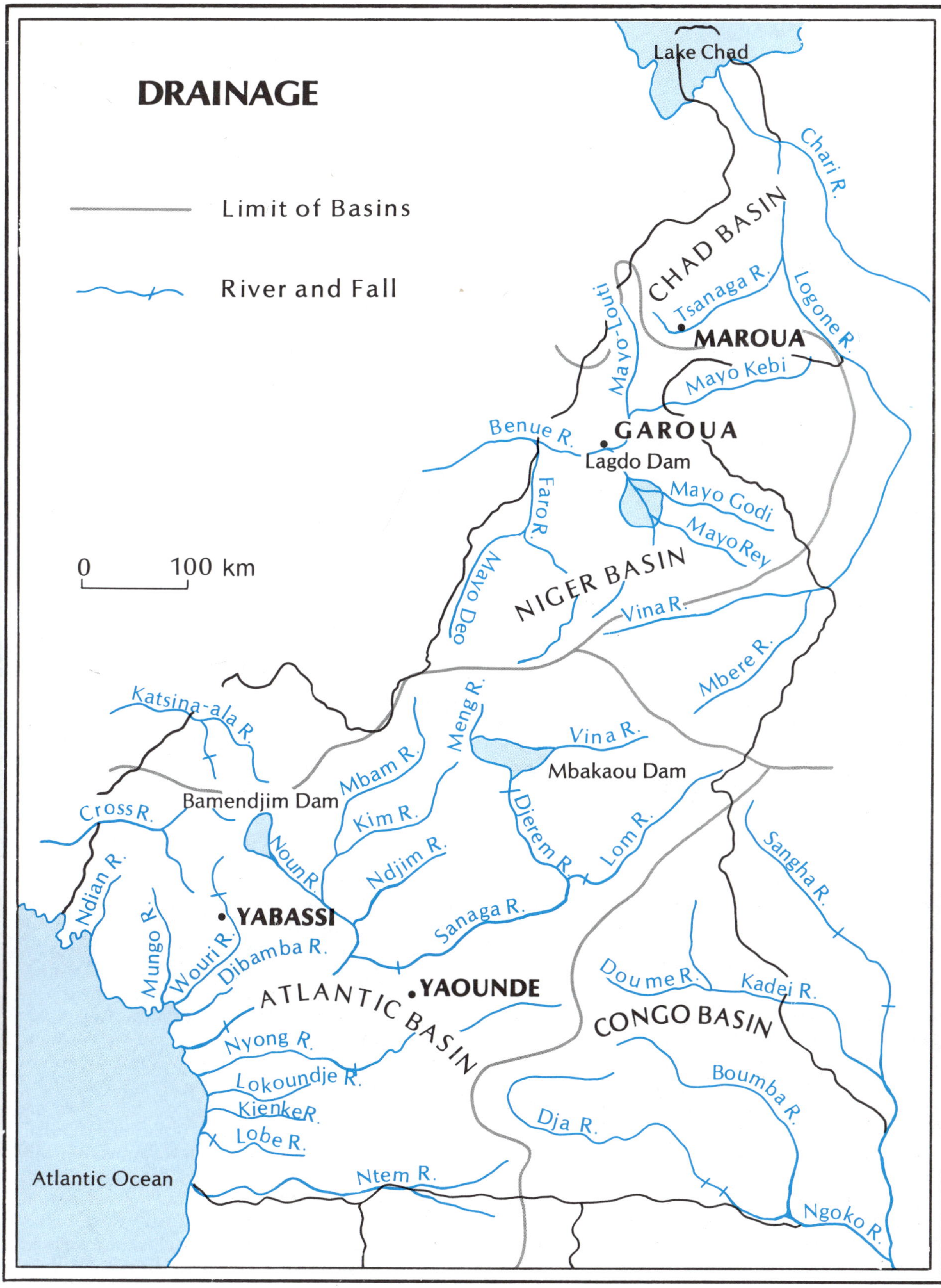
DRAINAGE
Limit of Basins
River and Fall
0 100 km
Lake Chad
Chari R.
CHAD BASIN
Tsanaga R.
Logone R.
Mayo-Louti
MAROUA
Mayo Kebi
Benue R.
GAROUA
Lagdo Dam
Faro R.
Mayo Godi
Mayo Rey
Mayo Deo
NIGER BASIN
Vina R.
Mbere R.
Katsina-ala R.
Meng R.
Vina R.
Mbakaou Dam
Mbam R.
Bamendjim Dam
Cross R.
Kim R.
Noun R.
Ndjim R.
Djerem R.
Lom R.
Sangha R.
Ndian R.
Mungo R.
YABASSI
Wouri R.
Dibamba R.
Sanaga R.
YAOUNDE
ATLANTIC BASIN
Doume R.
Kadei R.
CONGO BASIN
Nyong R.
Lokoundje R.
Kienke R.
Lobe R.
Boumba R.
Dja R.
Atlantic Ocean
Ntem R.
Ngoko R.

metres high. It receives two major tributaries. On the right flank this river receives the Dibombe which comes from the Nlonako hill at Nkongsamba and on the left flank it receives the Makoumbe which drains the western part of the Ndikinimiki plateau. From Yabassi to Douala it is navigable and at its estuary, where the Mungo and the Dibamba also enter the sea, there is a zone occupied by sand bars, mud and mangroves. It is from this estuary (Rio-dos-Cameroes: River of Prawns) that the country had its name.

The Dibamba: It is a very short river, measuring just about 150 km from its source at the Banen mountains situated to the south-west of Ndikinimiki. It serves mostly for the floating of timber logs.

iii) *The Rivers of the Southern part of the Coast:*

The Nyong River is about 800 kilometres long and originates from the central plateau east of Abong Mbang. It flows on a gentle slope especially between Mbalmayo and Makak. It also has a very large valley which is always submerged by floods during the rainy season. The gentle slope and large valley, are factors that favour the existence of marshes which are also a breeding ground for mosquitoes. This is exactly the situation along this river. Below the town of Mbalmayo, the course presents a series of falls and rapids for a distance of about three kilometres. From Dehane to the sea the river is navigable.

The Lokoundje which is some 185 kilometres long rises from the Nkolgam massif and flows through Lolodorf. Its course has many falls. It flows into the sea in an estuary.

The Kienke: It measures 100 km long and flows mostly on bare rocks. The slope of this river is very gentle but the course is interrupted by many rapids. It descends the Southern Plateau to the ocean in a series of rapids and the last one, four metres high, is just a few metres upstream from the town of Kribi.

The Lobe is 80 kilometres long, and like the Kienke it has many waterfalls because of the accidented relief. It is worth noting here that this area forms the western end of the Southern Plateau and rivers necessarily have to tumble down.

The Ntem (or Campo): This river measures about some 400 kilometres and takes its rise from Gabon. It has a drainage basin of 31,000 km^2. Around Maka the Ntem splits into two arms —the Bongola to the north and the Campo to the south. These arms encircle the Campo plain and then meet again before reaching its mouth. The river tumbles down in rapids and falls as it descends from the Plateau.

B) THE CONGO BASIN:

The rivers of this basin take their rise from the eastern part of the Adamawa in Cameroon but drain into River Sangha which is an important tributary of the Congo River. This river flows through Brazzaville and empties itself into the sea at Matadi in the Republic of Congo. The two main tributaries of the Sangha are the Kadei and Ngoko.

The Kadei River takes its rise from the south-east of Garoua-Boulai, flows southwards and then meets the Doume. From this confluence it flows east into the Central African Republic where it meets the Mbere to form the Sangha.

The Ngoko which forms the boundary between Cameroon and Congo for a distance of about 120 kilometres is formed of two rivers — the Dja and Boumba which both meet at Moloundou. The Boumba rises from west of Yokadouma and the Dja from the south-east of Abong Mbang. The Ngoko and the Sangha both form a navigable section from Moloundou to Brazzaville.

C) THE NIGER BASIN:

The Niger River is not a Cameroonian river but a big river in Africa emptying itself into the Atlantic ocean in Nigeria. The Niger has some of its affluents in Cameroon — the Benue and Katsina. The Katsina river is swelled by the waters of the Mentchum river.

The Benue River: The Benue river with a length of 1,400 kilometres, has only a section of 350 kilometres in Cameroon. It rises from the northern slopes of the Adamawa. Its main tributaries are Mayo Kebi and the Faro, and less important ones are Mayo Godi, Mayo Rey, Mayo Farda —all of them descending from the Adamawa. At its confluence with the Faro it flows out of the country to Nigeria. The Benue is void of falls and its course is navigable up to Garoua in the rainy season. It provides a short way from the ocean to Chad.

D) THE CHAD BASIN:

This is the least important of all the basins and formed mainly by the Logone River and its tributaries. The Logone is formed by the meeting of the Vina of the north (or Bini) and the Mbere which both rise from the north-east of Adamawa. From Bongor to Kousseri the Logone serves as a boundary between Cameroon and Chad. It is a river on a plain, dragging down a lot of floods in its course in the rainy season which result from the rains on the Adamawa. In the dry season these floods disappear just to allow some channels of water inserted between sheets of sand in its valley. In the rainy season the Logone is navigable for a greater part of its course.

CAPTURE OF RIVER LOGONE BY MAYO—KEBI

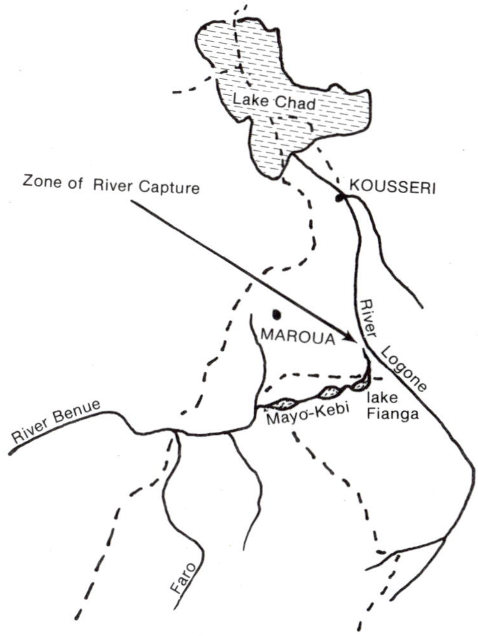

The floods of the rainy season are significant because of an important phenomenon they can produce in future, that of "river capture."

During the rainy season, and more especially in times of floods, water passes from the Logone valley over the narrow piece of land to the valley of the Mayo Kebi. Each time this happens its channel increases and the piece of land is lowered through the scraping force of erosion so that there is the possibility in the future of the Logone waters moving into the direction of Mayo Kebi.

More important even, is the fact that the Mayo Kebi is having a type of headward cutting into the narrow piece of land that separates the two valleys. Should this action continue, the Mayo Kebi will eventually flow into the course of the Logone and divert the waters of the Logone into its course.

It is for this reason that many geographers see that the future of Lake Chad is very bleak. Should the Mayo Kebi succeed to capture the waters of the Logone, Lake Chad would shrink and even disappear completely because its volume depends greatly on the Logone River.

CONCLUSION

Cameroon has rivers which belong to the equatorial and tropical zones and these rivers are generally regular in flow. They may reduce in volume in the dry season but still maintain their flow. Their basins are averagely wide.

The principal rivers of Cameroon take their rise from the Adamawa plateau. Apart from the Benue, the Logone and Chari, and the Wouri which are navigable, especially in the rainy season, the rest of the rivers are imbued with rapids and falls which hinder navigation.

On the other hand, these falls are of great importance as regards the generation of hydro-electric power. The Sanaga is a major example of a useful river for hydro-electric power generation. Two mighty hydro-electric stations are found on it, one at Edea and the other at Song-Loulou.

LAKES

Besides the numerous rivers in Cameroon, the territory is also endowed with a good number of lakes of varying shapes and sizes.

By 'lake' here, we are considering all the vast masses of water standing on the surface of the territory which are nearly or completely enclosed by land.

These lakes are as a result of several causes which range from natural to artificial. Each one can be studied following the causes which led to its formation.

a) Lakes created by volcanic activity:

Such lakes are caused by the accumulation of water in the craters of extinct volcanoes and are commonly known as *crater lakes*. They are often found at the tops of volcanic hills. The crater lakes of the territory include almost all the lakes on the volcanic range of Cameroon. The main ones are Lake Barombi in Kumba, Lake Baleng near Bafoussam, Lake Tison at Ngaoundere, Lakes Nyos, Oku, Awing, Wum and Befang in the North-West Province and the two beautiful Manenguba lakes in the South-West Province. There are many other ones which have not been mentioned. Such lakes are usually surrounded by very steep borders.

Volcanic eruptions can at times emit lava which flows down to solidify at the end of the flow channel, creating a type of dam that can cause the accumulation of water.

b) Lakes created by tectonic movements:

Tectonic movements within the crust can throw the surface into a series of undulations with large depressions and hills. Such depressions eventually get filled up with water to create what we generally term *tectonic* or *depressional lakes*. In Cameroon examples of such lakes are Lake Ejagham, south-

Lake Awing: a crater lake on the western highlands.

The Mefou Lake: an immense man-made lake.

west of Mamfe; Lake Ossa, west of Edea and Lake Chad in north Cameroon.

c) Lakes created by the accumulation of sand and silt:

Quite often on the Coastal Lowlands and on very flat basins, rivers deposit most of their load brought down from the hills. The load, which is composed mainly of sand or silt, may rise to block and hold back portions of water which form standing pools that can be called lakes, or more appropriately, *lagoons* when they appear at the coastal regions.

This is the situation with the Douala and Ndian basins lying on the coastal zone of Cameroon where numerous lagoons are found. Far into the hinterland the very low depressions can have marshes which contain such pools of water. They disappear or reduce in size in the dry season and increase with the arrival of floods from heavy rains. Such waters can be found on the vast plain of the north drained by the Logone river.

d) Artificial or Man-made Lakes:

Man has been induced by several reasons to build dams across rivers in order to hold back enormous volumes of water upstream. Some of these reasons include the generation of hydro-electricity; the supply of enough water for irrigation; the supply of drinking water and the creation of fishing and recreational centres. Examples of such artificial lakes include the Bamendjim Lake on the Noun River; the Mbakaou Lake on the Djerem river; the Edea Dam on the Sanaga River,the Mefou Lake on River Mefou in Yaounde.

On arrival at the town of Yaounde, one becomes alarmed at the presence of numerous lakes that cover some areas of the town. These lakes, especially the large ones, contain fish and all along, kids of the city are found with hooks and baits catching fish. On the big municipal lake, fishing is even done using dug-out canoes, but up till date lakes in this town have not been known to furnish fish of reasonable size.

Some of these lakes are also being used for resort. However, this use is fast losing ground because many hotels are increasingly being equipped with swimming pools.

The Lakes of Yaounde are obviously man-made. The aerial photographs of the town dating as far back as to 1951 show only wide valleys with practically no lakes. It was sufficient to dam carefully the course of the streams and raise the dam to any height then water will accumulate upstream. Some of these lakes in Yaounde are simply large pools of standing water which increase in size with the arrival of heavy floods and they contain a lot of rotting vegetation.

The case of the Mefou Lake found at the outskirts of Yaounde is of particular interest. It is the largest water mass of human creation around the Yaounde plateau. Unlike the other lakes of the town which needed unskilled labour for their creation, the Mefou dam employeed engineers for its construction.

LAKE CHAD

Lake Chad, which is found at the extreme north of the country, is situated at an altitude of about 240 metres. The lake is a remnant of a far larger inland sea which must have extended southward to the base of the Marghi and Mandara hills and far towards the Tupuri marsh. It is believed by some scholars that this sea discharged itself through the Benue into the Atlantic and that the connection was lost by subsidence.

The lake, which is shared by Cameroon, Chad, Niger and Nigeria, covers an area of 24,000 km^2 but the Republic of Cameroon owns only a small portion of it.

It is an inland drainage lake which receives fresh waters from the south and even though it loses much water through infiltration and evaporation, its waters are not very salty. The slightly salty waters are in the north where no fresh waters come in.

Even though Cameroon has only a small portion of this lake, its existence depends almost entirely on the waters that come from Cameroon. Its main sources of water are located in the south.

After meeting the Logone river, the Chari increases its volume and then flows into the lake. It is worth emphasizing that the Chari is the lake's principal supplier of water. Rushing down from the Bautchi plateau in Nigeria, is a small river which is clearly less important compared to the Chari river. Additionally, one could count the many seasonal streams which flow down from the Mandara Mountains. They are useful only during the rainy season in increasing the volume of the lake.

Although this lake is extensive, it is very shallow and varies considerably in size between the dry and wet seasons. The variations in size are not only noticeable in recent times, but there is enough evidence that they were even more serious in historic times, especially between the pluvial and interpluvial periods which occurred as a result of climatic changes in this part of the world. For example, at a certain time Bahr-el-Ghazal to the north-east of the lake in the Republic of Chad, used to be one of its major sources of water, especially during the humid periods. During the second half of the 19th century when drought conditions became very drastic this stream dried up completely and Lake Chad reduced greatly in size.

By 1955 when the river resumed its flow due to heavy and prolonged rains, the lake also increased in size.

The variations in the level of water are easily noticed on the northern banks because this section is shallow. It is only in the southern and western parts that navigation is very easy because the depth of water is greater than in the north, and there is no tall vegetation in the south as is the case with the northern section. The depth of water in the middle of the dry season is just about six metres, but the lake becomes pliable even by heavy engine boats when the floods of the rainy season are at their peak.

THE FUTURE OF LAKE CHAD:

There is the likelihood that Lake Chad will reduce considerably or disappear completely, first of all, because of the probable capture of the head-waters of the Logone river by Mayo Kebi as earlier discussed. Secondly, this may also happen due to the fact that high temperatures and increasing aridity cause greater evaporation of the waters of the lake. Thirdly, large deposits of sand arriving from the Sahara desert into the lake may eventually fill it up.

Should this lake disappear or reduce much in size, the inhabitants of the region would also suffer. It is for this pending danger that the Lake Chad Basin Commission (L.C.B.C.) was created in 1964 at Ndjamena, Chad. The main concern of this Commission is to encourage the development of forestry in order to check the advance of the desert; the promotion of fisheries, agriculture, animal and water resources and also the means of communication in the Lake Chad Basin. Should these goals be attained, the social and economic life of the people using the basin waters will be greatly improved.

The Lake Chad Basin Commission is made up of the Republics of Cameroon, Chad, Niger and Nigeria. These are countries which share the lake.

PART THREE: HUMAN STUDIES

CHAPTER 8: THE PEOPLES AND PEOPLING OF CAMEROON

The peopling of the Cameroon territory took place mainly in a relatively recent time in history. The movement of populations from diverse directions began a few hundred years ago, and continued through the nineteenth century until it reached its present stage of stabilization and organization. This, in itself, does not mean that the territory was bare of human settlements before this time. We learn about the Pygmies as the oldest inhabitants of the territory, and also about the Sao who established a civilisation around Lake Chad in the fifth century. The civilisation is said to have attained its apogee in the 15th century. Apart from these two groups of people there is almost no evidence of any settlements or large-scale movements of people into the area in historic times. There are movements currently under way within the territory but they are essentially an adjustment of people moving from overpopulated and less prosperous regions to underpopulated and more prosperous ones.

In the course of the disorderly early migrations, most of the groups underwent various social changes which broke them up into numerous subgroups inhabiting different geographical portions of the territory.

As a result of continual fighting, hazardous and precarious living conditions, lack of fruitful contact with other more highly developed peoples, ethnic groups of related ancestry inhabiting isolated geographical areas, developed linguistic, social and economic life-styles that are different one from the other. However, because of economic and social development in recent years the different peoples have been able to come in contact with outsiders and have ceased to live in isolation. Thus, there is currently a mixing up of the different peoples of Cameroon. Despite the mixing up, we can still identify some ethnic groups and tribes with the geographical areas they originally inhabited and in which their numbers are concentrated. It will be wrong, however, to think that the number of the indigenous inhabitants of an area necessarily outstrips that of in-migrants from other parts of the country.

In the discussion that follows, the peoples of Cameroon will be divided into those in the north and those in the south, each of which will be further split into subgroups.

It is important to note some problems which arise from such a division. First, the division of the population into north and south has little bearing ethnically, because a good many tribes that now inhabit the Adamawa and some regions south of it (the so-called semi-Bantu, for example) migrated from the north, following pressures initiated by the 19th century Fulbe invasion of that region. This means, therefore, that some of these tribes in the south (or centre) are ethnically linked to the groups in the north. Second, the peoples of north Cameroon originated from very different and diverse backgrounds, and despite the high level of intermarriage and intermingling over the past numerous years, many of these peoples are still distinct in their physical appearance and local customs.

Again, there is often a vague division into a "Christian South" and a "Moslem North" of the peoples of Cameroon. None of these regions forms a homogeneous unit either culturally or ethnically and such simplistic divisions should be adopted with some restraint.

The only justification that warrants the adoption of the north-south division in this study appears to be geographical — a tropical grassland region in the north and an equatorial forest region in the South, each of which has exerted an enormous influence on the history and life styles of the people.

An attempt will be made to examine the ethnic groups of Cameroon, where they are settled and how they came to be settled in those particular portions of the territory. By *ethnic group* here, we mean a large number of people who, as a result of their shared cultural traits and high level of mutual interaction have come to regard themselves, and to be regarded, as a cultural unity. Within ethnic groups are subgroups or human communities formed of the descendants of common ancestors, to which in certain cases are added allies by marriage, clients or servants. Such a wide group is known by different names, and may generally be called a *tribe*. The tribe itself is split up into smaller units, known as clans, and each clan in turn is composed of families.

In broad terms the peoples of Cameroon have traditionally been divided into:

a) The Bantu and Semi-Bantu in the South, and

b) The Sudanese and Fulbe in the North.

MAJOR ETHNIC GROUPS IN THE SOUTH

The predominant groups south of the Adamawa are the Bantu and Semi-Bantu. Besides these two large groups, there is a distinct minority group known as the Pygmies.

THE PYGMIES

They are believed to be the country's oldest ethnic group and are distinguishable from the other groups by their small size (averagely 1.5 metres in height) and usually looking like children thirteen or fourteen years old. Most of them have a light-brown complexion and a few of them are very dark. They are also timid and are always frightened by the presence of any stranger.

A typical male Pygmy has very long arms in relation to his legs. The broad nose is placed at the centre of a very short face, the eyes are also large and prominent.

There are presently about 5,000 Pygmies in Cameroon who are scattered almost everywhere in the south-eastern forest and around the Kribi region. Those who inhabit the dense forest are found around Yokadouma, Ebolowa, Lomie, Moloundou and Messamena. The Babinga who form the largest of the Pygmy groups in this area live along the frontiers of Cameroon with Gabon, Congo and the Central African Republic. The second group, which prefers the coastal forest to the interior of the forests, is found around Kribi, especially in the localities of Bipindi and Lolodorf.

The Pygmies do not practise a sedentary way of life. They wander in the thick Equatorial Rain forest as hunters, trappers and gatherers of wild fruits. They use poisoned arrows and spears for hunting. The animals they kill are eaten or exchanged for agricultural produce and superior metal tools of their more economically advanced neighbours, the Bantu.

As regards the Pygmy dwelling units, they are usually a collection of twigs and leaves arranged in such a way to form small huts in the forests.

The government of Cameroon today encourages the Pygmies to live in small villages and adopt more modern ways of living, and also to send their children to school for education. Some favourable response is being registered, though very limited. As a matter of fact, the typical life style described above is fast disappearing because of an increasing tendency of intermingling and intermarriage between the Pygmies and their dominant Bantu neighbours. This usually leads to the assimilation of the Pygmies into the Bantu groups. The Pygmies seem to be disappearing more through intermarriage and assimilation than through extinction.

THE BANTU

The central and southern parts of Cameroon are presently being inhabited by the Bantu. The name "Bantu" is a linguistic rather than a racial term. Here, it is used for the people as well as the languages which exist around this equatorial area in Cameroon.

The Bantu of Cameroon are divided into two main groups: the north-western (or coastal) and the equatorial Bantu.

The north-western Bantu include the Bakweri, the Bassa, the Bakundu, the Balong, the Bamboko, the Douala, the Mbo and some numerically smaller groups found around these major ones.

The equatorial Bantu are: the Maka who are located in the east and have other subgroups; the Pahuin (Fang-Beti-Bulu) who are made up of many smaller groups.

The Bantu inhabit the neighbouring forest areas close to the Pygmies and also the Savanna areas found at the northern fringe of the forest. The Bantu of Cameroon have distinct political and social structures. The *family,* which comprises the husband, the wife or wives and many children, forms the smallest unit of the social organization. Many of these families who generally claim to have a common ancestor and live in one geographical area form a *village.* A number of villages constitute *a clan.* With regard to political organisation we notice that the Bantu in this southern part of Cameroon lack the traditionally powerful rulers as would be found in the north and north-western parts of the country among the so called Semi-Bantu societies.

Settlement Pattern and Dwelling Unit

The Bantu live in isolated villages in the forest and, in most cases, build along the road, forming a type of linear village. Examples of such linear villages in the South-West Province are Ikiliwindi, Mato, Mbalangi and Kwakwa.

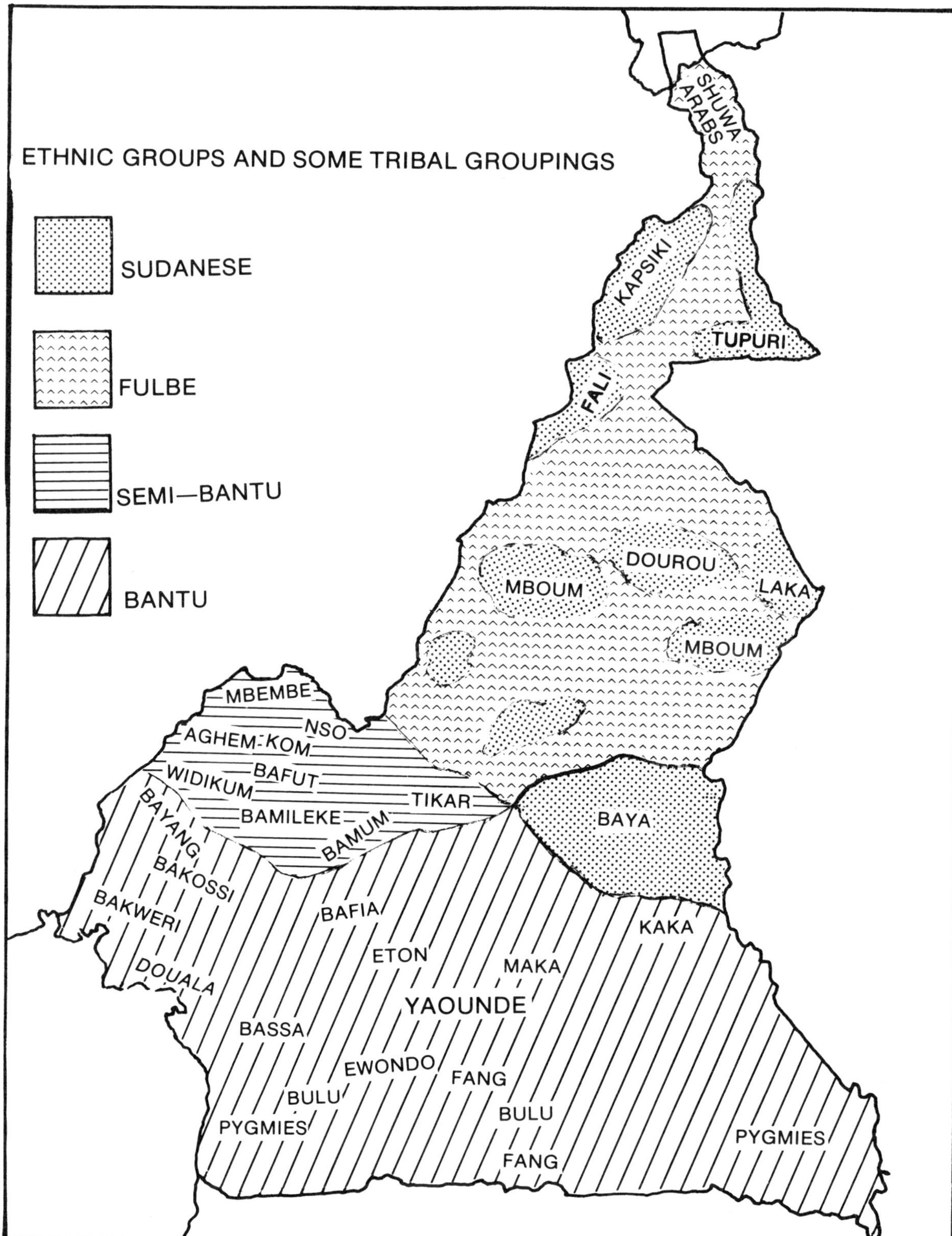
ETHNIC GROUPS AND SOME TRIBAL GROUPINGS
SUDANESE
FULBE
SEMI—BANTU
BANTU
SHUWA ARABS
KAPSIKI
TUPURI
FALI
DOUROU
MBOUM
LAKA
MBOUM
MBEMBE
NSO
AGHEM-KOM
BAFUT
WIDIKUM
TIKAR
BAMILEKE
BAMUM
BAYANG
BAKOSSI
BAKWERI
BAYA
BAFIA
KAKA
ETON
MAKA
DOUALA
YAOUNDE
BASSA
EWONDO
FANG
BULU
PYGMIES
BULU
FANG
PYGMIES

The Bantu of Cameroon, like other peoples within the territory, have striven in various ways over the ages to provide themselves with adequate shelter. Even though the materials for traditional dwelling units are greatly influenced by the environment and the culture of the people, a lot of changes have been introduced following the introduction of modern building materials. Thus, we will have a variety of dwelling units built with such local materials as bamboo, wood, sticks, palm leaves and mud.

In their original form, the traditional Bantu houses are aligned along the road. They are rectangular in shape, and the walls are made of wood.

A very common and cheap type of wood generally termed *Caraboard* is used for the walls by poor folk. The more financially viable people use harder and more expensive wood. In some areas mud is used for plastering the walls. The roofs are in an inverted V-form and made of raffia palm leaves. These raffia palm roofs provide for more cool temperatures inside the dwelling units than the corrugated iron sheets, but they constantly need repairs.

MAJOR BANTU TRIBAL GROUPINGS

a) The Douala:

The Douala inhabit the estuarine area of the Mungo, Wouri and Dibamba Rivers and their main centre of concentration is the zone occupied by the town of Douala.

Ardener (1956:17) collected a fairly detailed account of the traditions of Douala origin. The tradition holds that the ultimate ancestor of the Douala was named Mbongo. A descendant of his was Mbedi the father of the direct ancestor of the Douala who was called Ewale. A distant claim of their area of origin takes them to Bakota in the present territory of Gabon and Congo but the only fixed point on their migration is called Pitti on the Dibamba River.

It is believed by the Douala that at Pitti, a brother of Ewale called Dibongo, and the ancestor of the Limba, split off. The Douala remained at Pitti for a considerable period, and eventually migrated down the Dibamba River to its mouth and north-westwards to the east bank of the Wouri estuary, settling among the Bakoko branch of the Bassa. Later, the Douala occupied both sides of the estuary and drove the Bassa inland. The date of this settlement is not clear, but many believe that it was sometime between the 17th and 18th centuries.

At the beginning of the 19th century, King Bell was the most powerful among the Douala rulers who were competing with each other and trying to deal independently and directly with foreign traders. King Bell was of the sixth generation after Ewale. Bell lost the hegemony of the Douala in 1814 when the head of a collateral branch of the tribe set himself up as "King Akwa". After that time two prominent chiefs of the Douala, King Bell and King Akwa, were to be recognized. Later, other chiefs leading some of the lineages and sub-lineages emerged. The four major chiefdoms of the Douala presently in existence (Bell, Akwa, Deido and Bonaberi) represent the lineages of: *Bonanjo* ("children of Njo") of which *Bonapriso* ("children of Priso") is a collateral branch; *Bonaku* ("children of Akwa"); *Bonebela* ("children of Ebele-Deido") strictly a seceded collateral branch of the Bonaku, and *Bonaberi* (children of Bele") localised on the western side of the Wouri estuary. In the year 1884 the two Douala chiefs, Bell and Akwa, concluded annexation treaties with German businessmen who were already at the coast.

Many factors distinguish the Douala from other Cameroon tribal groups. These factors include their early and intensive contact with Europeans (traders, missionaries, and explorers) several years before colonisation. The Douala also established a hegemony as intermediaries in the early trade in slaves, palm oil and nuts, and ivory between Europeans and the peoples of the interior. During the colonial period and after, they quickly attained a high level of literacy. Richard Joseph (1972:15-17) maintains that the early attainment of literacy and the ability to vigorously pursue their interests unhindered by awe or fear of the whiteman are factors that have enhanced the position of the Douala among the peoples of Cameroon. They possessed an ability to use petitions, judicial procedures and every form of political pressure to resist what they considered to be the inequities and indignities of the German and French colonial rule.

A combination of these factors — early literacy, combativeness, and wealth acquired from the early coastal trade and landed property — makes the Douala a numerically small but politically and economically powerful group in Cameroon.

The Bakoko and Bassa

These two groups lie to the south and east of the Douala and to the west of the Pahouins. It is fairly difficult to differentiate these two tribes, especially in the northern portion because of the high level of intermingling and related ancestry.

Their oral tradition holds that they have the same ancestry (Mveng 1963:242). Like the other Bantu peoples they came into the Adamawa plateau from central Africa, then migrated

A Pygmy dwelling unit.

Rural dwelling units near Limbe.

Rural dwelling unit near Yaounde.

Residents of Sangmelima help a neighbour build his house. Building materials change as economic progress occurs.

towards the coast. This migration is placed sometime in the 17th and 18th centuries. The Bassa later moved inland leaving the coast to other migrants, notably the Douala who came after them.

The Bassa were later going to play a very important role in Cameroon's early nationalist movements. They are believed to be individualistic, with a segmentary and more diffused social structure. This spirit of individualism sets them opposed to all forms of authority whether of the colonial administration or their own chiefs, and has therefore resulted in the extreme dispersal of their population into small groups of isolated huts. As a matter of fact, after the 1955 uprisings, the French administration engaged in regrouping the Bassa into large, and more easily controlled villages as part of their military campaign against the Bassa *Maquis* (guerrilla fighters).

The Pahouins (Fang-Bulu-Beti)

The Pahouin are a component of the Equatorial Bantu. They occupy an area that extends from south of the Sanaga river in Cameroon to the Middle Congo, Gabon and Equatorial Guinea.

Alexandre and Binet (1958) distinguish between the Pahouin proper and those people that became assimilated (or *pahouinised)* as a result of their migrations southwards or contact with the Pahouin.

It is held that the Pahouin came into Cameroon in different waves from the north-east and then migrated southwards. The migrations, sometimes peaceful and sometimes violent, were caused directly by the Fulani, or possibly by pressure from other groups, for example, the Baboute and Mbum, who were fleeing from the Fulani. The newcomers were thus assimilated or pahouinised. The ancestors of the Pahouin migrated southwards crossing the Sanaga in different waves approximately between Nanga-Eboko and the Mbam confluence. The three subgroups of the Pahouin include:

The Beti in the north; the Fang in the south; the Bulu in the middle.

a) *The Beti*

Guyer (1980) states that the term "Beti" was originally used to designate status rather than ethnic identity, and as such applied to free-born people of the entire linguistic group, including those now known as Bulu and Fang. Today, it refers primarily to a section of the larger unit, which is also subdivided into local ethnic and dialect groups, each group being segmented into clans and lineages based on patrilineal descent.

The Beti are believed to have entered the country through a route which lay immediately north of the thick equatorial forest. The group consists of the relatively populous *Ewondo* of the Nyong-and-Sanaga region; the *Bane* of the Nyong-and Sanaga, Ntem, and Dja-et-Lobo region; the *Fang* of the Dja-et-Lobo and Ntem region, the *Mbida-Mbane* and *Mvog-Nyenge* in the east of Yaounde, and the three Eton tribes (Eton-Beti, Eton-Beloua and Beloua-Eton) mainly in the Saa and Obala region.

b) *The Fang*

They channelled themselves into Cameroon by way of the Lom valley. The Baya and the Mbum who were forcibly pushed south by the Fulani initiated a further migration of the Fang into the southern forest and even to Gabon, Congo and Rio Muni.

The Fang in Cameroon consist of —

The *Fang proper,* of the Ogooue valley in South Cameroon.

The *Ntumu* (or Ntoumou) in the Ambam region

The *Mvae,* divided into two — to the east and west of the Ntumu.

The Fang are more numerous across the Cameroon border in Gabon, Congo and Rio Muni.

c) *The Bulu*

The Bulu constitute slightly more than one-third of the Pahouin and live in the east and south of the Beti. Their subgroups include:

The *Bulu* of the region of Sangmelima, Kribi and Ebolowa

The *Zaman* of the Dja valley

The *Yengono, Yembama* and the *Yelinda* of the Nyong valley.

The *Yesum, Yebekanga, Yekebolo* and *Mvele.*

Alexandre and Binet list as those in the process of being *pahouinised* the *Manguissa, Yekaba, Bamvele, Evuzok, Batchanga* (Tsinga), *Omvang, Yetudi* of the Sanaga, and the Pygmy tribes of Bibaya, Babinga and Beye'ele who, in the past, lived a symbiotic life with their Pahouin neighbours.

It must be noted that almost all the people on the Cameroon territory migrated to their present sites from other places, especially from locations found outside the country.

THE SEMI-BANTU (GRASSFIELD BANTU)

The term *Semi-Bantu,* refers to those language groups and Bantu people found in the grassland zones of Central Cameroon who have received a considerable amount of Sudanic influence. Some scholars use the term, *Grassfield Bantu* to differentiate them from the Bantu of the forest region. The Semi-Bantu, therefore, are people of mixed races, living between the Bantu of the forest region in the south and the Sudanese of the grassland area in the north.

Settlement Pattern and Dwelling Unit

The settlement pattern of the Semi-Bantu is scattered. Family houses or compounds are dispersed all over the area. Surrounding most of the buildings are farmlands. In the Bamileke region there is a special kind of *enclosed landscape* where each family's farmland or plot is enclosed by fences or hedges which serve as boundary lines between one person's land and other people's land.

Towards the Bamenda highland area, this enclosure system disappears giving way to *open landscape* where plots are not fenced.

The traditional dwelling units of the Semi-Bantu, especially those of the Bamileke, have been celebrated as a real masterpiece of art, produced through communal labour using only local materials. There is a more ancient type of traditional house whose walls are made of neatly tied bamboo on which mud is applied. The framework of the roofs is of bamboo which is then covered with grass. Such houses are a lot fewer today and are restricted to the barns and secret society huts. The more recent and more common houses have walls made of sun-dried bricks.

MAJOR SEMI-BANTU GROUPINGS

Three main ethnic groups distinguished among the Semi-Bantu peoples, Bamileke, Bamum, and Tikar have very similar social and political institutions that they must have shared a long common history.

The Bamileke

The Bamileke who are very numerous compared to the other Semi-Bantu groups inhabit the western highlands of the West Province. They include essentially the peoples of Bafoussam, Bafang, Bawaju, Bagante, Bandjoun, Dschang and Mbouda.

It is not easy to establish detailed and accurate accounts of the traditions of origin for the Bamileke because different groups have diverse origins and different periods of arrival. It also appears that the Bamileke absorbed waves of population migrations that came from different directions, some from the north, others from the south-west and still others from the south-east. A powerful and prolific people, they were able to assimilate and impose their culture, and sometimes their language, on newcomers.

Evidence suggests that the bulk of the Bamileke descended from the Adamawa, crossed the Upper Mbam river to the area which is today occupied by some Tikar tribes. Pressure from the invading Fulani in the 17th century led to a series of migrations southwards, undertaken at different times by separate individual portions of the population.

The migrants travelled to the area now occupied by the Bamum and remained there long enough to establish villages. The Bamileke claim that they founded these villages in Bamum country: Nkoupit (Bapi), Folepan (Baleng), and Kounden.

Through a series of wars with the Bamum, they were chased out to the other side of the Noun River and with time they scattered into many groups which now form some of the chiefdoms.

The Bamileke are good agriculturalists who farm on the fertile volcanic soils in the region. They are also good at local crafts.

Social and Political Organisation

As earlier mentioned, the Bamileke are made up of many villages. Each of these villages has a rigid social and political organisation which, for long, has survived the forces of change. At the head of each village is a traditional ruler, the *fon,* who is assisted by many advisers. The prominent advisory body is the Council of Notables. The traditional ruler is the main custodian of the people's land and his major role is to administer justice to his subjects and preside over meetings.

The *fon* himself is a well respected and venerated personality. He is considered to have divine powers. He is the overall authority in his village. The powers of the *fon* are transferable to another person only through family inheritance. He selects his successor when he is still alive but this is kept secret until he dies.

The village itself is divided up into small wards and each of these wards has a head who is responsible to the *fon* and also carries instructions from the *fon* to his own ward for execution.

The Bamum

The present Bamum population is composed of people of two different backgrounds. There are the people of Sudanic origin who, some 300 years ago, broke away from the Tikar, settled at Rifum and then moved southwards. There are also those of Bamileke background who were found living in the present-day Bamum country and who were conquered and partially assimilated by the Bamum invaders.

The Bamum who are located in the Foumban area hold that their founder, Nchare, was the son of a Tikar chief. They resemble their neighbours, the Bamileke, in their art, social and political organisations.

The Bamum came into the area after conquering and sending away the earlier inhabitants who were the Bamileke and the Tikars. This conquest was achieved under their leader, Nchare Yen.

Semi-Bantu dwelling unit in Bamenda.

Semi-Bantu men in their traditional attire.

The Palace of a Bamileke Chief.

Since then one could count eighteen kings who have led and established the Bamum Dynasty.

They are:

1. Nchare Yen, 1394-1418
2. Ngouopou, 1418-1461
3. Monjou, 1461-1498
4. Mengap, 1498-1519
5. Ngouh I, 1519-1544
6. Fifen, 1544-1568
7. Ngouh II, 1568-1590
8. Ngapna, 1590-1629
9. Ngouloure, 1629-1672
10. Kouotou, 1672-1757
11. Mbwembwe, 1757-1814
12. Gbetnkom, 1814-1817
13. Mbeikouo, 1817-1818
14. Ngoohouo, 1818-1865
15. Ngoungoure, 30 minutes
16. Nsangou, 1865-1889
17. Njoya Ibrahim, 1889-1933
18. El-Haji Seidou, Njimoluh Njoya, 1933

Among them King Njoya Ibrahim was the greatest. He was chosen at the age of four by his father Nsangu who was a great warrior. The mother ruled as a regent though very much against the desire of Nsangu's children who staged a revolt that led to the killing of many people and burning of many houses. At fifteen, Njoya himself acceded to the throne. He wielded a lot of authority, had enough initiative and ambition. He introduced the moslem religion to the people and invented the Bamum writing. Njoya however, was deposed by the French administration in 1931 then taken to Yaounde where he later died in prison.

The Tikars (Tikari, Tikar)

It is believed that the Tikars came into the area near Garoua from Nigeria and were forced to move southwards to the plateau zone west of Ngaoundere. In the course of time they spread out in a south-west direction. Today the Tikars live mostly in the Mbam region and the northern half of the Bamenda grassfield area, especially in the Menchum, Mezam, Donga and Mantung Divisions.

The tradition of the Tikars holds that they migrated to their present sites from the Tibati-Banyo or Ndobo areas. The terms "Ndobo" and "Tikar" are loosely used to refer either to the area north of Foumban and just south of Banyo or to the people among whom the Tikar dynasts settled after their legendary exodus from the Mbum area on the Ngaoundere plateau.

From this Tibati-Banyo area a good portion of the Tikars moved and settled on the Ndop plain from where they scattered to found such *fondoms* as Bafut, Kom, Nsaw, Nsungli and many other smaller ones.

The social and political organisations of these peoples are almost the same as those of the Bamileke villages.

The Bafut:

Bafut is actually a composite group because the smaller groups of people who make it up did not migrate into the area under one leader. Neither did they come in at the same time.

The oral tradition of the Bafut brings them from Tibati-Banyo, the area in the upper reaches of the Mbam River, to Ndop plain. The migration from Ndop has been attributed to disputes of succession which resulted in a disintegration of the group. Some factions which split from this larger group moved westwards towards present day Bambili, Babanki, and Bafreng. One faction moved still further to settle on a stony hill called Mbebeli — one of the present day wards of the Bafut tribe. But these newcomers found some people who had earlier inhabited the small hill, so they settled with them peacefully after offering them gifts.

Another group, the Buwe Bukari, were already settled in the lower part of Bafut but the two early settlers seemed not to have established any contact with each other. The Buwe Bukari are believed to have found their way into the area through the Menchum valley. Other groups of migrants into the area include the Bawum who claim to have come from the uphills of Bamenda Station (i.e. around Mendankwe). The Mambu people who claim a Widikum origin passed through Mundum to Bafut; and finally were the Mankwi who came in from Njimuyaah.

Without examining the migratory legends of these migrants, it is important to note that none of the groups migrated under the name of Bafut as such. It was only when they settled and formed a single state or a confederacy of almost autonomous wards that they came to be known as Bafut. The only recognisable early settlement where the various migrants converged was Mbebeli and their ancestors are supposed to be the earliest settlers of present day Bafut.

The Kom (Beko, Bikom)

The Kom currently inhabit the open country in the mountainous heart of the Northwest Province. They constitute one of the chiefdoms believed to have been derived from the Tikar or Ndobo dynasties.

The royal legend of origin brings the ancestor of the royal clan from Ndobo to Bamessi where he settled with his people. Later, most of them were destroyed by a trick of the Bamessi chief. Under a new leader, Nandong, the remnant left Bamessi and made their way to Nkar to be joined there by the ancestors of today's Ndotitichia and Ijunasung clans. From Nkar they proceeded to Djottin on the western border of the Nsaw and from there they moved again to Ajung. Under the leadership of Jinabo I, they left Ajung and reached Laikom where they subsequently displaced the Andonalu clan. The latter moved the seat of their chiefdom to Achain.

The newly arrived Kom were later going to establish one of the major chiefdoms in today's northwest Province. The foundation of the dynasty at Laikom is closely associated with three clans — Ekwu, Achaf, and Itinelaa. The kingship became invested in one of the matrilineages of the Ekwu clan, but an Achaf lineage and an Itinelaa lineage provided the estoolers of kings. These three ruling families originally settled at Laikom from where they spread to other villages, including Mme, Kuk, Nyos, Fungom, and Kung.

The rise to prominence and the expansion of Kom is associated with one king, Yu, who reigned from 1860 to 1912. It appears that before Yu's succession the dynasty exercised its most effective control over Abasakom, the northern half of the kingdom, adjacent to the capital at Laikom.

The Kom defeated a number of villages and made them tributary: Nchang, Ake, Mejang, Mejung and Basaw. Having conquered these villages, the *fon* (or king) of Kom established his rule and secured tribal solidarity by marriage alliance. Today the heads of the five villages (Nchang, Ake, Mejang, Mejung and Basaw) plus those of Ajung as well as Besonaku (or *Bassin Oku,* an offshoot of the Oku sub-tribe in Nsaw) are the seven 'government' sub-chiefs of Kom. The seat of the paramount *fon* of Kom is Laikom on the high mountain terrain of the Kom hills.

The Nsaw (Nso, Banso)

The earliest known ancestor of the Nsaw group was Ngonso who is believed to have been the brother of King Nchare of Foumban. These two brothers are said to have moved from Rifen (Bankem) near Banyo about 200-300 years ago to the Mboh plain.

From Mboh Nchare and his followers migrated south, crossing the Mbam river and eventually founding the Bamum kingdom. Ngonso, for his part, moved to Koviferm where he and his followers met other families who had earlier settled on the piece of land. It was at Koviferm that Ngonso's group consolidated itself and was named, Nsaw (or Nso) after its leader. At this settlement they were attacked by Fulani, destroying them in their numbers. Ngonso escaped and took refuge at Tavissa where he later died.

Continued Fulani attacks forced the Nsaw people to abandon their Koviferm settlement and migrate to Kimbaw where they presently live.

After settling at Kimbaw, some 19 kilometres (12 miles) from Koviferm, rivalry between princes led to two Nsaw princes breaking away to found Mbiame and Oku fondoms in 1892.

It is worth indicating here that when the Nsaw people arrived at Kimbaw they met the people of Nkar whom they subdued under their rule. In their pursuit to have more people with them, they conquered other smaller villages: Djottin, Din, Lassin, Mbinnon, Nkor etc., but after defeating them they were allowed to live on their own and continue with their own traditional ways.

The Bali (Bali-Chamba)

It is believed that the Bali are part of the old Dindi Kingdom of the Adamawa region who migrated southwards, passing through Kontcha and Ngaoundere and entering the Banyo district where they fought and conquered the local inhabitants.

At Banyo they were nicknamed "Bani" because their soldiers used bows and arrows and were mounted on horses. Here, they split into two main factions. One of the factions under Gawolbe moved southwards towards the Bamum country, plundering and defeating those they met before crossing over the Noun River. This was about the year 1825.

In 1830, the Bali were involved in a serious battle with the Bafu-Fondong, near Dschang in which they lost their leader, Gawolbe II. The outcome of this battle was going to be felt by the whole Bali-Chamba group. It resulted in a split of the group into several smaller groups following rivalries for succession among the princes. Seven sub-groups which resulted from the disintegrating Bali could then be counted and each of them was led by a prince. Each prince then became the founder of the small chiefdoms of Bali-Gham, Bali-Kumbat, Bali-Nyonga, Bali-Gasho, Bali-Gangsin, Bali-Muti and Bali-Kontan.

Some of these chiefdoms were later subdued and incorporated into the stronger ones. Very significant is Bali-Nyonga, which was founded by one of the seven princes, Nyongpasi (or Fonyonga I). He led his group to Kufom where he died (c. 1855-60) and was succeeded by his son Galega I, who later moved his capital about six miles (9.6 km) to a new site, Ntanka, now known as Bali.

Galega transformed Bali-Nyonga from a band of raiders and refugees into an effective small state. He was the ruling chief when Zintgraff arrived in 1889. Galega gave Zintgraff a friendly reception and they both had a treaty (blood pact) of protection in 1891. In 1896 he again concluded a labour-supply agreement with the plantation company Zintgraff managed. He later died in 1901 and was succeeded by his son, Fonyonga II (1901-1940). The main language of Bali-Nyonga is Mungaka which was widely diffused in the Bamenda Grassfield area, even among non-Bali peoples, owing to the fact that it was adopted by the Basel Mission after 1903 as the medium of instruction and evangelisation. In the eastern Bali-Chamba settlements, Mubako is the leading language among the numerically larger group of the Bali-Chamba.

The Widikum

The term 'Widikum' refers to the name of the place from where the people of Widikum are said to have originated. This place is situated at the confluence of the Momo and Egwefi Rivers in the southern part of Momo Division. It also appears that early British colonialists used the term 'Widikum' to refer to the forest people who migrated northwards from the Mamfe Division into the Bamenda area in order to differentiate them from their neighbours, the Tikar and the Bali-Chamba.

It is, however, interesting to note that the Widikum people differ in their languages and claims to their ancestors. From the linguistic aspect we find that there are two distinct groups. The first group is made up of people whose languages (or dialects) are mutually intelligible and they have oral traditions which trace their origin to one common ancestor. This group consists of the people of Menemo, Moghamo, Ngie, Oshie and Widikum, almost all the people of Momo Division. They believe that two brothers, Iyirikum and Ongiekum, were the ancestors of the Widikum and Ngie respectively; and that from the beginning they lived at Dudum near Widikum. From the names of the two brothers which end with 'Kum' one is forced to believe that their father was Kum.

Following a quarrel between the two brothers there occurred a permanent division. Ongiekum took his supporters and migrated to a small place called Andegechung where he settled and bore many children. His children later became the ancestors of the Moghamo, Menemo, Oshie, etc.

The second group of people who claim to be Widikum are those whose languages are not mutually intelligible and whose ancestors do not correspond to that of the first group. The people of this group are scattered all over the North-West Province. They include the people of Ngwaw clan (except Oshie), the Bajong, the Mundum, the Akum, the Mbelewi (Awing), the Pinyin, some wards in Bafut (Mambu, Bugiri, Mankaa and Mbekum), Beba-Befang, Modelle, Badji, Batamo, Mukuru, and the Koromonjang.

MAJOR ETHNIC GROUPS OF THE NORTH

While accepting that the Bantu and Semi-Bantu are the dominant ethnic groups in the south and central parts of Cameroon, we could also say that the Sudanese and the Fulbe are the main groups in the north. A minority group in the north includes the Shuwa Arabs.

THE SUDANESE

The Sudanese form the original population of the north. We distinguish among these Sudanese the Paleo-Sudanese and the neo-Sudanese.

The Paleo-Sudanese were the first occupants of the north, who after some time, were chased out by the neo-Sudanese and Fulbe to particular defensive highland sites. Today, the Paleo-Sudanese occupy the lands stretching from the Mandara and Alantika mountains to the southern border of the Benue basin. The smaller groups that make up this ethnic group are the Mafa (matakam), Kapsiki, Guidar, Guiziga, Tupuri etc. The neo-Sudanese are in the northern plains, especially in the Logone basin. Those who belong to this group of neo-Sudanese are the Massa, Musgum, Kotokos, etc.

A Musgum mud hut.

Clustered settlements of the Mafa on the Mandara mountains.

The neo-Sudanese arrived after the Paleo-Sudanese and settled on the lowlands, especially the region of the Logone River. They came both from the East and West. It is important to note that some of the neo-Sudanese have remained very indifferent to Islam and to such, the name 'Kirdi' is given, which means the same as 'Pagan'. Kirdi, therefore, is used for the Tupuri, Matakam, Musgum, Bisiga, Fali, Kapsiki, Daba, Massa, Mundang.

The Kirdi hold fast to their traditional way of life so that modern lifestyles are just penetrating their societies today. They are strongly opposed to the very islamised Fulbe who in the course of spreading Islam through wars, conquered and seized their land. The Kirdi have no precise socio-political organisations.

Settlement Pattern and Dwelling Unit

It is noticed that in the northern part of Cameroon the traditional dwelling units and types of settlement vary from one locality to the other. On the Mandara mounts there is clustering of houses. Each clustered settlement is surrounded by a built-up wall which is mainly of rocks held together by soil and the roofs of the houses are of thatches. They are the typical Mafa settlements on the Mandara.

Around Pouss the clusters of huts are made completely of dried-up mud. They are the pure traditional Musgum dwelling unit.

THE FULBE (Fulani)

Any discussion about the peoples of Cameroon, especially those in the central and northern parts, has to be connected, at least in part, to the Fulbe because the pressure they exerted in the north engendered the waves of movement which affected even parts of South Cameroon. But the question arises, who are the Fulbe and what is their origin?

It is believed that the Fulbe migrated into Cameroon from the Senegal-Mauritania region in West Africa, bringing along with them the Moslem faith. The Fulbe are to be found not only in Cameroon but in Senegal, Mali, Mauritania, Guinea, Sierra Leone, Upper Volta, Niger, Nigeria, Chad and Sudan.

The Fulbe occupy mainly the northern part of Cameroon and are believed to have moved into this area from Nigeria in different waves. The first wave of nomadic animal-raising Fulbe arrived and settled peacefully in north Cameroon as far back to the 15th century. They came in slowly through diverse routes. These early Fulbe settlers were only partially Islamized. They intermingled and intermarried with the indigenous people.

The next wave of Fulbe immigration initiated by the revolt of Othman dan Fodio in northern Nigeria in the 18th century was going to alter, in a dramatic way, the course of events and the social organisation of North Cameroon.

Othman dan Fodio, called *Shehu* (chief) by his followers was by birth, education, and temperament, an embodiment of the Moslem teacher and philosopher. He was born in the Kingdom of Gobir in 1754 and received a thorough training from a series of learned men, gained a good knowledge of the Koran, a mastery of the Moslem law, rhetoric and classical islamic history. In short, he was a complete practising moslem.

At the age of twenty,Othman began to teach at Degel, his home in Gobir, and soon was travelling and preaching extensively on the need to end pagan practices and unsanctioned misgovernment which were common in the Hausaland of Gobir. Othman advocated a theocracy, rallied supporters who, within a period of ten years (1800-1810) overcame a reasonable portion of north Nigeria and Cameroon, and created the Sokoto empire which extended from the Niger to the Logone river and from the borders of the Sahara down to the Sanaga headwaters. The name of this empire was derived from that of its capital. The empire was divided into several provinces and at the head of each province was an *emir.*

Modibo Adama, a lieutenant of Othman dan Fodio, carried out the *Jihad* (or holy war) in the area of northern Nigeria across Yola to northern Cameroon. He was made *emir* of the Adamawa province (or emirate) with his seat at Yola. The subdivisions of his emirate were Yola itself, Maroua, Garoua, and Ngaoundere. The present-day region of Adamawa derived its name from Adama, its first emir.

It was this socio-religious movement that was going to lead to a domination of the original inhabitants of north Cameroon and even to a migration southwards of those who resisted the influence of the Fulbe.

By the mid-19th century the Fulbe hegemony was already established in the whole of northern Cameroon with each ethnic group placed under the authority of a *Lamido;* the only exceptions being the Kotoko and Mandara communities, and the portions of the Logone valley occupied by Shuwa Arabs.

The first half of the 20th century saw another wave of the Fulbe who settled in the north-western highlands and who are commonly referred to as, Fulani. The typical Fulani are tall and slim, with narrow faces, and have a clear complexion and wavy hair.

A typical Bororo hut on the hills of Fundong in the North-west Province.

Two groups of Fulani can be distinguished:

a) The town-Fulani (or settled-Fulani)

b) The Pastoral Fulani (or cattle-Fulani).

Although the Fulani are said to be nomadic, not all of them practise that way of life. The town-Fulani have lost most of their animal wealth, inter-married with non-Fulani groups and therefore lead a sedentary way of life. Because they live in close contact with non-Fulani groups they have developed a mixed culture. These town Fulani receive better education and more modern ways of life. They are found mostly in the towns located in the Diamare, Benue and Adamawa areas.

Only a relatively small proportion of the Fulani remain as pastoral or cattle Fulani, practising the nomadic life-style. These are the cattle-Fulani, or "Bororo". There exist interesting characteristics about the life-styles of the cattle Fulani. They are less willing to give up their centuries-old traditional ways. They remain strictly aloof, proud of their racial characteristics and customs. They have maintained their racial and linguistic purity by not intermarrying with non-Fulani groups. These cattle-Fulani care little about religion and nothing for power, but live mainly for their cattle, goats, and sheep. They move from one green pasture to the other at will, paying some tribute to the reigning chiefs or a grazing tax to those chiefs in whose lands they graze their cattle, but owing allegiance to none.

Numerically, they are relatively very few, but scattered all over, especially from the Diamare to the western and Bamenda highlands. They are fairly tall and lean. Light-skinned, with long, thin noses, they are rather Arab-looking. Their slow, long striding walk covers large distances relatively quickly, and they never seem to tire.

Mode of Living

They are seasonally nomadic, depending on the state of the pasture. There are two types of cattle (nomadic) Fulani: those who have no centre to which to return, but are always moving, setting up large encampments for various times and seasons; and those who have a centre where the aged and others in the family remain, and to which the herdsmen *(gainakos)* return from time to time. Such a permanent base is known as *ruga.* The women who carry milk in large shallow gourds for sale in towns and villages usually come from such centres. Most Ardos (Fulani chiefs) establish permanent bases.

For the highly nomadic type, the dwelling units are usually temporary ones and consist of dome-like huts, constructed with forest twigs and poles stuck into the ground and bent inwards to join at the top. The twigs are then tied together and covered with grass.

THE SHUWA ARABS

The Shuwa Arabs migrated into Cameroon from the Upper Sudan some over 300 years ago. They are of Hamitic origin and live mostly around Lake Chad in the extreme north, especially in the fertile areas fringing the Sahara desert. They have a clear complexion and sometimes intermarry with the Kotoko and Bornuans. The Hamites are a human race believed to have come all the way from Asia Minor.

CHAPTER 9: POPULATION AND DEMOGRAPHY

Population Growth Before the 1976 Census

Demographic studies on Cameroon, for long, had been neglected. This could be attributed in part to the lack of interest on the part of the colonialists who occupied the territory for more than one half of a century. Until 1976, when the first detailed population census was held in Cameroon, most of the figures on the size of the population which date as far back as to the period before the First World War were based on estimates. Despite the discrepancies surrounding the estimates, one would nevertheless remark that the population of Cameroon has been rising steadily since the arrival of the Germans and perhaps with interruptions only during the First and Second World Wars.

According to a United Nations Population Report on Cameroon, the earliest estimates date as far back to 1912 when Cameroon was still a German protectorate. From the report, the total population size at the time was 2,649,000 with about 2,000,000 living in what became in 1916 the French Cameroon and the rest in British Cameroon. The evolution over the years as shown by many records of the colonialists proved unreliable and of poor quality as reflected in the sharp fluctuations in the growth rates.

On the basis of better census findings conducted by the United Nations, the population of Cameroon was realized to be much higher than what the colonial documents stated. For example, the 1960 figure was revised upward by 1.4 million. That is, in 1960, the population of Cameroon was 1.4 million higher than 4.3 million contained in colonial reports. Revising the estimates back to 1950 the U.N. established the following population figures for the whole nation:

Year	Population in '000
1950	4,955
1955	5,273
1960	5,681
1965	6,177
1970	6,781
1975	7,528
1980	8,444

Source: U.N. Fund for Pop. Activities: Report of Mission on Needs Assessment for Population Assistance-United Republic of Cameroon. NY. Sept. 1980

After independence in 1960, the government found it increasingly necessary to have valid data concerning its population for the formulation of policies and execution of plans. It was for these reasons that between 1960 and 1966 "quasi-censuses" were conducted mainly by administrators for administrative purposes.

1. The first one lasted from January to June 1960 and it covered that part of the then North Province lying north of the Benue River.
2. The second one, conducted from April to August 1961 covered that part of the North Province located south of the Benue River, including the Adamawa Plateau region. The above two censuses sampled 1,400,000 persons.
3. The third one, from March to November 1962 was for the East Province and the then Centre-South Province approximately. It sampled about 1,200,000 persons.
4. The fourth census, from January 1964 to February 1965 was in the former West Cameroon State, and sampled about 1,000,000 persons.
5. The fifth one in 1965, was limited to the Bamileke country and its surroundings, sampling over 1,100,000 persons.

These "censuses" as described above could not provide enough information for the whole nation, firstly, because they were conducted over a very long time period. Secondly, the level of ignorance on the part of the masses was relatively very high. People could not understand the reasons for certain questions and would resent them as an intrusion on their privacy; thirdly, scepticism in some cases led to the belief that counting of people was associated with attempts to collect taxes.

The April 1976 population census was the first of its kind to be conducted over the whole territory within a short period of time. Unlike the previous ones, it took account of different kinds of variables that are necessary for a detailed demographic analysis.

Recent Population Trends

There are three ways by which a population grows or declines in size — births, deaths and migrations. A community or nation can gain population only through the fertility of its inhabitants or by immigration; it can also lose population only through the deaths among its residents or by emigration.

From the population figures of the past years, it would be realized that there has been a steady increase in the population of Cameroon and according to projections made by Cameroon demographers the population of the country is bound to increase still further for the rest of the century.

Projected Growth of Total Population

Year	Population
1981	8,965,700
1982	9,205,200
1983	9,491,000
1984	9,790,900
1985	10,106,600
1986	10,446,400
1987	10,857,100
1988	11,180,600
1989	11,540,300
1990	11,899,600
1991	12,243,700

Source: *Situation et Tendances de la Population Camerounaise.* Ministere du Plan et de l'Amenagement du Territoire. Direction de la Planification. Yaounde, Novembre 1985, P. 15.

Birth Rate

The birth rate is usually expressed as the number of babies born per thousand people per year. The total number of births during the year is divided by the estimated population at the midpoint of the period. This may be visualized as:

$$\frac{\text{Total Number of Births in Cameroon, 1976}}{\text{Total Population of Cameroon, 1976}} \times 1000 = \text{Cameroon's birth rate for 1976}$$

The 1976 census estimated a crude birth rate of 45 per thousand (45‰) for the entire nation. However, the rate varies from one region to another. For example, on the Western Highlands the birth rate is over 50‰ but among the Moslems of North Cameroon it is only 29‰.

Fertility Rate

Fertility rate is the number of live births per 1,000 women of childbearing age (15 to 44) in a given year. It is a more refined measure of births than the birth rate, which is based on the total population of all ages and both sexes, rather than only on women likely to give birth. Fertility rate is obtained by dividing the total number of live births for the specified period by the number of women of childbearing age, multiplied by 1000.

$$\frac{\text{Total number of live births in Cameroon, 1976}}{\text{Number of women of chilbearing age (15 to 44 yrs)}} \times 1000 =$$

$$\frac{\text{249,465 children of less than one year}}{\text{1,637,800 women of 15 to 44 yrs}} \times 1000 = 152\text{‰}$$

Infertility

Infertility was not well analyzed from the 1976 population census figures. However, the demographic survey in 1962-1964 for the South-East Region revealed that sterility was not an uncommon problem in Cameroon. The survey found that 23 percent of women aged 50-54 had never had children.

In 1975 a World Health Organization (WHO) Survey carried out in the East Province found that primary infertility affected 16 percent of the women and secondary infertility another 26 percent. Prominent among the causes seem to be gonorrhea (in women) and filariasis (in men), infections due to deliveries and abortions, goitre, genital tuberculosis, and certain practices appear to be other causes.

Death Rate (Mortality Rate)

The death rate (also called crude death rate) is the number of deaths per 1,000 population in a given year. In other words, it is the number of deaths within a specified time period divided by the total population within that time period, multiplied by 1000. Thus,

$$\frac{\text{Total number of deaths for Cameroon, 1976}}{\text{Total Population of Cameroon, 1976}} \times 1000 = \text{Cameroon's death rate for 1976}$$

According to the 1976 population census, the death rate for Cameroon was 20.4‰. This means that for every one thousand people there were 20.4 deaths (or 20 deaths).

Even though deaths occur everywhere, the rate of occurrence varies greatly between one region and another, between urban and rural settings, and between one age group and another.

Regionally, death rates in the southern part of the country are much lower (17‰) than those in the northern part of the country (40‰ on the Mandara Mts.). This is as a result of improved medical services and better equipped health institutions, cleaner sanitary conditions among the citizens in the south than those in the north where there is retarded economic development. The water problem in north Cameroon may also be the cause of many water-borne diseases. Sinking of deep wells, the provision of more health centres, the education of the people and a change of life style will help improve the deplorable living conditions in this northern part of the country.

Low levels of education and poor living standards have also been responsible for many deaths in most remote or rural areas. In the 1960's it was estimated that the death rate in rural areas was 26‰ as compared to 14.5‰ in urban areas. This large difference results partly from the medical and social services that are more plentiful in urban centres and accepted by the urban people (which is not always the case in rural areas). Moreover, the age structure of the population is much younger in urban than rural areas. Differences in death rate among the different age groups are also significant — usually many more deaths among infants of less than one year old and the very elderly above seventy years old than among youths.

Infant Mortality Rate

Infant mortality rate is the number of babies out of each thousand born alive who die before reaching the age of one; that is,

$$\frac{\text{No. of deaths below age one in Cameroon, 1976}}{\text{Number of live births in Cameroon, 1976}} \times 1000 = \text{Cameroon's infant Motality rate for 1976}$$

In Cameroon, infant mortality rates are the highest among low income families and in economically backward areas. Low family income, overcrowding, lack of sanitation, illiteracy and exposure to poisonous substances make it likely that a child will develop a disease and die within the first year of life. The 1976 infant mortality rate stood at 157 per thousand.

General mortality rate has fallen in recent years as a result of improved medical care, better living conditions and the elimination or reduction of certain infectious diseases such as smallpox and cholera. Despite these improvements infant mortality and maternal mortality are still high enough to call for more medical attention. Other diseases such as malaria and schistosomiasis are still rampant, and chronic malnutrition affects a good portion of the rural population.

Growth Rate:

The growth rate is the rate at which a population is increasing (or decreasing) in a given year due to natural increase *and* net migration, expressed as a percentage of the base population.

The growth rate takes into account all components of population growth: births, deaths, and migration.

$$\frac{\text{Births 1976 - Deaths 1976} \pm \text{Net migration 1976}}{\text{Total Population mid 1976}} \times 100$$

The growth rate can also be calculated from natural increase and net migration rates:

Rate of natural increase ± Net migration rate.

Demographers express growth rate as a percent annual increase, that is, not as rate per thousand but as a rate per hundred. The growth rate from the 1976 census stood at 2.46 percent. The growth in Cameroon's population is due essentially to excess births over deaths, with migratory movements having a very neglible impact on the total population count. These excess births over deaths can be attributed to the following reasons:

1. Improvements in medical facilities and general socio-economic development.
2. Countrywide innoculation and vaccination campaigns beginning from the 1950's resulting in the control and even eradication of such endemic diseases as cholera, measles, malaria, sleeping sickness and smallpox.
3. The establishment of many hospitals, dispensaries and health care centres both by the government and voluntary agencies in various parts of the country.
4. Great increases in locally trained medical personnel at various levels ranging from doctors down to nursing aides.
5. The provision of extensive periodic child immunization campaigns against common infectious diseases.
6. Regular radio, press and rural animation programmes to educate the masses in the rural areas on personal and environmental hygiene, disease prevention and nutrition.
7. Closer collaboration between traditional and modern trained doctors.

Life Expectancy

Life expectancy is the number of years that one can expect to live, usually estimated at birth. The rate of infant mortality is regarded as a very sensitive indicator of health and economic development. Infant mortality rate is very important in determining life expectancy because at birth (among babies) most deaths occur, especially in less developed countries with poor and inadequate medical facilities. This explains why life expectancy levels in less developed countries are usually very low.

The 1986 life expectancy stood at 50.08 years for males and 53.97 years for females. That means that for the whole country the life expectancy was 52.53 years.

POPULATION PYRAMID FOR CAMEROON 1976 : FIVE—YEAR AGE GROUP

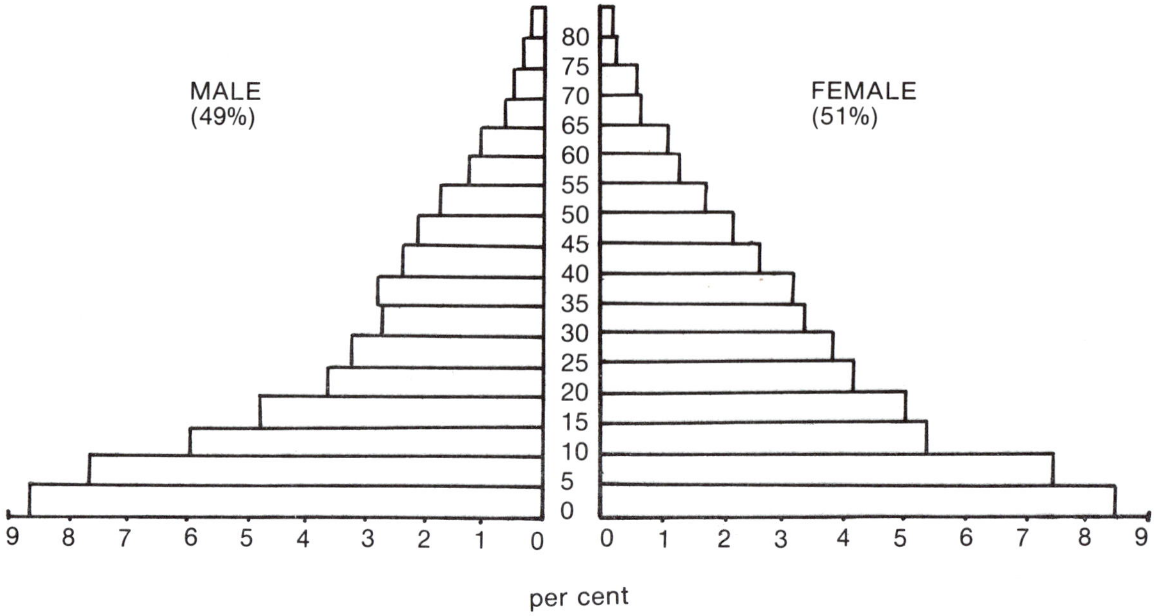

POPULATION STRUCTURE

Age Distribution:

The age distribution of a population refers to the number of people in each age category, say, birth to age ten, eleven to twenty, twenty-one to thirty and so forth. Age and sex distribution are graphically presented in the form of *Population Pyramids* as in the figure above for the Republic of Cameroon in 1976. A population pyramid is a graph which shows the percentages of males in each age category of the population on the left side of a vertical line and percentages of females in the same age category on the right side of the line. Less developed societies will have differently shaped population pyramids with many children and few elderly.

Cameroon has a very young population. The 1976 census found that children of less than 15 years represented 43.4 percent of the total population, adults aged 15 to 59 years represented 50.8 percent while the elderly, 60 and above represented only 5.8 percent. This is typical of less developed countries where both fertility and mortality rates are high, but where there has been a recent marked decrease in mortality rate as a result of advances in medicine which tend to benefit the young ones more.

Sex Ratio (Sex Distribution)

The sex ratio of a population is defined as the number of males per 100 females. A sex ratio of 100 thus represents a population in which there are equal numbers of males and females. A sex ratio over 100 represents a population that contains more males than females.

There are no major age-sex distribution differences in the population of Cameroon. The observed figures in 1976 revealed a sex ratio of 95.9 for the whole country, meaning that for every 100 females in Cameroon there were 95.9 males. This ratio, however, varies depending on the age group — higher for age groups less than 15 years and lower for age groups between 15 and 64 years. Sex ratio also varies according to geographical regions. There are out-migration regions, such as the western highlands, where in most cases, the males are breadwinners of the families. They leave their families behind and move out to the littoral or plantation zones in search of jobs. The Littoral and Southwestern regions on the other hand are in-migration zones, receiving mostly males so that the sex distribution here is reversed: a higher proportion of males here than in the western region. The deficit of males in Cameroon can be attributed to the following factors:

1. Under enumeration: It is likely that more males than females were omitted in the population

count. This is so because males, especially unmarried males in towns, tend to be more mobile than females and must have left home before the arrival of the enumerator on the day of the census. The under-enumeration is even more likely when the males do not possess a fixed residence and employment. Ignorance and scepticism as to the reasons of the count might have kept away many tax evaders.

2. Male deaths from political troubles: There were many adult male deaths before and immediately after independence (1960) resulting from political problems in some parts of the country. Even the recent political troubles under the Biya Administration in 1984 resulted in more male than female deaths.

NUPTIALITY

Nuptiality refers to marriage as a population phenomenon. It takes into account the rate at which marriage occurs, the characteristics of persons united in marriage, the dissolution of marriage (through divorce, separation, widowhood and annulment).

Marriage in Cameroon takes place early. Results of the 1976 census, analyzed by the U.N.O., show that the average age at first marriage among women was 18.8 years, and that 46 percent of those aged 15-19, and 80 percent of those aged 20-24 had married. Only 4 percent of women aged 40-44 were still single.

Marriage patterns, however, differ according to the milieu: marriage occurs earlier and celibacy is less common in rural than urban areas. For example, 50 percent of women aged 15-19 in rural areas had married as compared with 33 percent in urban areas. Similarly, whereas the average age at first marriage was 18.3 years in the rural areas, it was 21.1 years in Yaounde; and whereas in the rural areas 4 percent of women aged 40-44 had never married, the corresponding figure for Yaounde was 12 percent.

Forms of Marriage:

There are essentially two forms of marriage in Cameroon:

a) *Monogamy,* the marriage of one man to one woman

b) *Polygamy,* the marriage of one person to two or more partners of the opposite sex. Two forms of polygamy exist, namely:

I) Polygyny — This is the marriage of one man to two or more women.

II) Polyandry — This is the marriage of one woman to two or more men. Polyandry is rare in Cameroon as it is in the world at large, but it does exist in some parts of the world.

Monogamy in Cameroon, as in all countries of the world, is the most widespread form of marriage. Polygamy is also common and accepted as another legitimate form of marriage. In many western countries, however, polygamy is frowned upon and even looked upon as a sin, even though it was accepted by the church and state as late as the middle of the seventeenth century.

Like other population variables, there are differences in the forms of marriage between urban and rural settings.

Forms of Marriage (%)

	Total	Urban	Rural
Monogamy	76.4	81.4	74.6
Polygamy	23.6	18.6	25.4
	100	100	100

The table above indicates a number of things. It indicates that, out of every 100 marriages, 76.4% of them are monogamous and 23.6 are polygamous. In the urban setting, out of every 100 marriages 81.4% of them are monogamous and only 18.6% are polygamous. In the rural areas there are more monogamous marriages (74.6%) than there are polygamous marriages (25.4%). Finally, in both the rural and urban settings monogamy is the prevalent form of marriage.

Marital Status

Marital status refers to whether people are single, married, widowed or divorced. The marital status of a population, therefore, looks at the proportion of each of these groups in the total population.

	Singles	Married	Widows & Widowers	Divorced or Separated	Total
Males	39.5	55.8	2.1	2.6	100
Females	15.9	66.8	13.9	3.4	100

It is obvious that most singles are found among youths. Single adults either postpone marriage to some later date or have vowed to remain single for life, as in the case of monks, nuns, and priests.

Disparities exist between the marital status of males and that of females as seen on the table. There is a large disparity between the proportion of male and female singles. Possible reasons include the fact that young males are generally more reluctant than females to marry; and females mature faster and become ready for marriage much earlier than their male age mates.

With regards to widowhood, it was mentioned earlier in this chapter that there are usually more adult male than female deaths in Cameroon. This results in a higher proportion of widows (13.9%) than widowers (2.1%).

Concerning divorce, the disparity is not great, but the smaller proportion of divorced males found in the population could possibly be as a result of the fact that males more easily remarry after divorce than females.

MIGRATION

Migration is the movement of people into or out of a given geographic area. Migration is an important variable in the growth or decline of regional or national populations.

Two main categories of migratory movements are to be distinguished in Cameroon, namely;

1. Internal migrations
2. External migrations or international migrations

INTERNAL MIGRATIONS

Internal migrations are very intense in Cameroon. They are responsible for modifying the population size of the different geographical regions. There are currently a number of internal migratory movements which include migrations from one region of the country to another (inter-regional), rural exodus and transhumance.

a) *Inter-Regional Migrations:*

Inter-regional migrations are characterized by movements from areas with low economic opportunity to those which offer greater economic promise. Broadly speaking, inter-regional migrations take the form of three major movements: those movements originating in the western highlands, the west-east movements, and the north-south movements.

b) The movement originating in the Western Highlands

This is the most significant of all the internal migratory movements and it plays a major role in shaping the pattern of population distribution in the country. Out-migration from this region started as far back as the early part of the 20th century with the establishment of German plantations on the fertile volcanic soils below Mount Cameroon. Due to insufficient supply of local labour and the failure of the forced labour methods employed by the Germans on the Bakweri, there was a need to import workers from the hinterland and beyond. Thus , labour was brought in from the interior of the country.

Today this movement continues because of the out-migration of rural Bamileke people from an area that is heavily populated and from land that is already intensely cultivated. The western highland area of Cameroon, as a whole, has a high rural population density which acts as the source region from where people leave the countryside and stream down to the Littoral and South-West provinces. Some of the migrants seeking white collar and blue collar jobs move to the economic capital, Douala, and to Limbe, Tiko and Kumba.

Those who migrate as a result of demographic pressure on land settle on the rich agricultural region of the Mungo Division. Annually, vehicles from the Yabassi and CDC plantations travel all over the West and North-West Provinces recruiting workers who may migrate with their whole families or singly to these plantation areas.

The movement from the western highlands is especially facilitated by a well-surfaced road from Bamenda through Bafoussam to the Littoral and South-West provinces. It is estimated that four-fifths of the migrants from the western highlands move to this direction while a fifth move to Yaounde and its environs. With the recent improvement of the Bafoussam-Yaounde road, the movement to Yaounde would likely increase.

ii) *The West-East Movement*

The movement across the country affects the major cities of Limbe, Douala, Yaounde and Bertona but the most important wave is between Douala and Yaounde. The East Province, having been neglected in the past (especially in colonial times) is now witnessing some kind of economic development due to a number of factors which include: forestry projects, tobacco and coffee cultivation, the relative success of the ZAPI programme, the creation of the provincial administrative unit, and the increased trade resulting from all these activities. This increases the west-east movement.

iii) *The North-South Movement*

The last important movement is that originating from the north and leading to the south. It has several aspects: rural exodus towards the large cities of the south, with or without a temporary stay in the towns *en route* such as Maroua, Garoua and Ngaoundere; migration of whole families from the extreme north towards the area designated for rural development program by the North-East Benue Project; migration of herdsmen with their cattle from the Adamawa Province towards the Centre and East Provinces; the movement of job seekers from the northern part of the country towards the agro-industrial complexes in the south, particularly the sugar cane plantations in the Centre Province at Mbandjock.

The north-south movement is especially facilitated by the services of the Cameroon Airlines, the TransCameroon railway, and the easy car services operating in this part of the country. It is important to note that a corresponding movement in the opposite direction, that is, from south to north, is relatively less significant because economic opportunities are fewer in the north than in the south.

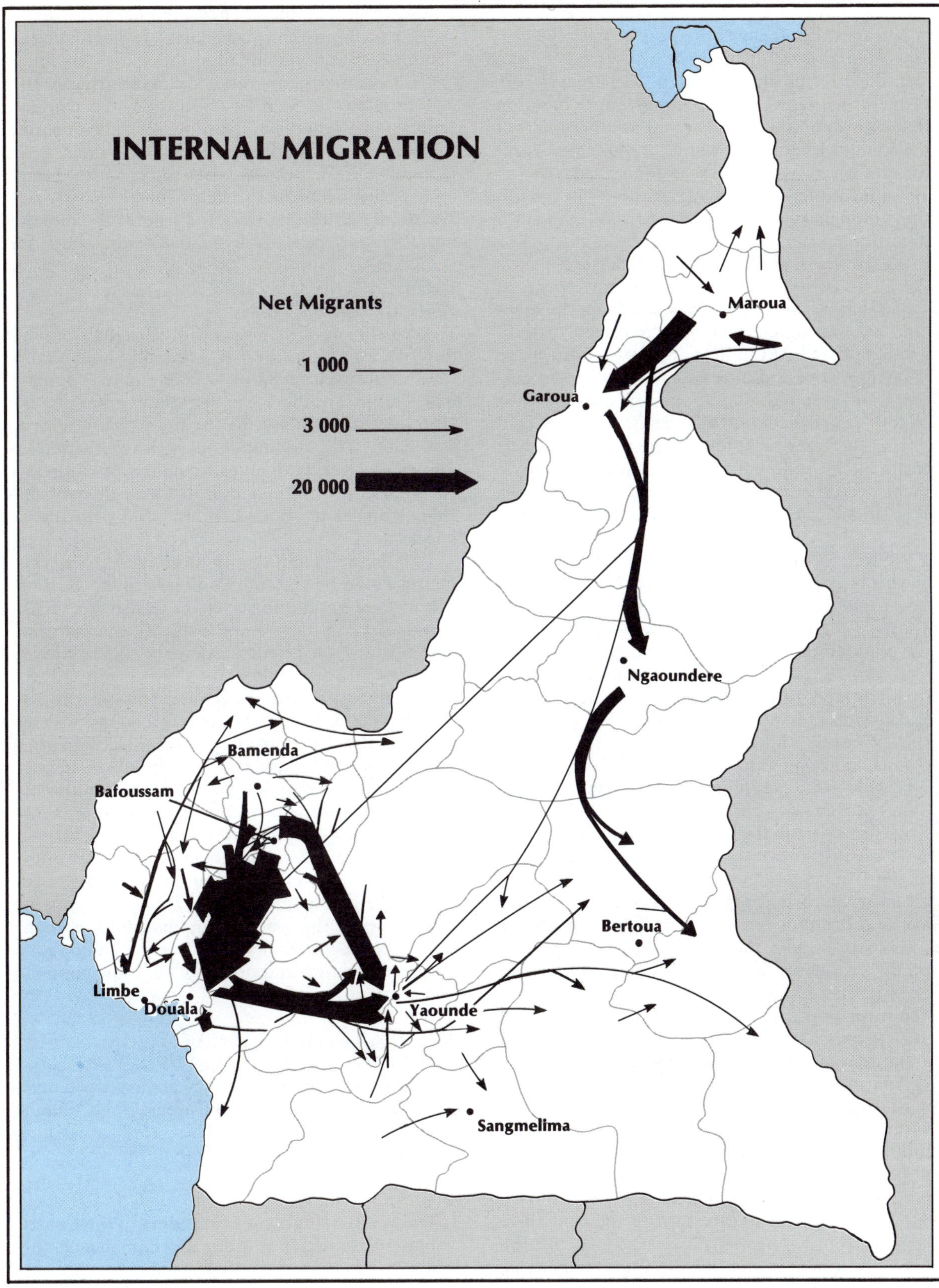
INTERNAL MIGRATION
Net Migrants
1 000
3 000
20 000
Maroua
Garoua
Ngaoundere
Bertoua
Bamenda
Bafoussam
Limbe
Douala
Yaounde
Sangmelima

Rural Exodus

Rural exodus is closely associated with the process of urbanization and the causes have been examined in the chapter titled "Urbanization in Cameroon". However, it is important to mention that rural exodus is greater in the southern part of the country where the towns offer more opportunities for economic progress in the industries, and also educational progress in the numerous institutions of higher learning.

Transhumance

Apart from the movements described above there also exist numerous seasonal movements of animal rearers with their livestock from the highland regions which have been deprived of their green pasture by bush fires and dry conditions to the depressions and river valleys where green pasture can be found. The movement down to the valleys takes place during the dry season, and during the rainy season the herdsmen drive their cattle back to the hill tops. This seasonal movement with cattle is practised on the Adamawa Plateau and on the western highlands.

EXTERNAL MIGRATIONS

This is what we may term, international migration, that is, migration from one country to another. It constitutes one of the ways by which the population of a nation may grow or decrease.

A look at the ethnic map of Cameroon shows that a number of ethnic groups lie astride each of the country's territorial boundaries. For example, the following ethnic groups in the north spread from Cameroon into Chad or Nigeria: Shuwa Arabs, Kotoko, Musgum, Massa, Tupuri, Mundang, Margui-Kapsiki, Fulbe and Fali. A similar situation exists in the south as well as in the east. This means, therefore, that population movements from one country to another cannot be accurately recorded, especially where the migrants avoid the motorized roads and instead follow bush paths where there are hardly any control points. In other areas, say, the eastern and western ends of the Adamawa Plateau there are numerous Mbororo populations who are mostly nomadic and follow their herds across the borders without any due regard for official control.

Apart from irregular cases as those cited above, it is still possible to have some figures to work with. The 1976 census found that foreign nationals totalled some 200,000 (about 3 percent of the population); of these about two thirds were Nigerians, whereas nationals of Chad, the Central African Republic, and France formed the next three largest groups. Nigerians, especially of the Ibo tribe, are found mainly in the English-speaking part of Cameroon and the two big towns of Douala and Yaounde. They are engaged essentially in trade and commerce and a few of them do fishing at the coastal areas.

Cameroonians emigrate to neighbouring African countries as well as to Europe, the United States and Russia. There are regular movements to countries of the former colonizers, France and Britain, but in recent years there have been increased movement to such countries as USA, USSR, Belgium and Italy for studies in courses not offered in local academic institutions.

POPULATION DENSITY AND DISTRIBUTION

According to the April 1976 census, the population of Cameroon stood at 7,663,246 inhabitants for a total surface area of 475,000 square kilometres. This means therefore, that the average population density at the time for the whole country was about 16 inhabitants per square kilometre. Ten years later, with an estimated population of 10,446,400 inhabitants, the population density for 1986 had risen to about 22 inhabitants per square kilometre.

To content ourselves with these average density figures could be very misleading because the distribution of population over the Cameroon territory is very uneven. Almost half the population is concentrated on about 20 percent of the whole area.

The highest population concentration is to be found in the Bamileke country where densities are between 100-200 persons per square kilometre. The region lying immediately north of the Benue River and covering the Mandara mountains is also very densely peopled with densities approaching 100 persons per square kilometre.

When the rural and urban populations are considered the uneven distribution of population becomes even more striking.

A look at the population distribution map shows that there are regions of very high population concentation and others of very low population concentration.

A. AREAS OF HIGH POPULATION CONCENTRATION

The population map shows at first sight that there are three outstanding areas of very high population concentration. These three areas put together hardly exceed one-fifth of the whole territory.

These areas are:

1. The Western highlands extending to the Coast:

This region extends from the Bamenda highlands down to the southern slopes of Mount Cameroon and to the coast. This is the most

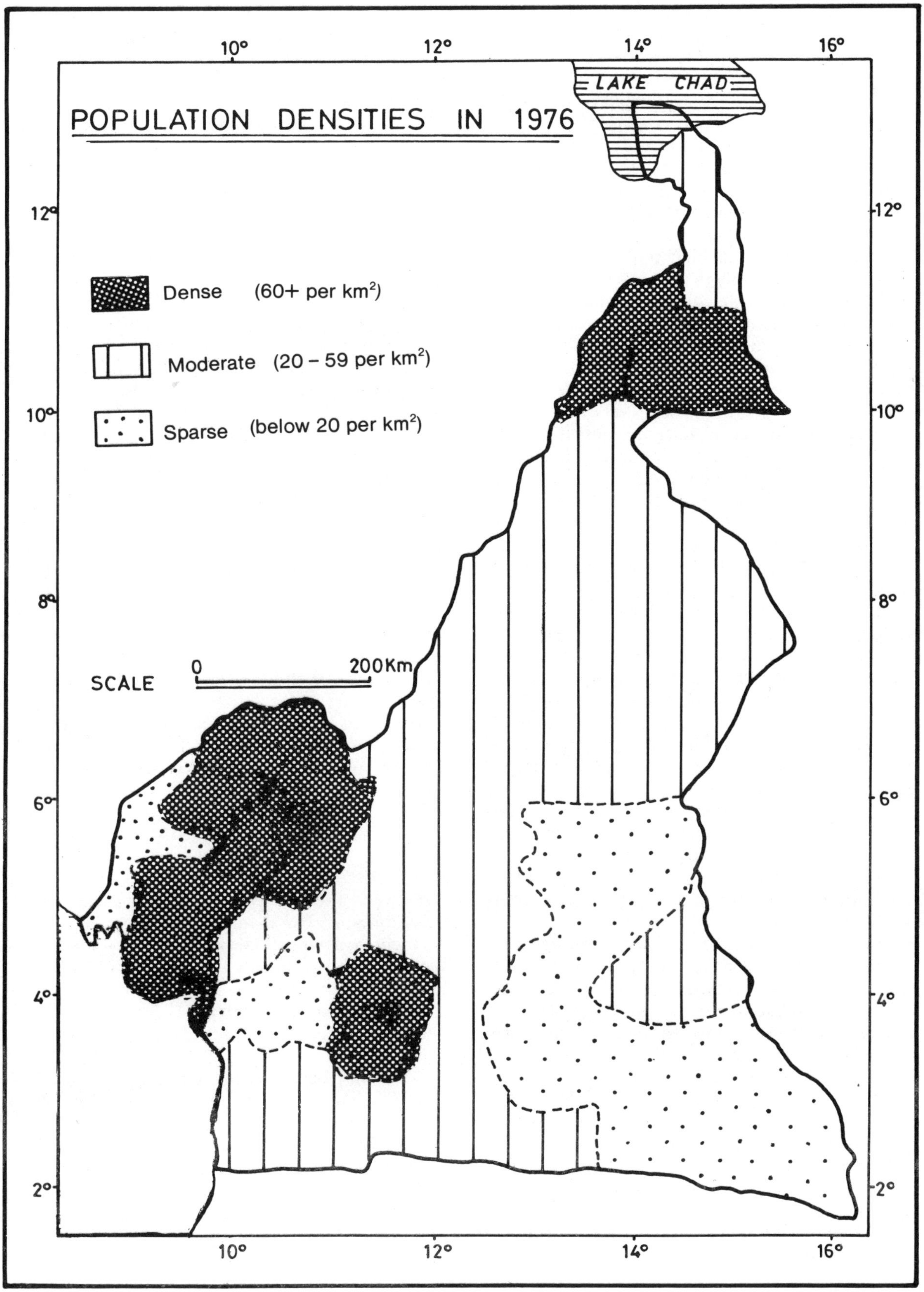

POPULATION DENSITIES IN 1976
LAKE CHAD
Dense (60+ per km²)
Moderate (20 – 59 per km²)
Sparse (below 20 per km²)
SCALE
0
200Km
10°
12°
14°
16°
12°
10°
8°
6°
4°
2°

important region of Cameroon as far as population concentration is concerned. It is in this region that the most important towns of Cameroon are concentrated. Notwithstanding the influence of towns, the region still has very high rural population densities. This, in particular, is the case with the Bamileke area. In the Bamileke lands the average densities are usually more than 100 inhabitants per km^2, especially in the localities surrounding the town of Bafoussam.

Elsewhere in this western region the densities are not at all low. The high density of the Wouri Division is due mainly to the presence of the most populous town of the country, Douala. The factors which influence the high population concentration in this region are as follows:

a) Fertile volcanic soils: The western highlands and their slopes are areas of many farms and plantations whose crops flourish very well because they grow on the fertile volcanic soils. Both food and cash crops do very well in this region thereby attracting many agriculturalists. The plains of Foumbot and the slopes of Mount Cameroon are clear examples of areas that are rich in food supply. These fertile areas have many banana, rubber, coffee and oil palm plantations, owned mainly by O.C.B. and C.D.C.

b) Climate: Another factor which attracts people to the highlands of the west is the fresh mountain climate. It is cool and almost free of disease carrying mosquitoes. The reverse is true of the low forest regions. This climate is congenial for human habitation and therefore attracts many people.

c) Infrastructural Development: Considering that the natural factors of climate and fertile soils already exist, the government has further equipped this region with relatively good means of transport such as roads and railways; communication network, such as, telephone, telex and radio.

All these facilities combine to improve living conditions, and job opportunities also become available.

These infrastructural facilities encourage a lot of rural exodus and movement from other parts of the country into the numerous towns of this sector, especially the commercial and industrial centre of Douala. The good road network here provides to the farmers a sure means of evacuating their crops from the farms to the towns.

2. The Western Part of the Southern plateau:

The second zone of high population concentration comprises the Bafia-Yaounde-Ebolowa region which is occupied by the Beti, Yambassa and Bafia people. This zone is comparatively smaller and less densely populated than the first zone. Here, the density ranges from 25-100 inhabitants per km^2 and the people settle in villages and small towns. The influence of the town in Yaounde is very great in attracting people for administrative, commerical and educational pruposes. We also reckon with an important historical event for the concentration of population here. The dense population of this region also resulted from the migrations of many tribes from the north, chased by moslem Fulani during the spread of Islam within the 18th and 19th centuries. In their southern escape, they halted after crossing the Mbam and Sanaga rivers. This was a good refuge area where they settled in small isolated groups.

3. The Mandara Mounts and the Eastern Plain

This zone includes the block of the Mandara Mounts, its southern and eastern borders. The three administrative divisions which occupy this region of the country have high population densities which generally exceed 45 inhabitants per km^2. The case of the Mandara Mounts is of particular interest. The northern part of these mounts, (average population density of over 50 inhabitants per km^2) has the highest population density in the north. The Koza district, situated north of Mokolo town has a density exceeding 100 inhabitants per km^2. These mountain areas are occupied by the Mafa, Moufou and other kirdi tribes who are of the paleo-sudanese group.

The other regions which come after this very densely peopled northern part include the rest of the Mandara, the southern part of Mayo Danay, N.E. of Garoua and the extreme north, near lake Chad.

The high population density in this region resulted from the fact that this was a zone where many migrating groups converged. The pushing of the original settlers (paleo-sudanese) to the small refuge zone on the Mandara Mounts explains the high population concentration in this zone.

B. SPARSELY POPULATED ZONES

Between these heavily populated areas there exist zones of very sparse population, with some of them being practically uninhabited. Among these sparsely populated areas the main ones include:

a) The Adamawa highlands

b) The south-eastern part of the Benue depression

c) The northern part of the Southern Plateau

d) The south-eastern part of the Southern Plateau

The Adamawa, the south-eastern part of the Benue depression and the northern part of the southern plateau were formerly inhabited, but as time went by these areas were deserted by their inhabitants. The Tikars, the Beti, the Bulu and the

Fang who occupied these regions later migrated to the southern and western parts of the country.

As regards the south-eastern portion of the southern plateau, we note that this zone had always been avoided by migrating populations. Migrants have always preferred the western part of the country towards the sea. The south-eastern part of the Southern Plateau is primarily very inaccessible due to thick evergreen forest coupled with a less conducive climate and many disease-carrying mosquitoes. There was also a complete absence of any infrastructural installations. All these factors made this region not to have been inhabited at all, at least in modern times.

THE POPULATION DEBATE

Many Cameroonians are probably not aware that a large portion of the inhabitants of less developed countries (LDCs) live in abject poverty almost throughout their life-time. Some may not even imagine what it means to go for days without food just because of the abundant food found in Cameroon. Yet, many of such poor countries either do nothing about population increase or pursue policies which encourage population growth.

Failure to check population growth and pro-natalist policies adopted in the less developed countries have sparked a lot of concern in western industrialized countries. The industrialized countries call for a control of population growth in the less developed world as the essential means of improving the living standards of the people. Members of the less developed world disagree, holding that rapid population growth is not as serious a problem as the 'developed world' would make it sound.

The debate was quite evident at the first World Population Conference held in Bucharest in 1974. Officials from the less developed countries raised a number of strong arguments to show that rapid population growth is not the real problem. They held that "population growth is a false issue deliberately created by dominant rich-country agencies and institutions to keep the less developed countries in their underdeveloped, dependent condition," and that "for many less developed countries and regions, population growth is, in fact, desirable."

According to the argument, the real problem is, underdevelopment, which to a great extent has been caused by the dominant countries through the depletion of the world's limited resources for use in their own countries at the expense of the rest of the world. The American economist, Michael P. Todaro, points out that the average North American (or European) consumer uses up, directly and indirectly, almost sixteen times as much of the world's food, energy, and mineral resources as his counterpart in the less developed world.

It is sometimes argued that the rising affluence, extravagance, and greed in rich countries and among rich people in poor countries should be the major concern and not population growth.

Nowhere do such charges hold true than in Africa where, for more than three hundred years, beginning with the period of slavery and colonialism to imperialism, the material resources of the continent were enormously shifted to the western world and its population reduced almost to the brink of extinction.

The overconcern in the rich nations coupled with their aggressive population control programs necessarily have to be viewed with suspicion, especially when we consider that they themselves passed through a period of rapid population growth that accelerated their own development process.

A look at the 1984 population figures of some industrialized countries, even those with limited natural resources, shows that they are more populated than less developed countries. Many less developed countries are, in fact, endowed with both agricultural and mineral resources.

Country	Population '000	Average Density (per Km^2)	GNP Per Capita (U.S.$*)
U.S.A.	236,681	25	16,270
Japan	120,018	318	10,120
United Kingdom	55,624	228	8,970
France	54,935	100	10,400
West Germany	61,181	246	10,672
Switzerland	6,442	156	16,340
Zaire	32,084	14	160
Cameroon	9,467	20	834
Chad	4,901	4	59
Ethiopia	35,420	29	115
Nigeria	92,037	100	777

Source: United Nations Demographic Yearbook, 1984. New York, 1986
* *1986 Britannica World Data Annual.* Encyclopaedia Britannica, Inc. Chicago, 1986

From the table above, many observations could be made both between the less developed and more developed countries, and between the less developed countries themselves.

a) The high population concentration in the more developed countries does not constitute an economic burden for the nations because of their level of development and their grip on world resources through multinational corporations, financial institutions etc. On the other hand, Zaire, which is endowed with natural resources that are destined to serve American and European populations, has a relatively small population whose income per capita is lower than that of most European countries that have very limited natural resources. The critical question here is: Who should cut its population, the more developed or the less developed nations?

b) Among the less developed countries, there is no substantial difference in the levels of income, say, Nigeria and the other countries, despite the high population concentration in Nigeria, which, of course, is linked to the level of economic development.

There is actually little or no proof to show that the standards of living in Cameroon with a population of about 5.5 million in 1960 were higher than in 1986 when the population was 10.5 million. Rather, standards have improved because of the economic development that accompanied the population growth. It has to be noted that within this time no population control programs were actually implemented.

Neo-Marxist writers view population control efforts by rich nations and their allied international agencies as racist or genocidal attempts to reduce the size of the poor, largely non-white populations of the world who may, some day, pose a serious threat to the welfare of the rich, predominantly white societies. European nations are currently worried about the declining birth rates in their various countries. For example, in January 1983, Francois Mitterrand "declared that France's low birth rate was one of its major problems". His minister of Family Affairs, Georgina DuFoix, also stated in an interview that *"denatalite* puts at risk France's place in Western civilization". The French public also agreed with their officials that "if the birth rate continued to fall, France's standing in the world would be undermined".* Yet, the same European nations that are worried about a decline in their population, think it necessary that other nations have to check population growth.

Whatever the arguments, however, one remarks, with some amount of worry that such population figures recorded in the People's Republic of China, 950,000,000 for 1979, and 684,000,000 in India for 1980, need to be addressed. Population policies should be devised on an individual country basis and not on a global scale. In almost all African nations, the governments need to still approach their optimum population levels. For Africa, as a whole, and Cameroon in particular, the problem now is not population size *per se*. Rather, attention should be directed towards the causes of such population-related problems as, low levels of living, malnutrition, ill health, poverty, and rural exodus, resulting from the underdeveloped economies and a maldistribution of the inhabitants over the national territory in relation to where resources are plentiful. Some areas, such as the southern and eastern regions of Cameroon are underpopulated in terms of available potential resources. In fact, Cameroon offers one of the clear examples where enormous resources are still untapped.

*(*The Wall Street Journal,* June 20, 1984, p. 31)

In conclusion, it could be said that population growth is not the primary cause of poor living conditions of the less developed world. The causes are within national and international economic and social structures, operating from colonial to the present time. Natural hazards, such as drought and epidemics have contributed also, especially in those areas which lie close to the Sahara desert.

Influence of Economic and Social Development on Population Growth

It has been said by many that the most effective method to control population growth is to attack poverty and low standards of living.

Poor families will continue to have high birth rates because, without adequate incomes, employment, good health, education and social services, the only reliance for future security would be their children. Where infant mortality is high there is a tendency to have as many children as possible to make up for any loss through death. Education and good knowledge of childbearing would enable youths to postpone childbearing to a later date when they so desire. Higher academic and professional education, for example, tends to keep many people in learning institutions, thereby avoiding early marriage.

When women become better educated, and tend to earn a large share of household income, their traditional child rearing activities change, leading to a reduced rate of childbearing. Education and increases in income have been known to be very important in reducing the size of the family, because they enable parents to adopt good child-nutrition programs, which in turn, reduce infant mortality.

The status effect of increases in incomes raises the relative desire for material goods, especially among the lower class of citizens whose financial constraints made the purchase of durable and luxurious goods almost impossible. In such cases, the desire would be to maintain that already acquired status, or even go higher. The fear that more children not accompanied by higher incomes leads to 'stepping down' to "inferior goods" may force families to desire fewer children.

Family planning programs that advocate limiting the family size and using contraceptives, have often failed when they are forced on poor and uneducated people.

POPULATION GROWTH AND ECONOMIC DEVELOPMENT

In 1798, the Reverend Thomas Malthus put forward a theory of the relationship between population growth and economic development. The content of his *Essay on the Principle of Population* can be simplified and summed up in one statement that, *population tends to outrun the means of subsistence.* He maintained that human beings are bound to increase in numbers because of a powerful natural urge, the attraction between the sexes. Malthus explained that population increases *geometrically* (i.e. 1,2,4,8,16,32, etc.) whereas the resources to meet people's needs increased only *arithmetically* (i.e. 1,2,3,4,5,6 etc.). By this, Malthus meant that if a population could double in, say, twenty-five years, it would double again in the next twenty-five years, and so on indefinitely. Food production and other resources, on the other hand, increased at a much slower rate. As a result, human population, if not checked, will continue to increase rapidly without any limit and outstrip resources needed for human survival.

Malthus' belief of a slower rate of food increase was based on the *Law of diminishing returns.* This law states that: *if more and more men are employed on a given area of land, beyond a certain point, even though total output may continue to expand, average output per worker will shrink.*

Also important to this law is the notion that the amount of good farm land to be farmed in the world is limited. Once all the undeveloped regions of the earth have been settled and cultivated really intensively, any further attempts to increase food production will become less and less effective because they will bring into operation the *law of diminishing returns.*

Despite advances in agricultural technology, at some point, even in the distant future, the possibility of raising food production to cope with the massive population increases will remain dim.

According to Malthus, a balance between population increase and food production could only be achieved when there are natural or human calamities, or when people engage consciously in "moral restraint" and limit the numbers of their offspring. By "moral restraint" he apparently meant pre-marital chastity, late marriage, and sexual abstinence within marriage. Thus, Malthus was advocating some kind of birth control measures.

Malthus' views were shocking revelations about the fate of the human race in the face of declining resources. His pessimistic views did not turn out as he thought in Europe during the nineteenth century when many European countries at the time, were experiencing vast population increases. This was so because resources were massively being transferred from colonial possessions overseas to meet the population increases in Europe. Moreover, there were improved techniques of production that provided sufficient supplies of resources to cope with the population increases.

Even though Malthus might have been correct to state that the quantity of land was fixed, he was not aware that the quality could greatly be improved.

However, today the gloomy predictions made by Malthus still threaten the whole world, especially the so called less developed world. This is true of the countries which lie in drought-stricken areas.

Another flaw of Malthus' postulation is that it does not take account of the fact that national rates of population increase are directly related to the level of economic development (at least in the western industrial nations where the relationship has been well documented).

The relationship between population increase and economic development can be understood by examining *the theory of demographic transition* which is discussed below.

THE THEORY OF DEMOGRAPHIC TRANSITION

The theory of *Demographic Transition* is the most widely accepted theory explaining population change over time. It states that a population's fertility and mortality will both decline from high to low levels as a result of economic and social development. The decline in mortality usually precedes the decline in fertility, resulting in high poulation growth during the transition period. This is so, primarily because people in urban, industrialized societies possess well equipped medical centres and voluntarily limit the size of their families. In other words, the theory shows the relationship that exists between birth rates, death rates and the process of economic and social development. It tries to explain why contemporary industrialized nations have more or less passed through the same "stages" of modern population history. The demographic transition appears to proceed through three basic stages:

Stage One is the situation found in all traditional societies, that is, before economic modernization takes place. At this time, there are very high birth rates and very high death rates especially among infants. As a result, population growth is slow and in most cases remains fairly stable.

Stage Two is the situation found in all societies in their early stages of economic development and when this development is associated with improved public health methods, better diets, higher incomes, etc. At this time, the birth rate remains the way it was but the death rate drops sharply as a result of the improved living standards, thus raising life expectancy from under 40 years to over 60 years. Consequently, population grows very rapidly as the gap between birth rates and death rates widens. Stage two thus marks the beginning of the demographic transition, (i.e. the transition from stable or slow-growing population to rapidly increasing numbers).

Stage Three is the situation found in advanced industrial societies. The birth rate drops drastically as large families are to be seen as a burden, and the death rate (especially infant mortality) remains extremely low because of the existing medical services and good living conditions. Falling birth rates coupled with low death rates result in fairly stable population size with a growth rate almost at zero.

One may ask if this theory really works. At best, it describes what has happened in Western Europe from pre-industrial to the present. However, it should be realized that in many countries, the government could start specific population policies at any stage of economic development that will substantially increase or decrease the growth rate of the population. Moreover, cultures differ greatly. In some societies, the desire for children has little or nothing to do with the ability to raise them. In many African societies, children are wanted for their own sake and not as an economic asset as western writers tend to believe.

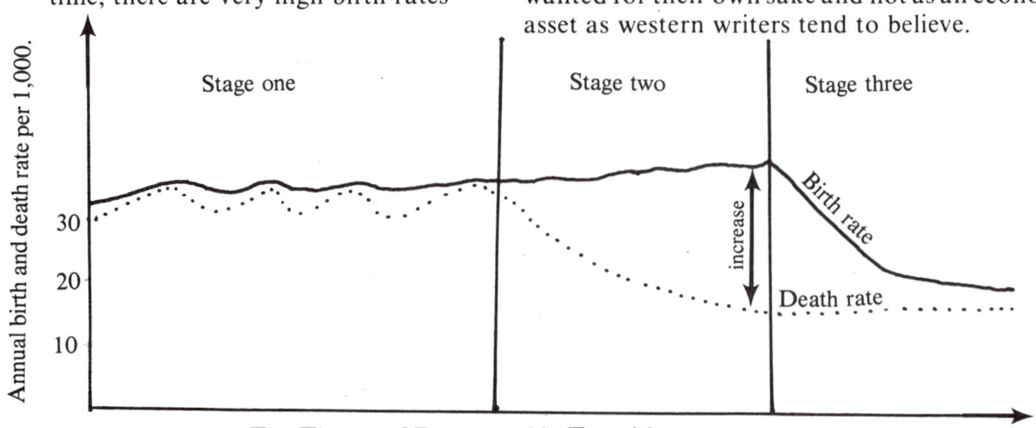

The Theory of Demographic Transition.

CHAPTER 10
URBANIZATION IN CAMEROON

What is Urbanization?

Urbanization is the increase in the proportion of the population living in the urban areas — the process of people moving to cities or other densely settled areas.

Urbanization involves two aspects:

1. The movement of people from rural to urban places where they engage in primarily non-rural activities; and

2. The change in life-style from rural to urban with associated values, attitudes and behaviour. The term 'urbanism' has often been used to refer to a way of life in which the traditional life-styles associated with village life are replaced by modern ways of life.

The important factors in the first aspect are population density, size and economic functions; the important factors in the second aspect are psychological and behavioral. These two aspects are interdependent. Urbanization is closely tied to modernization and industrialization, machine technology and division of labour.

What Then Is An Urban Centre?

It is fairly difficult to establish a clear distinction between a rural population and an urban population because what might appear to be a rural setting in one country may be considered an urban centre in the other. To put it differently, how many people must live in a concentrated area and engage in primarily non-rural activities to constitute an urban place? Minimum population concentration figures vary from country to country. Thus, 2,000 in France, Israel and Argentina; 2,500 in the United States, Mexico and Venezuela, 5,000 in Belgium and Ghana, and 10,000 in Greece. By Cameroonian considerations, an urban centre is any agglomeration of over 5,000 people engaged mainly in non-rural activities.

We may talk of an urbanized area, which is the *built-up area,* where buildings, roads, and other essential urban land uses predominate.

RAPIDITY OF URBANIZATION IN CAMEROON

Despite the fact that many more Cameroonians reside in rural areas than in towns, one nevertheless notices that the population of urban residents has continued to increase from one year to the other. By 1967 urban residents in Cameroon represented slightly over 18 percent of the total population, in 1976 they represented 28.5 percent and according to projections made by the ministry of Plan and Regional Development, by 1991 urban residents will comprise 40.35 percent of the total population.

Urban Population as percentage of Total Population 1967-1991

Year	%
1967	18%
1976	28.52%
1984	34.33%
1985	35.13%
1986	36.14%
1987	36.96%
1988	37.28%
1989	38.59%
1990	39.42%
1991	40.35%

Source: *Situation et Tendances de la Pop. Camerounaise P. 17. Y'de, Nov. 1985*

The growth of urban centres in Cameroon, especially since independence, has come about as a result of many factors, namely:

1. An accelerated rate of population growth and a large-scale migration from the countryside of rural residents to towns for jobs and services.

2. The creation of a strong agricultural base capable of producing food surpluses that can support both the rural and the urban population.

3. An efficient transportation system enabling rural residents to travel easily to towns.

EVOLUTION OF CAMEROON'S URBAN SYSTEM

The large-scale movement to urban centres of rural residents is a fairly recent phenomenon. It was not the case in the precolonial or colonial times, but since independence the growth rate of many towns has greatly increased. This, however, does not mean that there were no urban centres in Cameroon before independence. The Cameroon urban system is largely the product of three periods of town formation, namely: precolonial, colonial and postcolonial (or independence).

Many precolonial towns existed in the northern part of the country, built up around the residences of powerful traditional leaders (Lamibe) and around trade centres. Such

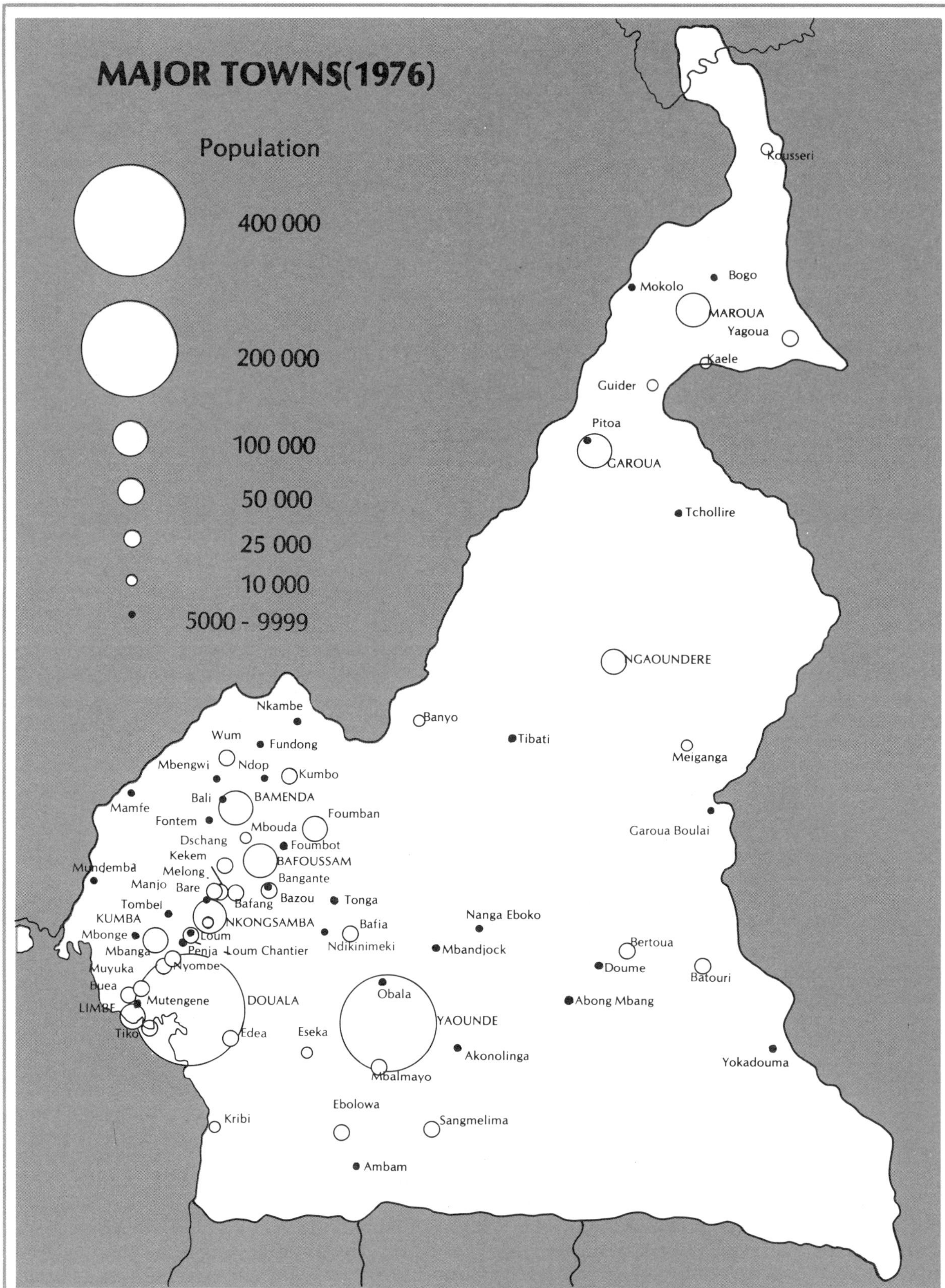
MAJOR TOWNS(1976)
Population
400 000
200 000
100 000
50 000
25 000
10 000
5000 - 9999
Kousseri
Bogo
Mokolo
MAROUA
Yagoua
Kaele
Guider
Pitoa
GAROUA
Tchollire
NGAOUNDERE
Banyo
Tibati
Meiganga
Garoua Boulai
Nkambe
Wum
Fundong
Mbengwi
Ndop
Kumbo
Mamfe
Bali
BAMENDA
Fontem
Mbouda
Foumban
Dschang
Foumbot
Kekem
BAFOUSSAM
Mundemba
Melong
Bangante
Manjo
Bare
Bazou
Tonga
Tombel
Bafang
KUMBA
NKONGSAMBA
Bafia
Mbonge
Loum
Ndikinimeki
Nanga Eboko
Mbanga
Penja
Loum Chantier
Mbandjock
Bertoua
Muyuka
Nyombe
Doume
Batouri
Buea
Obala
LIMBE
Mutengene
DOUALA
Abong Mbang
YAOUNDE
Tiko
Edea
Eseka
Akonolinga
Yokadouma
Mbalmayo
Ebolowa
Kribi
Sangmelima
Ambam

towns which were centres of *lamidats* included Ngaoundere (founded in 1836), Rey Bouba (founded 1804), Garoua (1938) and Foumban — centre of the famous Bamum Kingdom. In the South, Douala was already a major trading centre before colonization ever began. Some of the towns gained more significance with the arrival of colonial rulers and other new towns were created by the colonialists for administrative purposes. Colonialist, for the most part, did not create new settlements. Main centres of tribal settlements were later made *chefs-lieux* of *departements*. Such centres as Dschang, Buea, Bamenda, Garoua, Yaounde and many others, became so significant during colonial times that they served as nuclei, attracting people from the surrounding countryside.

GROWTH OF TOWNS WITH MORE THAN 5000 INHABITANTS IN CAMEROON, 1967-76

	Urban pop. ('000) 1967	1976	% annual change
1. Douala	219,843	458,429	12.1
2. Yaounde	151,237	313,706	11.9
3. Nkongsamba	45,278	79,343	8.4
4. Garoua	28,974	69,285	16.7
5. Maroua	32,333	67,187	16.1
6. Bamenda	25,900	67,184	17.7
7. Bafoussam	35,274	62,239	8.5
8. Kumba	37,065	58,469	6.4
9. Limbe	24,900	51,600	11.9
10. Ngaoundere	20,277	42,038	11.9
11. Foumban	22,863	39,425	8.0
12. Bafang	19,635	26,660	4.0
13. Edea	18,392	25,493	4.3
14. Loum	14,280	24,276	7.8
15. Tiko	13,185	23,411	8.6
16. Mbanga	17,099	23,153	3.9
17. Dschang	17,000	22,171	3.4
18. Mbalmayo	15,195	22,106	4.6
19. Guider	8,591	21,011	16.1
20. Bafia	14,664	19,295	3.2
21. Meiganga	10,681	19,174	8.8
22. Bertoua	7,458	18,460	16.4
23. Batouri	7,917	18,332	14.6
24. Ebolowa	21,385	18,277	-1.6
25. Wum	11,000	16,385	5.4
26. Manjo	12,499	15,435	2.6
27. Mbouda	7,555	15,098	11.1
28. Sangmelima	7,321	14,843	11.5
29. Muyuka	8,656	14,696	7.8
30. Yagoua	12,022	14,686	2.5
31. Kousseri	5,762	14,589	17.0
32. Banyo	4,448	13,612	22.9
33. Kumbo	8,311	12,603	5.7
34. Kaele	7,252	11,811	7.0
35. Kribi	5,178	11,286	13.1
36. Kekem	6,610	10,894	7.2
37. Bangante	10,080	10,556	0.5
38. Eseka	7,088	10,358	5.1
39. Foumbot	11,997	9,886	-2.0
40. Ndikinimiki	4,025	9,674	15.6
41. Tibati	5,089	9,509	9.7
42. Mamfe	11,729	9,082	-2.5
43. Tombel	6,262	9,010	4.9
44. Dizangue	300	9,007	322.5
45. Mbanjock	700	8,900	130.2
46. Obala	5,183	8,141	6.3
47. Akonolinga	6,095	8,063	5.4
48. Tchollire	2,426	7,755	24.4
49. Mbengwi	1,388	7,560	49.4
50. Tonga	7,288	7,518	0.4
51. Nkambe	4,724	6,826	4.9
52. Abong-Mbang	2,723	6,689	16.2
53. Fontem	9,162	6,343	-3.4
54. Nanga Eboko	9,656	6,201	-4.0
55. Bazou	9,846	6,129	-4.2
56. Yokadouma	4,876	6,007	2.6
57. Mokolo	3,450	5,794	7.5
58. Tignere	2,606	5,788	13.6
59. Doume	2,198	5,299	15.7
60. Batibo	5,460	4,311	-2.3
61. Bamendjou	12,166	3,144	-8.3

Source: Tandap, L. "Urbanization and Population Redistribution Trends in Cameroon" P. 144.

The most dramatic period of change in Cameroon's urban growth came after independence when new administrative units were created with *chief centres* within each Province, Division, and Sub-Division. The creation of these chief centres was often followed by the provision of attractive socio-cultural facilities (e.g. schools, colleges, police and gendarmerie centres). Job opportunities which attract people are thus created in such centres both in transformation industries and government offices.

Another important event which led to a relatively high rate of urbanization in the Bamileke lands and in the Sanaga-Maritime region, was the unrest of the 1950's and early 1960's caused by political upheavals in the country. Troubles and insecurity in the countryside drove the populations of the most affected zones to the towns. With the return of peace, most of those who had been rooted out of their homes in the countryside and who had lost all their wealth preferred to settle permanently in towns. It is partly for this reason that the rate of urbanization in the Bamileke territory was very high between 1958 and 1965. In the Sanaga-Maritime region there was the formation of the *Zone de Pacification (ZOPAC)* along the main roads between Douala and Yaounde. The Mbida government regrouped some ten to fifteen thousand civilians. The populations were resettled within stockades and supplied with alarm signals and weapons. The purpose of ZOPAC was to destroy the shelters of *Maquisards* in the forests and also deprive them of access to innocent civilians. ZOPAC enabled the political authorities to systematically hunt out and kill those considered to be terrorists. Those who were resettled formed small towns which have continued to expand in size.

DISTRIBUTION OF URBAN CENTRES

The distribution of urban centres in Cameroon depends essentially on the population density of the area, the available means of transportation, the level of economic activity and, of course, the dynamism of the local inhabitants.

Most of the urban centres of Cameroon are concentrated along a line on the south-western part of the country, starting from Limbe and Douala at the coast to Nkambe and Kumbo in the 'grassfields' zone. This is the main urban belt of the country and it corresponds to the zone of the highest population concentration. Meanwhile, west of this belt, along the Cameroon-Nigeria border, there are only a few urban settlements.

The rest of Cameroon towns are dispersed all over the territory; many of such isolated towns are on the South Cameroon Plateau and Northern Plain. The south-eastern part of Cameroon is least affected by urbanization because of relative inaccessibility, low population density, and an almost complete absence of any major economic activity.

PROJECTED CHANGE IN URBANIZATION RATE BY PROVINCE (1976-1991)

	Rate of Urbanization								
PROVINCE	1976	1984	1985	1986	1987	1988	1989	1990	1991
1. ADAMAWA	25.08	22.43	22.54	22.64	22.75	22.86	22.96	23.07	23.17
2. CENTRE	27.62	47.82	48.92	50.02	51.12	52.22	53.32	54.42	55.51
3. EAST	20.66	23.31	23.65	23.99	24.33	24.67	25.02	25.36	25.71
4. FAR NORTH	8.91	10.07	10.18	10.29	10.40	10.51	10.63	10.74	10.85
5. LITTORAL	75.13	79.70	80.26	80.81	81.34	81.87	82.38	82.88	83.37
6. NORTH	20.99	17.58	17.63	17.69	17.74	17.80	17.85	17.91	17.97
7. NORTH-WEST	14.97	16.81	17.08	17.35	17.62	17.90	18.18	18.46	18.74
8. WEST	12.43	26.43	26.84	27.26	27.69	28.11	28.54	28.97	29.41
9. SOUTH	17.62	24.52	25.14	25.77	26.41	27.06	27.72	28.39	29.07
10. SOUTH-WEST	32.28	43.25	43.97	44.68	45.39	40.11	46.82	47.53	48.24
CAMEROON	28.52	34.33	35.13	36.14	36.96	37.28	38.59	39.42	40.35

Source: Situation et Tendances de la Pop. Camerounaise. Y'de, Nov. 1985. P. 17

The rate of urbanization is highest in the littoral, followed by the Centre and then the South-West Provinces. The least rates are found in the Far North, North-West and the Adamawa Provinces. Apart from economic and political reasons, it would be realized that most of the least urbanized Provinces are those in grassland areas where the tendency is for the people to adopt a dispersed form of settlement on the hills rather than the grouped form of settlement to be found in the forest region.

URBAN HIERARCHY

The cities of a geographical area, if properly arranged according to size and function, will fall into some kind of hierarchical order. This notion of hierarchy shows the relationship among the cities in a particular area and elaborates on the concept of Central Place theory which assumes that places of varying sizes perform a variety of services for different hinterlands. In every nation, therefore, the cities could be arranged and grouped into categories according to the functions they perform. Thus, we may have a *national metropolis* or *first-order city* serving as the political, economic and cultural headquarters of the whole nation. A *regional-metropolis* or *second-order* city has its functions spread over a particular region within a nation. For example, Bafoussam may be said to be the regional metropolis of the Bamileke territory. Other cities with reduced size and function occupy a lower position on the hierarchical order.

URBAN PRIMACY

With the exception of a few countries, most African countries are characterized by massive population concentration in the largest city and its environs. That largest city stands out as the *primate city*.

The concept of *urban primacy* looks at the tendency for countries to have one city that is at least twice the size of the second city, and combines the functions of the port city, political capital, cultural centre and commercial centre. Such cities as Dakar, Conakry, Abidjan and Cairo, for example, are many times larger than the second-ranking city in their countries and contain a significant proportion of the national population and as much as half of the productive capital of the nation.

What one meets in Cameroon is a situation of *dual primacy,* where two large cities, Douala and Yaounde, share the major urban functions of the country. These two cities, Douala, the economic capital, and Yaounde, the political capital, are clearly at the head of Cameroon's urban hierarchy. Douala, the economic capital, contains the headquarters of companies and is the centre of the country's basic industries. It offers more job opportunities than Yaounde, most of which require little or no school training. In this way it has the advantage of attracting many more illiterate rural migrants than Yaounde. Unlike Douala, Yaounde stands as the intellectual centre of the nation attracting essentially literate migrants from the countryside to attend its numerous schools. Below these two cities the other towns reduce gradually both in size and function in their regional settings. This seemingly conscious distribution policy of urban centres over the national territory seeks to avoid the problems associated with urban primacy.

Only the two towns of Douala and Yaounde with growth rates about 10 percent per year, have made some giant strides ahead of the others. They form the two *first-order* towns of Cameroon. Other towns such as, Bafoussam, Bamenda, Garoua, and Nkongsamba are also growing fast, and could be termed intermediate-size or *second-order* towns. Such towns as Ngaoundere, Foumban, and Limbe belong to another category, the *third-order* towns. These categories are not static because the growth rates of some towns, for example, Buea and Mamfe, have tended to stagnate, while others (e.g. Bamenda) have increased over the years.

URBAN PROBLEMS IN CAMEROON

The development of cities has always proved a mixed blessing. Although cities are an acknowledgement of progress, they have problems, many of which have seemed to increase rather than diminish over time. These problems can be grouped as follows:

1. Inadequate Urban Route Infrastructure and Traffic Congestion:

Transportation problems concern mainly the congestion of pedestrians, bicycles and cars. Such congested traffic occurs especially in Douala and Yaounde where the pedestrian, cyclist or motorist tries to jostle his way through to go before the other. The streets are narrow and poorly maintained, many containing potholes. In many cases, there are no traffic lights, and where they exist frequent power black-outs occur. There are few or no parking spaces for cars so that many motorists park at random by the road-side, thereby narrowing the passage way. There is inadequate and sometimes no mass transporttion system. Urban mass transportation (SOTUC) buses are often overcrowded.

2. Inadequate Urban Services:

Planning agencies, health and welfare services, educational facilities (e.g. public libraries), fire protection and recreational facilities are either inadequate or non-existent in many Cameroonian towns. As a matter of fact, buildings and built-up markets in many towns (e.g. Bamenda, Tiko, Kumba) have been ravaged by fires within the past recent years.

3. Unemployment and Underemployment:

The migration of rural residents to towns since independence for jobs and better standards of living has caused enormous problems of unemployment and underemployment. Despite the shortage of jobs many youths continue to come and others are still kept in urban centres by the hope that some day things will get better. Many of these rural in-migrants often lack the rudimentary skills needed for industrial work; they do not possess high rates of literacy, and are otherwise ill-prepared for city living.

Another stimulus to city in-migration is education. Many of the secondary schools and other institutions of higher learning are located in the towns. The youths who attend these schools get accustomed to better conditions of living — good roads, electricity, piped-water and health facilities. After their studies, their main objective is to live and work in the town in preference to the village where the facilities mentioned above are non-existent. Unfortunately, not many of such youths get employed.

4. Poor and Inadequate Housing Conditions:

Urbanization which occurs without adequate industrialization, sufficient formal employment or secure wages, has condemned many urban residents to poor-quality housing.

The unemployed and underemployed cannot afford to buy or rent the decent type of housing constructed by the public and private sectors. Consequently, slum settlements have mushroomed in sections of all the major cities and towns and the residents are constantly being afflicted by diseases and are frequently being displaced in times of urban renewal processes. Lighting, sanitation, cooking facilities and telephone services are minimal in the populous sections of Cameroonian towns. In fact, many Cameroonian slum dwellers simply do not care and are not aware of health dangers they are exposed to in the filthy, degraded and overcrowded housing conditions under which they live.

5. Social Adjustment Problems:

The in-migrant is typically from a homogeneous rural society. In the city, he is confronted with a variety of issues and people of diverse backgrounds with whom he has to deal daily. Unlike in the village where his transactions were limited to his relatives and fellow villagers, in the city he has to learn to communicate in both English and French, he has to learn to fill out documents and write applications for jobs, he has to read road signs and cope with the congested traffic. He has to adapt to a new money economy — paying taxi fares for trips he would have done on foot, he must adjust to regular working hours and adopt new forms of recreation. All this constitutes a new way of life altogether, and this change from village to urban life-styles does not come easily to everyone, often resulting in emotional stress and a feeling of social rejection.

6. Crime:

It is essentially when an in-migrant fails to adjust properly to urban life and when the hopes of 'making it' are lost that other negative aspects of life arise. These may include alcoholism, armed burglary, drug addiction, gambling, murder, prostitution and stealing. As a matter of fact, the rate of alcohol consumption in Cameroon is one of the highest in Africa. Prostitution in an environment of inadequate medical facilities complicates and increases the incidence of communicable diseases. Even though the jobless town resident continues to dream of success in some later period, his real goal is simply to make ends meet from one day to another no matter whether he begs or steals.

7. Pollution:

In the cities and towns the tendency for air, water, soil and sound to be polluted is usually greater than in the village or countryside.

a. *Air Pollution:*

Air pollution is the presence in the atmosphere of harmful gases, liquids and solids. Air pollution in towns results mainly from the emission into the atmosphere of different types of gases from industries and households, from the exhaust pipes of motor cars, planes, trains and motorcycles. Such pollution is often referred to in urban areas as *Smog*. The presence of different types of air pollutants in the atmosphere is known to produce eye and respiratory problems in animals as well as in human beings.

b. *Water Pollution:*

Water is used for drinking, washing, industrial processes, irrigation, recreation and many other related purposes. Most of the wastes produced by human beings are disposed of in bodies of water —rivers, lakes and seas. Even those wastes disposed of in the air and on land are eventually washed down to waterways. When human population is concentrated in towns or cities, the bodies of water located near such settlements become easily polluted.

Urban waters are polluted by wastes from homes and commercial establishments, by water used in factories then discharged into rivers, lakes or seas (where fish become contaminated and killed). Polluted water has become filthy and impure for drinking, and can be the cause of many diseases, such as cholera.

In the towns of north Cameroon where water supply is limited, wells are sunk and in most cases their waters are impure because they are shallow.

c. *Sound Pollution:*

Unwanted sound or very loud sound is what we call noise. The major sources of noise to which town residents are exposed include, transportation noises from cars, trucks, motorcycles, airplanes; construction and demolition noises from builders and carpenters especially in the large cities; noises from musical appliances in bars and record-selling shops concentrated in the populous sections of the towns. Excessive sound levels or noises of any type can lead to hearing loss, annoyance and distraction.

d. *Solid and liquid dumps on land:*

Solid and liquid dumps on land, especially on city streets, can cause serious health problems. Solid wastes include refuse dumped near residential areas, and debris abandoned on streets. In fact, poor and inadequate drainage systems complicate matters by causing bad-smelling air from decaying refuse.

Industries are known to produce many chemcial wastes, some of which may be very toxic. When such wastes are dumped near residential areas they can produce negative results both on plants and living things. The concentration of industries on a relatively small piece of land as in Bonaberi or Douala, can cause many environmental hazards.

In many highly industrialized countries where pollution has become a serious problem (examples are acid rain and toxic waste dumps in the United States of America and the Scandinavian countries), laws have been enacted to clean up, and regulate the pollution of the environment. In Cameroon where the foreign industries are only out to make profit, little or no attempt is being made to spend money or clean up the contaminated lands and waters.

8. Irresponsible Municipal Authorities:

Urban administration, and more precisely municipal authorities, have so far failed to demonstrate an understanding of the needs of the urban poor and the gap between the elite and the masses keeps widening. Municipal authorities have either shown that they are irresponsible or lack the basic knowledge of what to do. They have not been able to develop policies to teach residents that littering is an offence which carries a penalty. Indeed, nobody seems to care about littering. Living under unsanitary conditions seems to be a way of life. The people seem to be completely cut off from any decision-making process that affects their lives.

The layout of streets and sidewalks; the naming of streets; the control of building appearance, size and shape; the separation of incompatible land use types (i.e. zoning), and many other town planning activities are either lacking or come late after the people have set up their settlements.

9. Urban Sprawl:

Most Cameroonian households, especially those in the middle and upper-income groups, have never really reconciled themselves to living in small apartments, let alone tall buildings. The dream has always been to own a spacious, single housing unit with enough yard. This has led many individuals to construct dwelling units, each detached from the other, and the consequence is urban sprawl. By *urban sprawl* we mean, the expansion of towns across vast areas, the uncontrolled spread of houses at the urban fringe, usually implying the taking up of more agricultural lands for residence, and increasing transportation problems from one end of the city to the other.

10. Inadequate Urban Food Supply:

As indicated above, urban sprawl means taking up agricultural lands. It follows, therefore, that urban sprawl results in a shortage of urban food supply. In the past, many crops, vegetables, fruits and nuts were produced close to big towns. Now, most towns are dependent on distant food sources and the problem is bound to increase, given the growing number of urban residents.

URBAN STRUCTURE AND DEVELOPMENT

A majority of western urban centres reveal the existence of various specialized areas following different land uses. Thus, we find residential areas, industrial areas, commercial areas, and the like. These specialized areas are created as a result of certain natural values offered by the land itself or as a result of planning by municipal authorities.

With the growth of specialized areas (or zones) different urban centres reveal a structure that is closely associated with the development process. Research on the internal structure of the city (or urban centre) started in the early part of the 20th century in north America and three classical models of urban structure and development were developed by American urban geographers.

These models include:

1. The Concentric-zone model,
2. The Sector model
3. The Multiple-nuclei model.

THE CONCENTRIC-ZONE MODEL.

In the early 1920s, Ernest Burgess, a sociologist at the University of Chicago, developed a model of internal city structure and urban growth: the *Concentric-zone model.* It must be stressed here that Burgess was not concerned with the static structure of the city, but with how industrial cities change over time; that is, how cities develop

Concentric Zone Model

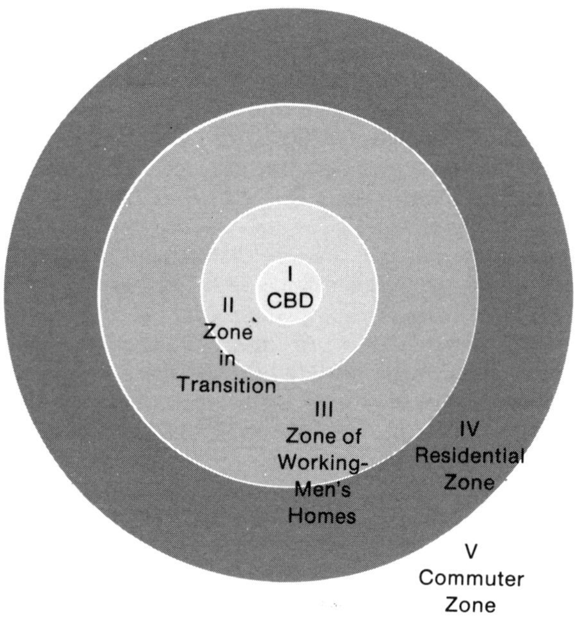

Sector Model

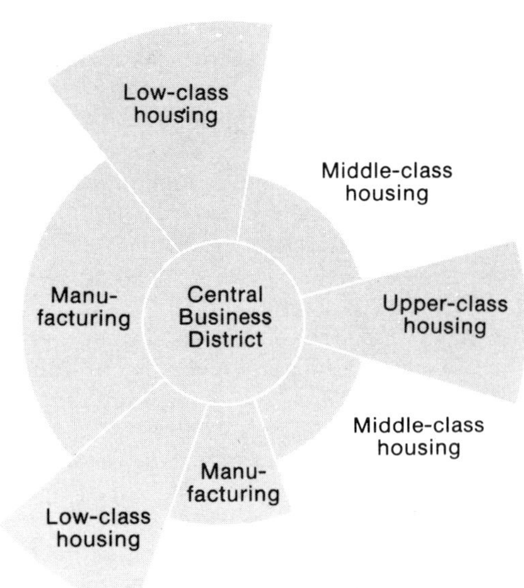

Multiple Nuclei Model

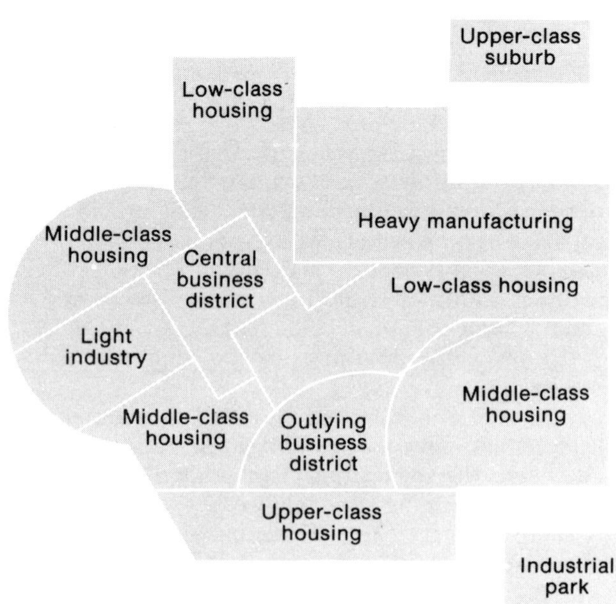

spatially. He saw the city of Chicago as a series of concentric circular zones of typical combinations of land use. The city, according to him, contained five successive zones, each of which tended to be expanding and encroaching upon the further outlying zones. Burgess then suggested that a city grows radially in a series of concentric zones or rings. The five zones he distinguished were as follows:

Zone 1: The Central Business District (CBD). This is the centre of business and civic life. It is the centre of retail trade, with major department stores, theatres, hotels, banks, central offices of economic, political, legal and civic leaders. This typical view could be seen, at least, in the 1920s, but a lot has changed since then. For example, due to transport facilities, most of the shopping centres in American cities have been moved away from the downtown location, but the demand for office space in downtown locations is still very high.

Zone 2: The zone in Transition. Residential areas in this zone deteriorate as a result of encroaching new businesses and industries from zone one. This zone also contains all factory complexes, many of which were created in the 19th century. Here, one finds residential slums, rooming houses, cheap hotels and other rundown dwellings. In the United States of America, this is the area of high crime rate and social disorganization.

Zone 3: The zone of independent workingmen's homes. This zone contains dwellings for blue-collar workers as well as shops designed to meet their needs.

Zone 4: The zone of "better residences". This is the zone of single-family houses, where white-collar workers reside and shopping centres develop. In the city of Chicago this was the zone of the great middle-class, small businessmen, professional people, salesmen. However, even in the 1920s other kinds of land-use patterns had started coming into the area.

Zone 5: The Commuter Zone. Members of the upper-middle class inhabit this zone. Thanks to the dependable commuter railroads and ownership of private automobiles which enabled the husband and wife to leave in the morning for the city and return in the evening after work. Beyond this zone is agricultural land or the hinterland of the city.

As a result of growth and expansion, the city must necessarily be organized spatially, and each zone extends its spatial area by encroaching on the next outer zone.

It must be borne in mind that Burgess was only proposing an "ideal type" of what American cities would look like if other factors did not intervene — but of course, many factors do intervene. He himself noted that topographic features such as hills, valleys, rivers, and artificial barriers which also include city plans and transportation systems, can intervene to introduce distortions into the zonal pattern.

Criticisms of Burgess came in the form of alternative models to the concentric-zone model.

SECTOR MODEL

The second model was formulated by Homer Hoyt in 1939. He was an American economist who worked for the Federal Housing Administration (FHA). Hoyt's model differed from that of Burgess. According to Hoyt, the growth of cities occurred from a slow outward expansion from the core of the city in the same direction resulting in the formation of sectors with similar land uses. Hoyt arrived at this conclusion after studying data from 142 American cities in which he observed that certain physical aspects of the city, such as transportation routes, unusual scenic beauty, or geographic barriers, modify the shape and the growth pattern of cities. Those areas best suited for industry or trade because of their location expand in one sector, residential areas in another sector.

Homer Hoyt added the following logical arguments.

1. Similar activities group together for mutual advantages, as in the case of the central business district.

2. Activities that are not similar are mutually detrimental or incompatible with one another. For example, it is unlikely that high-income or high-status residential areas will locate close to heavy industry.

THE MULTIPLE-NUCLEI MODEL

The third model of urban growth and structure is the *multiple-nuclei* model formulated by Harris and Ullman in 1945. Unlike the concentric-zone and sector models, the multiple-nuclei approach assumes that the city has, not just one, but several nuclei or centres, each of which is devoted to a different activity. Similar land uses tend to cluster. Thus, one nucleus may be devoted entirely to banking activities, another to manufacturing, another to government buildings, and so forth. Specialized facilities within a given area attract those who need and support it, and the city becomes differentiated.

Evaluation of the Models

To begin with, all the research work and the models that were developed concentrated only on American cities, and no universal pattern will be obtained from observations made only in one country. It has been observed that great differences in urban structure exist between industrial and non-industrial societies, between capitalist and communist societies, and even between European and North American cities. For example, it has been observed that the typical pattern of growth of non-industrial cities outside North America shows a situation where upper-class and upper-middle-class groups occupy the city proper and poor in-migrants settle in the periphery or in squatter shanty towns. This is to be contrasted with the North American situation where the upper-class and upper-middle-class (as seen in Burgess's model) reside far out of the city proper.

THE CENTRAL PLACE THEORY:

The Central Place Theory provides a good way of understanding the role of the city as a service centre. As early as 1826, Von Thunen, a German geographer, stated that under ideal conditions of a completely uniform land area surface a city would develop in the centre of the land area. It was this concept that Walter Christaller was going to use to formulate his famous Central Place Theory.

The essential argument raised by Christaller was that a central place (city) would provide goods and services to those living in the surrounding region, commonly referred to as hinterland. When viewed as the focal point of the region, the city becomes the supplier of wholesale and retail goods, services, and administrative functions, plus market-oriented manufacturing to the surrounding countryside or influence area. Walter Christaller, an economic geographer for that matter, also argued that a certain amount of productive land is necessary to support the urban place, and that the larger the city the larger its supporting land or hinterland. This implies, therefore, that the surrounding region and the city are interdependent in economic terms. It was Christaller who first recognized this economic relationship between cities and their hinterlands. Services directed toward the surrounding hinterlands are termed "Central functions" and places performing them are "Central Places".

According to the initial version of the theory as earlier stated by Von Thunen, each of the hinterlands surrounding a city would be circular in nature, with the city in the centre, but it was later suggested that the best shape that would fill all the spaces would instead be hexagonal.

Other aspects of the theory as outlined by Christaller are as follows:

1. Centrality: According to Christaller, the central location of a place is more important than the size and population. The term "central place" is used because to perform efficiently its basic function of service centre providing goods and services for the surrounding area, a city locates at the centre of minimum travel of its influence area.
2. The centrality of a city is an indication to which it is such a service centre; the greater the centrality of the place, the higher its "order." That is, the greater its centrality, the more influence it has upon its surrounding areas.
3. Higher-order places tend to offer more goods, have more establishments and kinds of businesses, larger populations, tributary areas and tributary populations, do greater volumes of business, and are more widely spaced out than lower-order places. Thus, we find small hamlets performing a few simple functions; small-sized towns, medium-sized and big-sized towns performing functions of different scales. As would be observed anywhere in the world, small settlements and towns serve as shopping and market centres for the nearby area, while major towns provide more complex services such as wholesaling, large-scale banking, insurance and the like to a much larger area.
4. Low-order places provide only low-order goods, such as foodstuffs, to low-order influence areas; these low-order goods are generally necessities requiring frequent purchasing with little consumer travel, and sold by low-order establishments. Conversely, high-order places provide not only low-order goods but also high-order goods sold by high-order establishments. These high-order goods are generally such manufactured goods as television sets and appliances, for which the consumer is willing to travel longer distances, although less frequently.
5. Central places can be arranged in a hierarchy comprising groups of centres. Centres of each higher-order group perform all the functions of lower-order centres plus some central functions that differentiate them from and set them above the lower-order centres.

Evaluation of the Central Place Theory

From the theory, there is little doubt that population size alone is not a true measure of the importance of a place. It is the central place functions that make the city important.

The notion of tributary, hinterland or spheres of influence that are characteristic of each urban area are well outlined. It is equally explicit in the indication that each city is intimately tied to the surrounding area in its functions.

Criticisms:

Christaller's theory, however, does not apply to all cities. The theory does not work with cities with specialized functions, i.e., performing one service, such as, mining, industry or recreation, for large areas. It applies mainly to trade centres.

The theory best works only under the conditions of a flat geographical surface with uniformly distributed population and a relatively unspecialized economy.

The pattern of hexagonal areas of influence in the ideal model are not found in real life. The real world has all sorts of barriers — topographical (e.g. mountains, lakes) and socio-political (e.g. territorial boundaries between nations) which alter the hexagonal pattern of exchange between a central place and its hinterlands. Moreover, the relation does not only exist between a city and its hinterland but also between one city and another (or many others) thereby making it difficult to delimit spheres of influence. For example, Yaounde and Douala interact with each other and are linked in their functions. Douala serves Yaounde in a good number of ways. Yaounde also serves Douala in other functions. Thus, it is difficult to say which of these two big cities in Cameroon constitutes the sphere of influence of the other. For the fact that the cities of a nation or region interact and are linked to each other in diverse ways it is preferable to talk of an urban system.

It is important to note that *geographical distance* is not really relevant in measuring the sphere of influence. What is more relevant is *economic distance.* Economic distance takes into account such variables as the costs of travel and shipping and socio-political arrangements which could seriously facilitate or hinder a city from rendering its central functions.

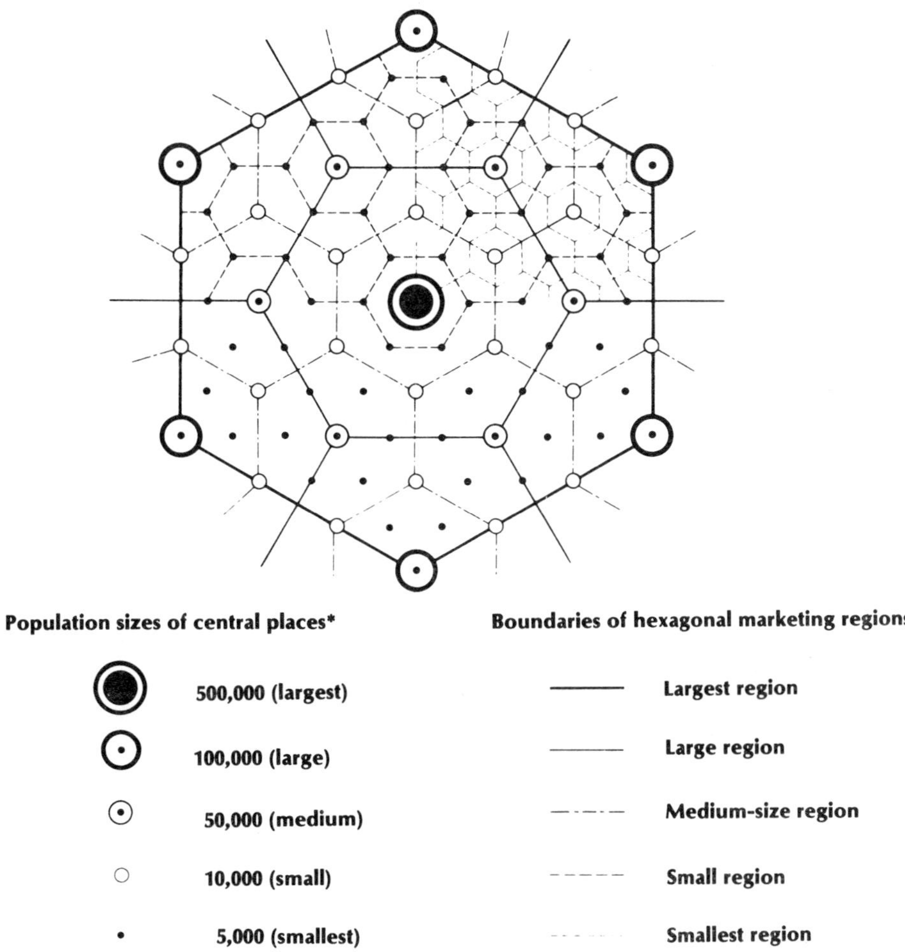

*The population sizes of central places have been chosen arbitrarily for illustrative purposes.

MARKETING REGIONS IN A SYSTEM OF CENTRAL PLACES.
(Source: Walter Christaller, *Central Places in Southern Germany* [Englewood Cliffs, N.J. : Prentice-Hall, 1966], p. 66. Copyright © 1966 by Carlisle W. Baskin. Reprinted by permission.)

PART FOUR: ECONOMIC STUDIES

CHAPTER 11: AGRICULTURE

Agriculture in Cameroon, like in most less developed countries, forms the main economic activity of the population. In Cameroon, over 80% of the population is involved in agricultural activities, and agricultural products account for 40% of the gross domestic product and for 70% of Cameroon's total exports in terms of value according to the 1976 figures.

The Cameroon government has realised the importance of agriculture and has therefore given enough attention to it. Through government agents, some resources have been placed at the disposal of farmers to enable them to increase the output and quality of their crops. Such resources include, the provision of farm loans (FONADER LOANS), pesticides, prizes for best farms, agricultural shows, etc. There are also some programmes, especially the "Green Revolution" launched in 1973, which aim at increasing agricultural productivity. Co-operative and Produce Marketing Organizations are all aimed at facilitating the activities of farmers.

TRADITIONAL AGRICULTURE:

General Characteristics of Traditional Agriculture

1) The agricultural undertakings by the Cameroonian farmer are usually on a small scale. The farms are generally small.
2) The agricultural technology is low, employing mostly hand tools such as cutlasses and hoes.
3) Though the plots are small, they are often scattered from one another.
4) Manure is not an important aspect in the farmer's programme.
5) Bush fallowing is an important characteristic of traditional agriculture in Cameroon (i.e. farming on one piece of land and after the yields start decreasing, the plot is abandoned for thick vegetation to grow so that better yields could be obtained from the next farming).
6) There is a lack of full incorporation of livestock farming and cultivation of crops.
7) Most of the crops are locally consumed i.e. for subsistence.
8) The yields are generally low since the act of manuring is never considered.
9) Each ridge is usually an agglomeration of different types of crops.
10) Most farmers own separate farms for subsistence crops and commercial crops.

The diversity in feeding habits among ethnic groups and natural conditions result in the creation of three main zones of traditional agriculture.

i) THE EQUATORIAL FOREST ZONE

The Bantu who occupy the southern part of the country practise a type of shifting-field cultivation. Shifting-field cultivation is an agricultural system which is characterised by a rotation of fields rather than crops, by preliminary clearing by "slash and burn", and by short periods of cropping alternating with long fallow periods. Only human labour is used; there are usually no draught animals — manuring is not practised, and the chief tools are cutlasses and hoes. The people clear a patch of woodland or forest, sow and harvest their crops year by year until decreasing yields show that the land is losing its fertility, and then without difficulty move into another area and clear new farms.

Method:

Usually in the dry season, after choosing the piece of land for the farm, it is then cleared down using mainly rudimentary tools such as cutlasses, hoes and axes. Sizeable trees are felled and when the vegetation dries up, fire is set on it and after that, tilling takes place. Typical crops here are tubers, such as, cassava, yams, cocoyams and there are also a lot of plantains and bananas.

In some very fertile areas, for example, the slopes of the Kupe Mountain, some of the inhabitants do not even till the soil. After clearing, the grass decays and cocoyam tubers and plantain suckers are just infused into the soil. After some months the weeds are removed while the crops are allowed to grow.

As regards seasonal plants such as cereals — groundnuts, maize, etc. — they do not do very well in this zone. The problem here is that cereals can grow quite well but will not ripen fast due to insufficient sunlight and too much humidity. The inhabitants therefore cultivate the cereals on less humid spots.

The yields on the piece of land reduce as years

go by for about three years and after that, they move to another new forest and allow the already cultivated zone on fallow for about ten years.

In this forest zone, some tribes have staple foodstuffs and therefore give priority to the cultivation of such crops. For example, in the South-West Province, the Bakundu are more inclined to plantains; the Bakossi to cocoyams; and the Banyang people eat cassava in its various forms.

On the other side of the Mungo, the Bassa and Bakoko depend mostly on cocoyams while the Maka and Fang depend on plantains. A lot has been written on the dangers and limitations associated with the shifting-field type of agriculture. The system depends on sufficient land being available so that adequate fallowing can take place. Especially where population density is high or where custom restricts the cultivable area, the system can lead to soil exhaustion, loss of fertility (especially by destruction of humus in the upper layers of the soil) and soil erosion. Shifting-field cultivation may also be responsible for the disappearance and degradation of forests. Again, this type of cultivation has been held to constitute a major limitation to the spread of permanent tree crops — oil palms, mango trees, pear trees and many other valuable fruit trees are burnt down by fires.

The system, however, should not be condemned out of hand because Cameroonian farmers have realised some advantages which go with it. Shifting-field cultivation as an agricultural system requires less labour and tools to produce a specified amount of food than does the system of permanent-field cultivation. Some farmers have suggested that the system provides a way of dealing with the severe weed problem. Burning destroys weeds and controls the growth of weed seeds for some time. The system is also suitable where population density is low enough to permit sufficiently long fallow periods so that soil fertility can be regained. In fact, it has always been the best form of land use in the early stages of settlement where population density is low. It would therefore be unwise to outlaw the practice without showing rural cultivators an alternative, and without showing them how to develop a more intensive type of land use in their social and natural environment.

ii) THE SUDAN SAVANNA ZONE

This includes mostly the Bamileke and Bamum zones, and then the north-west highlands which are generally referred to as the Bamenda grassfield zone.

This area falls within the Sudan Climate where there are two precise seasons — the dry and rainy seasons.

On the Bamileke and Bamum plateau, a type of intensive cultivation is practised due to the high population concentration. Land under fallow may not be allowed to last up to ten years. Maize is the main crop which supports the fast growing population. The land may remain uncultivated for only two or three years.

The ridges are aligned across the slope and on the same ridge maize is usually mixed with cocoyams, plantains, beans and groundnuts.

In some areas, especially around the Bamenda grassfield zone, the cleared grass is arranged on the furrows and covered with soil and after that fire is set on it. The grass burns inside the ridge and forms manure which accounts for heavy yields during the first year. After the first year, harvests for the subsequent years start decreasing. With time the farm is abandoned to create room for more vegetation to cover the area again.

This method is, however, being discouraged because the burning destroys many microbes in the soil.

iii) THE SAHEL ZONE:

More northerly as one enters the sahel zone from the Adamawa and as one enters the less fertile zones, tubers thrive with difficulty and give way to cereals. Here, we have only one agricultural season which begins with rains and in the end the harvest is done. Planification on food consumption is undertaken so that the food will last till the next harvest season. Maize dominates around the Adamawa and as one moves further north, it only comes next to millet which is the main crop, and cultivated almost everywhere. Swamp rice is also found on both sides of the Logone river.

The farming systems in the north vary according to the locality. Shifting-field cultivation, however, dominates in most areas but the system of the mountain dwellers is of particular interest.

The people who settle on the hill tops practise a type of terrace farming on the slopes of the hills and mountains. Each terrace is well supported by an alignment of small stones. This is particularly the case with the Mafa and Mofou on the Mandara Mountains.

THE CULTIVATION OF LOCAL FOOD CROPS:

Agricultural production in traditional societies is mainly geared at producing what is needed to nourish the society. This means that the supply of the necessary foodstuffs should go to feed the population of the society and to serve in their social needs such as marriage, death, birth, and sacrificial ceremonies.

Women in the grassland area prepare a field for planting.

Terrace farming around Mora in north Cameroon.

We have to note here that there are no such communities still existing today where production is only for self-sufficiency. Remnants of such human groups in Cameroon are the pygmies and the inhabitants of the Mandara and Alantika mountains.

With the expansion of urban centres, families do not only produce for self-sufficiency but to sell the surpluses in the neighbouring towns. As a result of this, families cultivate large areas and even use some little manure in order to sell the excesses and be able to afford manufactured goods which they themselves cannot produce.

The principal food crops of Cameroon are:

A) TUBERS:

Cassava: Cassava grows almost everywhere in the country, preferably under a hot and humid climate. It is thus found in the southern part of Cameroon especially in the Centre, Littoral, South and South-West Provinces, but as one moves to the north it disappears. It occupies the soil for up to eight months. Its nutritive value is, however, low but it is widely eaten in the country.

Cocoyams, Yams and Potatoes: They are some of the widely eaten root crops in Cameroon. Cocoyams are intensively cultivated in the rich volcanic soils of Bakossi, Bakweri and Foumbot. These crops grow well in the humid zones of the south. The only tuber which does not favour very humid conditions is Irish potatoes which can be found in great quantities in the Bamileke lands, Santa, Bambili and the whole of Bui Division.

B) CEREALS:

Their main domain is the savanna zone of the country. They are often intercropped with tubers on the same ridges.

Maize: This cereal needs physically good soils which are rich in humus but at the same time it avoids very humid soils. Maize is found all over in the country but wide-spread on the western highlands. In the North-West province it forms the staple food of many tribes such as the Bali, Nso and Kom.

Millet: There are two main varieties of millet found in Cameroon — the small or fine millet, and the sorghum or coarse millet.

Both types do not do well in very humid areas. This means that millet is completely absent in the southern part of the country which is very humid. The small or fine millet prefers light sandy and less humid soils, whereas the sorghum does well on fairly wet soils. The fine type of millet is more nutritive than the sorghum. Millet is the main crop in north Cameroon since its ideal growing conditions are found there.

Rice: The ideal conditions for rice are met in alluvial valleys or depressions. Rice is cultivated both on hills and down in the valleys where swampy conditions exist. It really does well where water is irrigated into the plots. Rice could be grown all over the territory where the favourable conditions are present but at present it is found in many isolated plains and valleys, for example, in Nanga Eboko, along the Logone river; in the Ndop and Mbo plains; along the Menchum and Noun valleys. Experimentation is being undertaken in many other places.

Groundnuts: Groundnuts are produced all over the territory in varying quantities by individual families. The best areas for groundnuts are where rainfall is just about 450 mm a year and there are light and soft soils. The north, therefore, offers the most favourable environment and is the region that produces the greatest quantity of groundnuts. They, however, avoid the swampy coastal areas. They are produced in the south, centre, west and east of the country but all the harvest is consumed by the population. Groundnuts are popular for use as a munching food when freshly harvested. Only the northern region exports decorticated groundnuts and it is this region alone which is responsible for the inclusion of groundnuts among Cameroon's cash crops. For the year 1975/76 this region produced 56,800 tons of groundnuts.

C) OTHERS:

Plantains: This is one of the main foodstuffs that feeds the urban population of the country. Plantains are cultivated in the forest zone and on the fertile western highlands. They are also found in areas which are naturally fertile, such as, Bambui, Bambili, Kom, and Bafmen, but their main environment is the forest zone. Plantains are convenient foods because, once planted, a crop can be expected each year for several years from descendants of the same "father" sucker. The plantain is always eaten cooked and the leaves are used as cover, plates and wrappings.

Wheat: Wheat is not widespread over the territory. At the moment, it is cultivated near Ngaoundere at Wassande by SODEBLE.

Wheat does not do well in very damp areas but requires enough moisture for the early growing period. Its growing areas should have an annual rainfall of 35-40 cm. For the grains to ripen enough, sunshine is needed but the temperature conditions should not be too hot. The soils should be rich in nitrogen, well drained and clayey enough to support the plant.

Sugar Cane: It is mostly planted around the compounds to benefit from refuse of the kitchens. However, since 1965 it has been cultivated on a plantation scale at Mbandjock, some 95 km from Yaounde town. These agro-industrial complexes — SOSUCAM set up in 1968 and CAMSUCO

Women husking maize.

Millet Farm in North Cameroon.

(Cameroon Sugar Company) set up in 1975 — both cultivate and process sugar cane destined for local consumption.

Beans: Beans are vastly produced in the western and northern parts of the country. They are also produced on a smaller scale in the other provinces. The collection from the country side for sale in the big towns is well organised and undertaken in certain localities by cooperative societies.

Oil Palms: Very few of these palm trees are planted by human beings. Most of them grow on their own and are spread by rodents, such as, squirrels and mice. Vast expanses of palm trees are found in the forest areas, especially in Manyu, Meme and Ndian divisions.

Many of these oil palms can also be noticed in Widikum and Bafang. From them palm oil, kernels, local cotton and wine are obtained but the oil proceeds are not as high as in the case of the plantation palms. It is important to note that the preferred drink in the southern part of the country is palm wine made from sap drawn from the tree. Some tribes fell the tree before drawing the sap inspite of regulations which forbid the felling of palm trees for that purpose.

It could be reasonably assumed that the local food resources of Cameroon would be adequate to support the population if the numerous hectares of unused agricultural land were cultivated; if the cattle from the north were all exploited for meat and milk; if transportation were improved to allow the domestic marketing of food crops; and if the share of capital investment allocated to the production of food crops were increased.

CASH CROP FARMING

It is difficult to make a clear distinction between "subsistence crop farming" and "cash crop farming" in Cameroon because there are almost no situations whereby the farmers do not use their crops for exchange.

Subsistence agriculture today in Cameroon may perhaps most usefully be viewed as an agricultural type which is devoted mainly to the production of food for local consumption but includes also a varying proportion of cash cropping. A critical factor, clearly, is the intention of the farmer. If the farmer merely sells his surplus, or sells some of it when he needs a bit of cash, then he is still a subsistence farmer. The distinction is further facilitated by the fact that some crops cannot be processed by the farmer for immediate consumption, so that their cultivation is solely for the sake of the market. Such crops include cocoa, coffee, rubber, and cotton. It is this form of agriculture that one refers to as "cash crop farming."

Commercial crops in Cameroon play a very important role in swelling the exports of the nation. These crops include mainly:

Cocoa: Cocoa was first introduced in Cameroon by the Germans in 1905. The cocoa tree is on the average, six to eight metres high, with a trunk diameter of about fifteen to twenty five cm. and it begins to produce after four or five years. This is the leading cash crop of the country both in tonnage and value. For example, in 1974-75, cocoa sold in all its forms accounted for about 36% of the country's sales.

Cameroon ranks fifth among world cocoa producing countries after Ghana, Nigeria, Brazil and Ivory Coast.

Major Cocoa Producers in the world (1975)

1) Ghana	396,000 tons
2) Brazil	230,000 tons
3) Nigeria	220,000 tons
4) Ivory Coast	210,000 tons
5) Cameroon	100,000 tons

This crop needs an average annual rainfall of over 1,000 mm, high and constant temperatures of about 30° C., and a good shade to keep the soil moist and also protect the plant from wind and the direct rays of the sun.

It does well on deep alluvial soils which retain enough moisture but drains freely. The atmospheric humidity at all seasons should be high. Apart from the former Cadbury and Fry Plantation in Kumba (which has now passed over to individual hands), most of the cocoa is cultivated in the framework of peasant farms. Cocoa is not cultivated in large capitalist plantations. Cocoa is not produced in the forest region. The major producing areas are the Centre, South, and South-West provinces at an average altitude of 700 m. Important localities in this region include Bengbis, Obala, Saa, Amban, and even the immediate outskirts of Yaounde town. The cultivation of this crop is gaining grounds now towards Bafia. The Centre and South regions account for over 70% of the total production of cocoa in Cameroon.

Other significant cocoa producing areas in the country include Kribi, Mbanga, Meme, Abong-Mbang and Akonolinga.

Cocoa in Cameroon suffers from, at least, three problems:

a) Blackpod disease and capsid pests.
b) Blowing down of the stems by serious storms.
c) The plants have always been trampled upon by elephants which also break down the branches.

For the fact that cocoa occupies an important place on the list of the country's exports, the government created the SODECAO (Societe de Developpement de Cacao), which is an organisa-

Cash Crops

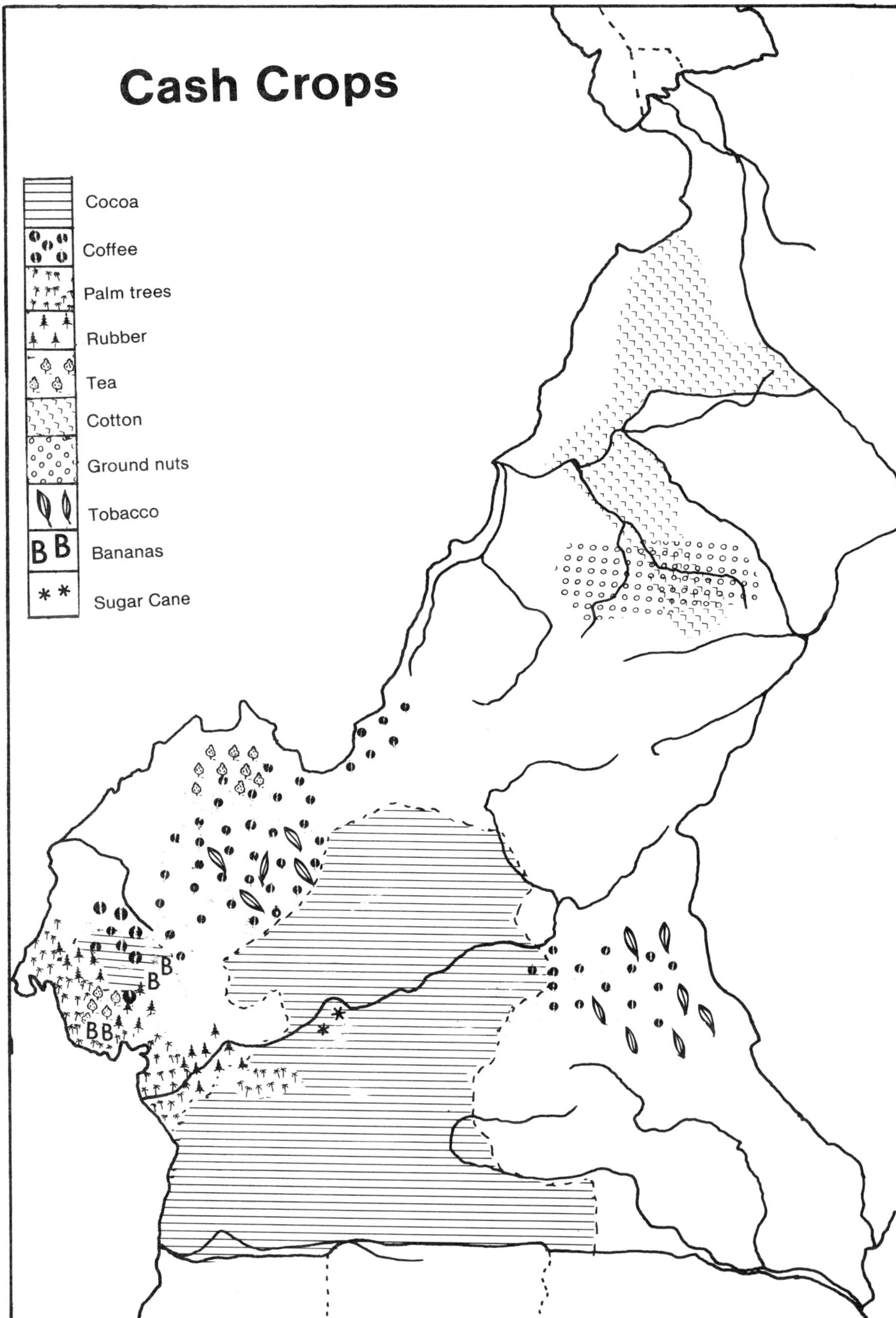
Cocoa
Coffee
Palm trees
Rubber
Tea
Cotton
Ground nuts
Tobacco
Bananas
Sugar Cane

Cocoa beans being sun-dried.

tion charged with the duty of stimulating farmers and improving the production of cocoa.

Coffee: It is the second crop after cocoa by value on the list of Cameroon's export crops. Coffee is believed to have been introduced in Cameroon in 1905. Two main varieties of coffee are cultivated: The Robusta Coffee and the Arabica Coffee.

a) *The Robusta Coffee:* It has broad, dark-green leaves. This variety thrives in the forest zone where humid and high temperature conditions are found. The soils should be fertile and well-drained. The Robusta Coffee finds its favourite conditions in fertile volcanic soils located at an altitude between 300-800 metres in the regions of Loum, Nkongsamba, Bare, Melong, Bafang, Mbo Plains, Bangwa, Kumba, and also in the eastern part of the country in the regions of Akonolinga, Ayos, Abong-Mbang, Bertoua, Batouri, and Yokadouma.

b) *The Arabica Coffee:* The zone of Arabica coffee is less extensive than that of Robusta Coffee. This variety is held to be higher in quality than the Robusta type. The Arabica coffee thrives on the mountain areas situated about 700 m with low temperatures. Because of these conditions, it is limited mostly to the western highlands, especially around Foumbot, Bafoussam, Dschang, Santa, Bui, Kom and Bafmen.

In the North-West Province the Produce Marketing Organisation (P.M.O.), through the intermediary of many cooperative societies, is responsible for the collection and sales of this crop. In the West Province the U.C.C.A.O. (Union des Cooperatives de Cafe Arabica de L'Ouest) takes care of the Arabica coffee produced by the indigenous farmers. The production of coffee by expatriate farmers is handled by COOPAGRO.

Flowering coffee plant.

The cultivation of Arabica coffee by the indigenous people is usually done in association with some local foodstuffs such as maize and beans. A new species of Arabica coffee commonly termed "Java" is under experimentation in these upland zones. This type is more high-yielding and more resistant to disease attack than the already existing Arabica coffee. There is a general strive towards replacing the already existing Arabica coffee with this new Arabica species.

Bananas: Bananas need certain conditions. They require much humidity, average temperatures of about 25° C. The volcanic and alluvial soils of lowlands are ideal conditions for bananas to do well. These conditions have given rise to their vast cultivation at Muea in the South-West Province by the Cameroon Development Corporation. Other very extensive farms for bananas are established by the O.C.B. (Organisation Camerounaise de la Banane) at Loum and Nyombe and other localities in the Mungo region.

Private individuals also have their own farms located mostly near those owned by these para-public organisations. Frequent disease attacks of the *Gros Michel* variety led to its replacement with the *Giant Cavendish* variety which is commonly known as *Poyo*. Bananas are mostly cultivated for commercial purposes by public organisations or large capitalist plantation owners because exportation needs special care and peasants cannot afford such care.

These corporations even use helicopters to spray the bananas — a thing which is almost impossible for peasants in Cameroon to do. Bananas are exported mainly to France and Germany.

Oil Palms: It was remarked earlier that most

Cultivation and Export of Banana by O.C.B.

Location	Area Planted	Exportations 1975/77	Exportations 1977/78
Loum	400 ha	7,941 tons	9,814 tons
Mpoula	235 ha	4,351 tons	4,017 tons
Bonandam	110 ha	2,546 tons	2,612 tons
Mbaka	100 ha	1,572 tons	1,312 tons
Tiko	120 ha	2,224 tons	1,653 tons
Nyombe	11o ha	4,095 tons	2,645 tons
Mantem	400 ha	5,266 tons	8,041 tons
	1475 ha	28,026 tons	30,094 tons

(Source: National Yearbook of Cameroon 1980)

Oil palm plantation in the Fako Division.

Cotton farm at Guider in North Cameroon.

Rubber Plantation at Tiko.

Ndu tea Estate:

palm trees in Cameroon are wild and planted by rats, mice and squirrels. Besides these wild palm trees whose products are used mostly for local consumption, we also have vast plantations of palms set up and exploited in a capitalist way, and products from such exploitations are commercialized externally.

The oil palm needs generally high temperatures, which should not be lower than 30° C. There should be high humidity all year round as is the case in the forest zone. Rainfall should be over 1300 mm and lasting not less than eight months. These conditions are present in the following areas which are consequently marked by great quantities of palm trees — almost the entire South-West province where the C.D.C. and PAMOL have created vast plantations. The Cameroon Development Corporation palm plantations are located at Tiko, Mutengene, Ekona, Mpundu, Sasse. The C.D.C. is still extending the palms to new areas. The PAMOL farms are found at Tiko, Ndian and Lobe.

At Mbongo and Eseka in the Littoral Province the SOPAME (Societe de Palmeraies de Mbongo et Eseka) is developing many palm plantations. Elsewhere in the forest zone the wild palms grow indiscriminately, notwithstanding the rate at which they are being cut down for wine tapping. This is noticed in Mamfe, Kumba, Dibombari, Eseka, Kribi, Akonolinga and Abong-Mbang, where the inhabitants fell the palms before tapping. Oil palms supply the populations with oil, wine, kernels, brooms and the stems are used for wood when they are old.

Rubber: Cameroon is the fourth greatest producer of rubber in Africa. This crop was introduced in the country by the Germans. The required conditions for growing rubber are almost similar to those for palm trees, namely: high temperatures, heavy rainfall, lowlands with deep, rich and well-drained soils. Four corporations are involved in the cultivation of this crop, namely: the C.D.C. and PAMOL, operating in the South-West Province, SAFACAM in the Littoral province at Dizangue, and HEVECAM in the Ocean Division of the South Province. The C.D.C. at present owns the largest farms for this crop.

Cotton: Cotton is one of the main cash crops of Cameroon and is cultivated mainly in the northern part of the territory. Unlike the crops of the south, cotton needs only 1200 mm of rain or moisture in the early growing season which begins in June. It needs plenty of sunshine during the growing period and a long dry season to cover the ripening and picking periods. It does very well on deep and well-drained clayey or sandy soils which retain just enough moisture.

The above conditions exist in the entire northern part of the country especially in the Diamare region. The only zone in the north that excludes cotton is the swampy area of the Longone and Shari rivers. Cotton is an annual crop which is planted in June and by the end of November or beginning of December picking is done. Abundant labour is usually needed during the picking season. The harvested cotton is ginned in many factories in North Cameroon at Kaele and Maroua, then sent to the textile industry (CICAM) at Garoua for spinning.

Tobacco: The cultivation of tobacco in Cameroon is practised in two main regions namely: the western highlands and the East Province. On the western highlands it is grown in the Bamileke and Bamum lands and consumed mostly in the country. It is sometimes cultivated in this region in association with other crops on the same plot. Drying methods after harvesting are not very well organised. It is often smoked in pipes locally and part is exported to the Bastos Company in Yaounde which manufactures cigarettes.

In the East Province, tobacco is produced in the localities of Batouri and Betare-Oya. Some of it is cultivated at Kribi in the South Province. The tobacco of the East province is of fairly high quality and part of it is therefore destined for export.

The cultivation of tobacco in Cameroon is undertaken by individual farmers under the supervision of the agents of the Cameroon Tobacco Company and the Bastos Company who buy and transform what the farmers produce. Since July 1978 the Cameroon Tobacco Company (SCT) has been fabricating the "SANAGA" cigars which are sold both within the country and to the western world.

Tea: Tea is almost entirely in the hands of the C.D.C. which is the principal producer in West Africa. The main producing areas are at Tole near the town of Buea in the South-West Province and at Ndu in the Donga and Mantung Division in the North-West Province. Currenty, the country's total output is estimated at 2,000 tons of dry tea per annum. Tea does well on the slopes of the western highlands. It needs rich and well-drained soils and, of course, an enormous labour force since the picking of the leaves must be done by hand. It does well only in areas of cool climate.

Much of the tea is consumed within the country and part is exported to Chad, Nigeria and other neighbouring countries. Presently it is being experimented at Dschang, Santa, Nkolbissong, Bafang, Melong and Otele, and an agro-industrial complex is being envisaged for Djuttisa near Dschang.

The aim of the Djuttisa project is to set up a

25-hectare selected tea plantation and also to build a factory which will provide about 1,000 tons of dry tea per annum.

Pepper: This is another crop that is exclusively in the hands of the C.D.C. This exportable type of pepper is grown mainly at Meanja and Bimbia. Unlike the local type of pepper, this variety grown by the C.D.C. bears in small clusters made up of small berries. Most of Cameroon's pepper is exported to France.

Pineapples: Pineapples that are exported are cultivated on the fertile soils of the Mungo division especially in the locality of Nyombe. Another area which produces exportable pineapples is Mbandjock. Large amounts of these pineapples are usually consumed locally.

Quiquina: The cultivation of this plant from which quinine is obtained, started in Cameroon in the locality of Dschang around 1959. A factory was set up in Dschang for the fabrication of quinine sulphate. Production work in this factory was interrupted for some years but since 1965 work has resumed in full swing.

MAJOR CASH CROPS

CROP	CORPORATION	REGION
Selected oil palms	CDC PAMOL SOCAPALM SAFACAM	South-West South-West Littoral Littoral
BANANAS (fruit)	CDC OCB Programme Private plantations	South-West Mungo
PINEAPPLES	Private plantations	Littoral & East
RUBBER	CDC PAMOL SAFACAM HEVECAM Project	South West South West Littoral Ocean
TEA	CDC	Tole and Ndu
TOBACCO	SCT(1) SACTA (Bastos)	East West
SUGAR CANE	SOSUCAM CAMSUCO	Centre at Mbandjock
COTTON	SODECOTON (took over from CFDT in 1974)	North (Private farms)

1)S.C.T.: Societe Camerounaise des Tabacs (Cameroon Tobacco Company).

It is worth noting that these companies are not involved in the same types of undertakings. Some of them, such as the C.D.C., PAMOL, SAFACAM, SOCAPALM, HEVECAM, SOSUCAM, CAMUSCO actually grow and process the crops but others, such as, the S.C.T., SACTA, and SODECOTON have only the role of stimulating the farmers and they also help to buy and transform what the farmers produce.

PLANTATION AGRICULTURE IN CAMEROON

It was effectively during the colonisation period that Europeans began to explore the tropical world. Their first interventions were mostly for commercial purposes. As regards agriculture, their interest initially was only to collect products from the interior of the countries but as time went on there was need to augment the quantity of products they needed and this led to the creation of plantations. By *plantations* here we are referring to the agricultural pieces of land which are being exploited on modern basis with modern techniques as opposed to the traditional types of farms already discussed.

Economically, most of the crops in the plantations are destined for export to the temperate countries which are in need of tropical products. Almost all the plantation crops in Cameroon, as in most tropical African countries, are not originally of the country. For example, cocoa grown all over in the southern part of the country originated from America, coffee from Ethiopia and Saudi Arabia. The establishment of these plantations has led to the creation of a network of roads, workshops and factories within the localities in which they are found, and thus, improving life in the rural areas.

PLANTATION TYPES IN CAMEROON

One can easily distinguish two types of plantations, namely:

a) The large capitalist plantations,

b) The small peasant plantations.

1) LARGE CAPITALIST PLANTATIONS

Characteristics: Some of these large plantations were formerly owned by Europeans, such as, the present plantations of the Cameroon Development Corporation. Today, almost all these large capitalist plantations in Cameroon are owned either by the government or corporations. Their main aim was to produce *en masse* for export. Most of these plantations usually cover vast areas of several hectares. Scientific monoculture is an important aspect of these plantations. The large estates here are generally specialized in the cultivation of a single crop and it is very rare to see two or more crops on one estate. The techniques of farming are scientifically advanced, with modern machinery for tree felling, land tilling, sowing, fertilisation and harvesting of the crops.

Quite often they employ a large labour force for such jobs as clearing and planting. They are usually equipped with a factory for preliminary processing of the products before export, e.g. the rubber factory at Tiko; sometimes they have agro-industrial complexes to bring the crops to finished products fit for consumption e.g. the Tole Tea factory and the Mondoni oil mill. The plots for various crops are usually well marked out geometrically, having well chosen crops with phytosanitary bases to care for them. These large capitalist plantations are also characterised by a special social structure, ranging from the administrators down to the seasonal labourers. Because of the social hierarchy, there also exist different lodging quarters for the different categories of workers, with the camps of the unskilled labourers being the least comfortable.

Besides the lodging facilities, medical and educational care of elementary level is also provided to the workers and their families.

LOCATION OF THESE PLANTATIONS

They are found mostly where the climate is conducive for the crops cultivated and also where the soil is very fertile. Their location is also influenced by the proximity to a good network of roads and rails, and near ports for the evacuation of their products. In Cameroon, therefore, the main plantation zones are:

a) The coastal region, especially on the slopes below Mount Cameroon and the Ndian Basin which are famous for oil palms, rubber, tea, and banana cultivation. Right up to the Meme Division, there are rubber plantations at Kumba and Tombel and cocoa plantations at Mungo-Mbongo and rubber plantations at Dizangue. It is within this coastal zone that we have most of the major plantation companies, such as, the C.D.C. with about 22,000 hectares, Pamol west of Tiko and at Lobe with about 9,000 hectares and SAFACAM at Dizangue.

b) The other less important plantations are on the western highlands and in the eastern section of the Southern Plateau. On the western highlands we have notably the Ndu Tea Estate, Santa coffee, and many Arabica coffee farms around Foumbot and Babajou owned by some rich foreign and Cameroonian individuals.

At Mbandjock the SOSUCAM sugar company (Societe Sucriere du Cameroun) exploits an area of over 4,000 hectares for sugar cane. At Batchenga and around Bertoua the Tobacco Company (Societe Camerounaise des Tabacs) has a large plantation, and some coffee plantations in the East Province and around the towns of Abong-Mbang, Batouri and Yokadouma. These large capitalist plantations account for a large portion of Cameroon's agricultural exports.

2) PEASANT PLANTATIONS

This is what we generally refer to as "peasant cash cropping." The crops involved here are usually commercial crops introduced to the peasants by foreigners. Because of lack of processing factories the peasants often send their products to special organisations for processing, for example, cotton, sugar cane.

It is for this reason that some peasant plantations are usually located near large plantations for easy sales of their produce to the agro-industrial complexes. This aspect is, however, absent with peasant plantations engaged in crops whose preliminary processing does not necessarily need a factory as in the case of cocoa and coffee.

Most of these peasant plantations cover a fairly small area of about one to two hectares with a limited labour force usually supplied by the family alone. The tools are often very rudimentary, comprising mostly cutlasses, hoes and sticks. The yields are never as heavy as in the case of the capitalist plantations.

The Main Peasant Plantation Zones:

— The western part of the Southern Plateau: This covers the Bassa, Ewondo and Bulu lands which are very famous for cocoa farms. This zone produces the greatest amount of cocoa in the country.

— The Bertoua and Yokadouma zone has tobacco, cocoa, and coffee plantations.

— The Diamare and Mayo Danai plains in north Cameroon have numerous cotton and rice peasant plantations.

— In the South-West Province, the Meme and Ndian divisions have a lot of cocoa and coffee peasant plantations, and in the North-West Province the Kom and Nso tribes have peasants who own large coffee plantations.

— The Ndop plain, the Mbaw plain in Donga and Mantung division, including the Menchum river valleys are zones of rice peasant plantations.

— The western high-plateau has coffee plantations and factories. Such coffee farms can also be found in the Mungo Division and in the surroundings of Loum, extending beyond Nkongsamba.

Problems of Agriculture in Cameroon

Farmers in Cameroon encounter a good number of problems. Some of them are as follows:

1. Climate

The equatorial climate is generally warm and wet, favouring the growth of insect pests. Some of these insects spread a lot of diseases which attack human beings, thereby making them less active and less productive. The heat in this region also quickens the decomposition of manure and the heavy precipitation is very active in leaching and washing off plant food.

2. Poverty

Most Cameroonian farmers have insufficient financial and material resources for the purchase of farm equipment and for the eradication of insect pests and diseases which attack their animals and crops. They also lack money to finance large-scale development projects such as irrigation and land reclamation. Many farmers lack the vehicles to transport their farm produce to the markets in the cities and as a result some of the produce rots and decays in the farms.

3. Transport

There is a limited number of good roads. The poor state of the few existing roads for transporting agricultural produce from the farms to the markets located in the urban centres forms a barrier to the farmer's progress. Even when vehicles are available, the roads are muddy and almost impassable during the rainy season.

4. Illiteracy and Conservatism

Most Cameroonian farmers are illiterate and conservative. They are slow and reluctant to adopt useful modern farming methods. They are also very ignorant about many useful farming techniques. Despite the existence of many government programmes and the presence of agricultural specialists many local farmers lack adequate information on how and when to plant particular crops, time of weeding, methods of controlling pests and other diseases, and how to use fertilizers and pesticides.

5. Land Tenure System

A growing population on a given amount of land means that the price of that piece of land has to go higher. In many villages where traditional privileges entitle some people to land, a relatively small percentage of the population owns a disproportionately large share of the productive land while most of the villagers own little or nothing. Moreover, much land is held by poor old people who cannot improve it and are unwilling to give it out for the simple reason of preserving it for their offspring.

Without land many peasants are deprived of the chance for economic betterment and improvement, especially in a rural environment where the almost single source of progress is agriculture.

The problem has led the government to pass a new land law. Under this law the government is the sole custodian of all land not put into effective use. Anyone wishing to acquire such land for any type of economic development has to apply directly to the government. Despite this, some individuals, especially in remote areas, still behave in a way designed to thwart the law.

6. Aging Plantations

Some crops which occupy the same piece of land for many years, such as, palm trees, cocoa and coffee are difficult to replace when they get old. Pruning may only serve for the time being, but it usually means much for the farmers to uproot the crops completely and plant new ones. This will mean staying for many years without an adequate source of income.

7. Low Prices

The poor selling prices which farmers receive for their food and cash crops can be very discouraging given the amount of energy expended. The individual farmers only accept the prices already determined by the government for commercial crops such as cocoa and coffee. Exposing these farmers to foreign exploiters can even be more devastating to the whole nation. Secondly, agricultural products do not compete favourably with manufactured goods. For these reasons the farmers are bound to remain poor, at least for the time being.

8. Destruction of Crops by Animals and Wind.

There is constant destruction of crops by animals, such as, elephants, monkeys, and even reared animals such as cattle, sheep and goats so that conflicts often arise between pastoralists and agriculturalists. Strong winds blowing across farms usually pull down many crops.

9. Inadequate Emphasis on Foodcrops

It has often been said that inhabitants of developing countries cannot feed themselves. Many causes, especially natural ones, have always been advanced to explain why they cannot feed themselves. The paradox, however, is that citizens of developing nations devote their energies to feed people in Europe and America and Cameroon is no exception. Through colonisation Cameroon, like other former colonies, was hooked up to the Metropoles and taught to specialise in agriculture in order to be able to earn foreign currency. The vast fertile lands were consequently taken up for the cultivation of cash crops and other plants which provided raw materials needed in Europe and America while the less fertile patches of land were left for food crops. The vast expanses of fertile lands occupied by the C.D.C. plantations were initially seized from the Bakweri natives by Germans for crops needed in Europe. Today, Limbe is one of the Cameroonian towns with very high food prices because the neighbouring lands are occupied by cash crop plantations.

Elsewhere cocoa and coffee occupy larger areas than plantains, cocoyams and maize. The farmers also devote more energy and time for these cash crops which will always be priced cheaper than the manufactured goods imported from abroad. As a result of this practice many farmers are deficient in manufactured goods and even in the supply of the basic food crops which they are capable of producing. The final consequence is poorly nourished and under-nourished families.

THE ROLE OF AGRICULTURE IN ECONOMIC DEVELOPMENT

1) The Source of Foreign Exchange:

The economy of Cameroon is essentially dependent on the export of unprocessed agricultural products. Such exports earn a lot of foreign exchange and also help to improve Cameroon's balance of trade with its trading partners.

2) Source of Savings for Capital:

A developed agricultural base is capable of feeding the nation's population. Money which otherwise would have been spent on the importation of food is saved up as capital to finance other investment projects.

3) Creation of Employment:

Agricultural development, especially in less developed countries like Cameroon, creates employment for rural residents, thereby reducing rural exodus.

4) Income Redistribution:

The tendency has always been for the government to decree salary increases for workers. By so doing, urban incomes rise because most of the civil servants and industrial workers reside in towns and are oriented towards an urban way of life,paying little attention to the farmer. With a developed agricultural base, rural farmers are capable of increasing their incomes by selling their products at reasonable prices to urban residents. The distribution of incomes becomes significantly more equal because farmers in Cameroon are able to sell to the urban workers. Farmers who form a large percentage of the poverty-stricken population see their incomes rise. In this way, rural households will not see themselves as struggling to subsidize the luxurious lifestyles of urban workers but as working to improve their financial and nutritional standards.

5) Raw Materials for Local Manufacturing Industries:

Some industries depend on agricultural raw materials. Food processing in Cameroon has become a manufacturing sector which is solely dependent on agriculture. The quality of the agricultural products, the quantity and the reliability of the supply has led to the creation of food processing industries. Tea factories, cocoa factories, sugar factories and palm mills are some of the food industries to be found in Cameroon. The manufacture of plastic buckets and motor tyres all depend on rubber produced in Cameroon. Agri culture has provided the initial basis of industrialization in Cameroon.

6) Food for Strategic Bargaining Purposes:

This aspect concerns the relationship that exists between a nation and multinational corporations. Multinational corporations can contribute to employment, can increase the stock of capital, can introduce new technology, and can increase tax revenue. They are capable of contributing to the development of less developed countries. However, from empirical evidence, multinationals are not philanthropists. Their interest has always been to minimize their contribution to development, so they will always try to cut the best deal for themselves. They will try to minimize wage levels, repatriate as many profits as possible, employ as few people as possible, and exploit as much wealth as possible.

The host country's interest, on the other hand, is to see that the multinationals make as much contribution as possible to development. In this case, the host country is in a bargaining process with a multinational corporation. To do this effectively, a country should make sure that it is not food-vulnerable and will then be able to say, no, and still survive.

There is nothing more critical to a country's bargaining position with a multinational corporation than the ability to feed its population. If a country has the capability of producing for local market, then with respect to this one strategic area of potential vulnerability, the country should ensure that it can feed itself with locally produced food. That ability to feed its own population is a valuable tool when the country starts bargaining with a multinational corporation. If a country is able to feed itself and does not need foreign exchange to do that job it has a greatly enhanced bargaining position *vis-a-vis* multinational corporations. It could refuse unfavorable terms without starving its people.

Ways of Improving agricultural Productivity

There are a number of ways to do this, but the ways will vary according to the region of the country because of the diversity in natural as well as human conditions. In general terms, the methods include:

1) Water Control: Increasing irrigation, improving the management, maintenance and design of existing irrigation systems, and investing in the sinking of more wells. Hand pumps and irrigation ditch maintenance could all play a major role. Water control will be most appropriate in North Cameroon where limited rainfall constitutes a major handicap to agriculture.

2) Other partial steps toward improving agricultural productivity include, the construction of roads, terracing, technological improvements, making credit terms easy for the small farmers, establishing institutions for marketing and distribution of improved seeds and fertilizers, and developing institutions for the effective dissemination of knowledge.

It might not be easy to undertake these measures to improve agriculture for several reasons. Most important is the fact that it is difficult for an urban elite to appreciate the importance of agriculture. Even if it did, it is oriented toward an urban way of life but being intellectually sophisticated in the rhetoric of agriculture during political rallies. Talk is cheap for those in authority. For the leadership to be successful it must have its fingers on agriculture. There should be no attempts to hold down food prices because that is the only way local farmers can increase their incomes. If the urban residents worry about food prices, they could create their own farms or invest some of their wealthy salaries in agriculture.

CHAPTER 12
ANIMAL RAISING IN CAMEROON

Importance of Livestock:

Livestock raising is particularly important because animals provide protein in diets. There are also many agricultural and industrial products associated with livestock raising, notably hides, skins, wool, and dairy products. Livestock is especially desirable where there is a short rainy season and farming operations have to be concentrated into very short time periods. Finally, and perhaps most importantly, by providing manure, livestock farming plays an important role in providing that change over from shifting-field to permanent field systems of cultivation.

Livestock raising is one of the activities in Cameroon which attracts government's attention to the rural areas. Due to the increasing part played by meat (or proteins in particular) in our diets, the government has stepped in to encourage more animal rearing to cope with the fast growing population and the generally improving standards of living. The various ways of government intervention will be discussed later.

FACTORS INFLUENCING LIVESTOCK RAISING IN CAMEROON

Cameroon has quite a good variety of animals and birds, some of which are suitable for certain areas and some are completely absent from other parts of the country. The factors which influence the raising of livestock in Cameroon are:

a) NATURAL CONDITIONS

Climate and vegetation: The equatorial climate has been realised to be bad for cattle rearing because it favours the spread of the tse-tse fly. This fly causes the cattle to be infected by diseases. On the other hand, the very dry sahel climate is also not favourable for cattle rearing because the areas with this climate have no grass for the cattle to feed on. The vegetation is always scorched by the burning heat of the sun. The ideal zone for cattle, therefore, is that of the humid tropical climate found in the northern part of the western highlands and on the Adamawa. It is very important to note here that the presence of pasture for grazing is more important than the influence of the tse-tse fly. The absence of cattle in the forest region is more as a result of no pastureland than the presence of the tse-tse fly. It is noticed that the savanna zones of Cameroon are always marked by the presence of cattle while the forest zones are conducive for other animals, such as, pigs which do not need much pasture.

b) HUMAN FACTORS

i) *Population density:* Heavily populated areas are always not very favourable for animal rearing (especially cattle). For example, the Bamileke used to rear cattle extensively but this has been greatly reduced by the increasing population densities. With the expansion of towns and farms cattle is carried further away.

ii) *Ethnic problems:* Some tribes have an interest in some animals but would detest the rearing of other types of animals. The Bororo are typical cattle rearers. They would always like to assess their wealth in terms of the number of cows each of them has. Due to local cultural ceremonies and sacred sacrifices some tribes prefer one type of animal to the others. For example, the fowl is very important for social ceremonies among the Kom people than any other animal or bird. Some groups would instead use the goat and others would like the pig. On the other hand, the Islamised populations do not rear pigs because according to their religion pigs are unclean.

The method of raising most of these animals by individuals is rather traditional and archaic. It is characterised by transhumance, nomadism and the tendency to only increase the number of animals for prestige purposes.

The different types of animals are:

a) **Cattle:** From the next table we see that the principal zones for cattle rearing are the Adamawa, the north and north-west. The cattle is mainly of two species, namely: the Zebu cows which represent about 95% of all the cows in Cameroon, and the Taurin which are just about 5% of all the cattle.

1) *The Zebu Cows:* The Zebu cows have a hump-back and are also of a comparatively higher beef-yield than the taurin cows. There are the zebu of the Adamawa and the zebu of the north.

i) *The Adamawa Zebu:* This type can again be divided into two sub-groups, the Peul zebu of the Adamawa is very rich in meat and is found in the localities of Ngaoundere and Banyo. It has a dark-brown colour which may also include white spots. The Bororo zebu on the other hand is not as rich in meat as the Peul zebu. This type has a brown colour and can trek very long distances. The white Akou type belongs to this second group.

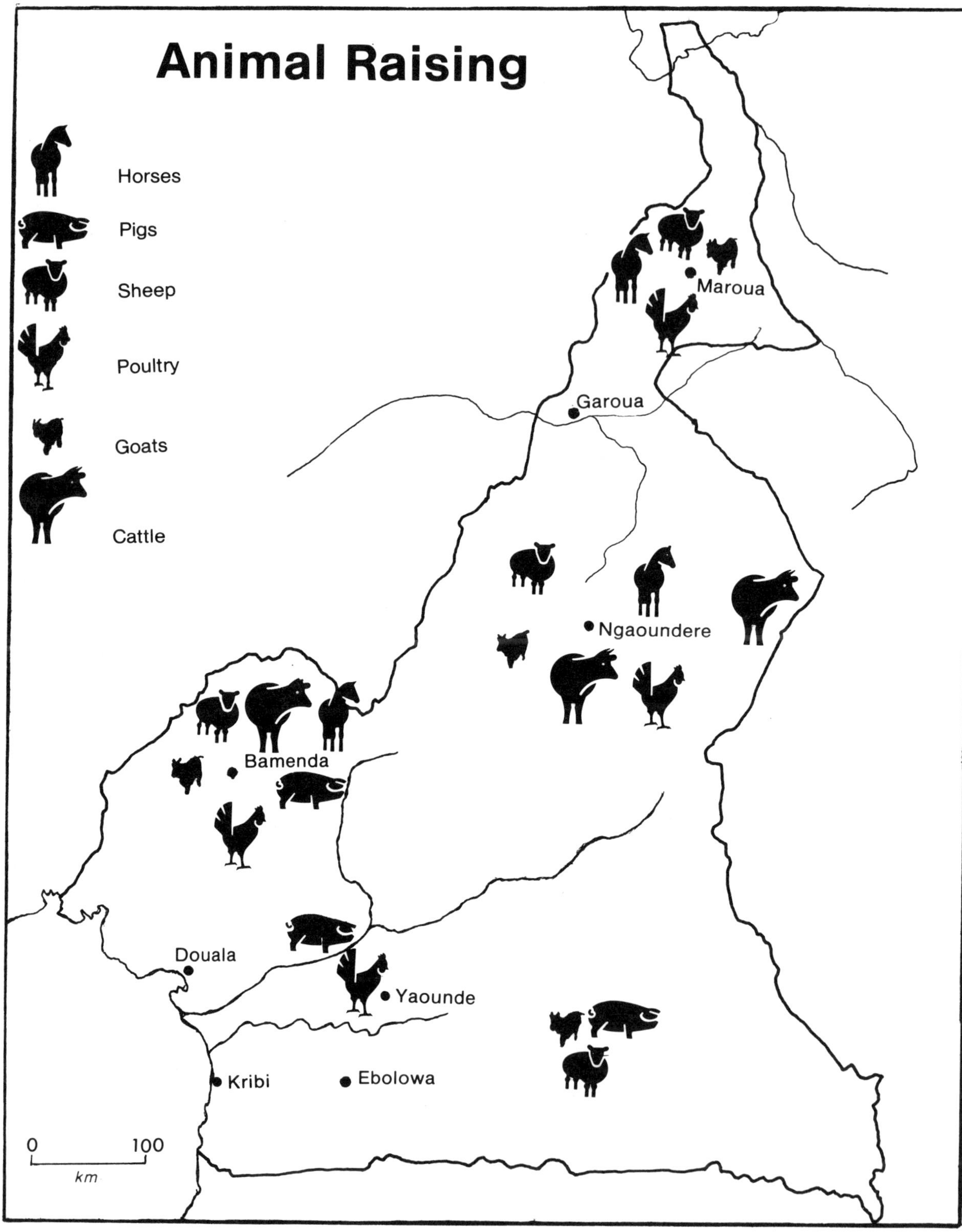
Animal Raising
Horses
Pigs
Sheep
Poultry
Goats
Cattle
Maroua
Garoua
Ngaoundere
Bamenda
Douala
Yaounde
Kribi
Ebolowa
0
100
km

Herdsmen from Sangmelima bringing cattle from the north.

ii) *The Zebu of the north:* This is a variety of the Peul zebu but the difference comes from its comparatively small size which is as a result of the less favourable environment and much cross-breeding.

2) *The Taurin Cows:* These are cows without hump-backs which are naturally smaller in size than the zebu, and are more resistant to the trypanosomiasis disease spread by tse-tse flies.

This species of cows forms a very small percentage (about 5%) of the cows in Cameroon.

In the taurin species we distinguish:

— The Rumsiki taurin in the Margui Wandala region

— The Ndama Taurin of the littoral, the west and the east

— The Muturu taurin of the Bakossi area

— The Namshi taurin of the region of poli

Goats and Sheep: Goats and sheep are raised almost everywhere in the country. Even the forest zone does not exclude them as in the case of cattle. An important fact still remains clear, that the north and the Adamawa are the leading zones for the raising of these two animals.

Next in importance are the western highlands. Elsewhere in the south only small quantities are raised because of little pasture. A look at the table will be more explanatory. On the whole, the goats and sheep in the north are different from those in the south of Cameroon. The Peul type of sheep in particular in the north is by far bigger in size and has more weight than the sheep of the south. The goats too, are much taller and bigger than those of the south. The goats and sheep of the north weigh averagely 40-50 kg whereas in the south they weigh between 30-35 kg.

Pigs: From the beginning the rearing of pigs was done through traditional methods but following the increasing demand for pork, selected species were imported, first of all, for higher supplies of pork and secondly, for crossbreeding with the local species.

Pigs are raised mostly in the western, the littoral, the eastern and the south-western parts of the country. It is worth noting that these are regions which are humid for a greater part of the year. The animal likes moist environments which also have much food and consequently, will not do well in the dry north. From the figures on the table we see that the western part of the country leads in the raising of pigs with more than half the number for the whole territory in 1978.

Another factor which accounts for the low quantity in the north is the tradition of the Muslim

THE TYPES OF ANIMALS:

The different types of livestock raised in Cameroon are cattle, sheep, goats, pigs, horses, asses and poultry.

The table below shows approximately the figures of these types of animals according to the various regions as on the 30th of June, 1979.

SITUATION OF LIVE-STOCK BY 30/6/78

ZONE	TYPES OF ANIMALS						
	CATTLE	SHEEP	GOATS	PIGS	HORSES	ASSES	POULTRY
NORTH & FAR NORTH	1,215,750	885,900	982,500	4,400	8,391	31,305	1,600,000
ADAMAWA	1,400,000	720,000	380,000	200	—	—	400,000
EAST	166,000	28,000	35,390	8,700	—	—	270,000
CENTRE & SOUTH	70,000	163,000	327,000	444,000	—	—	4,100,000
LITTORAL	1,300	23,150	32,800	17,400	140	—	441,800
WEST	155,000	247,600	493,620	982,500	1,017	130	3,092,800
NORTH-WEST	385,000	85,000	187,500	44,000	5,000	—	720,000
SOUTH-WEST	6,950	22,350	44,650	10,800	420	—	485,400
TOTAL	3,400,000	2,175,000	2,484,000	1,512,000	14,968	31,435	11,110,000

Source: Ministry of Livestock, Fisheries and Animal Husbandry.

REMARKS:

This table should not be taken to be very accurate because the figures were got through information from herdsmen who, for the most part were very sceptical as to why the data were being collected, and some of the data were obtained during prophylactic treatment of the animals.

religion which is against pigs.

Moslems regard pigs as unclean animals. The traditionally raised pigs are more numerous in the country than those raised under modern methods. In the western part of the country, a hybrid of the local and the imported varieties raised under modern techniques at Kounden, is more widespread. There are also many private piggeries all over the country.

Horses and Asses: The northern part of the country is the main region for these animals. The horse in particular is used by Fulani people. Were it not of the western and north-western highland zones which have some Fulani people who also own horses, one would be entirely correct to say that all these animals are found only in north Cameroon.

The lack of data on these animals in the Adamawa region is highly regretted, but for the fact that this region shelters many Fulani people it is believed that it also has a good number of horses and asses.

Poultry: Poultry are kept in all the regions of the country, but the West, Centre and South Provinces are in the lead. The poultry in Cameroon consist mainly of fowls reared both in the traditional and modern ways. The local fowls are fairly smaller in size than those bred and raised under modern conditions. In the traditional ways, the fowls for a greater part of the day move about in search of food, whereas in the modern poultry farms, the fowls are enclosed and provided with their feed for all the time.

Many citizens own individual poultry farms from which they obtain enormous supplies of eggs daily. The government also runs major poultry farms at Muyuka, Mvog-Betsi in Yaounde, Kounden in the West province, etc.

GOVERNMENT PARTICIPATION IN LIVESTOCK RAISING

In order to meet the growing demand for meat, the government intervenes in several ways.

1) *Improvement of rearing methods:* The actions undertaken to improve the rearing methods of

animals include the creation of modern ranches to demonstrate in a practical way, the modern methods of animal rearing.

In such ranches also, experiments of cross-breeding between imported and local species are performed. Such experiments are usually done under the control of the I.R.Z. (Institut de Recherche Zootechniques). Examples of such I.R.Z. centres are found at Bambui and Wakwa.

2) *The provision of sanitary care:* The aim of sanitary care is to fight against animal diseases. This is done in many ways. First of all, the Ministry of Livestock, Fisheries and Animal Industries has set up within every locality a veterinary dispensary which cares for animals and cures them of their diseases. The animals are often bathed to kill ticks and vaccinated or given some curative treatment to fight against such animal diseases as tripanosomiasis.

At times teams are equipped with drugs and sprays to fight out some disease-carrying insects in a particular zone; for example, the fight to control the tse-tse flies in the Benue and the Adamawa regions.

3) *Provision of trained personnel:* In order to effectively teach the animal rearers what to do, the country should have enough trained personnel. It is for this reason that veterinary doctors and agents are being trained. Veterinary schools exist both at Maroua in the Far North Province and at Jakiri in the North-West Province. There are more than 200 zoo-technical and veterinary centres in the country where these trained personnel work.

PROBLEMS OF LIVESTOCK RAISING IN CAMEROON

Livestock raising in Cameroon faces a number of handicaps. As earlier mentioned, many livestock rearers value animals for their own sake and count each one as a unit of wealth. Thus many cows of an inferior type will be valued more than fewer healthy cows each capable of providing large quantities of milk daily for many months after its calf is born. Often, the most important thing to the rearer is *number,* and each head of cattle, each sheep and each goat is thought of as a unit of wealth almost independently of anything it may produce. In other more advanced parts of the world animals are valued because they can change animal food like grass into human food like milk, from which the farmer can obtain butter, cheese and cream.

In areas where the desire for quantity and not quality becomes the essential concern, the size of the herds, in fact, can only be restricted by the conditions under which they live: animals are killed off by starvation, by the attack of wild animals and by disease. In recent past years drought has contributed to the elimination of hundreds of cattle in the northern part of the country.

A further problem of livestock raising concerns the location of markets. The main markets for cattle reared in the north and western highland regions are the southern towns of Douala, Yaounde, Kumba, Limbe and many of the smaller ones. Live cattle are sold to traders who drive them on the hoof, to the markets where they have to be slaughtered and sold. The effects of many kilometres of trekking to the south are indicated by the loss of weight and by the death rate of stock *en route.* It is estimated that animals on leaving the grasslands, normally weigh 450-500 kilograms and lose between 50 and 60 kilograms whilst on the journey. This loss of weight, accompanied by deterioration of quality of meat, is due to: a) lack of pasture and suitable drinking points along the route, particularly in the dry season, (b) the effects of insect bites and heat exhaustion, (c) the disturbing influence of the frequent necessity to drive stock off the narrow roads to allow vehicles to pass.

CHAPTER 13
MINERALS AND POWER

A) MINERALS:

In recent years prospection for minerals has taken many strides in Cameroon and even now it could still be said that the knowledge of Cameroon's underground is very limited. Despite the discovery of some of the minerals the level of exploitation is also minimal.

Petroleum or crude oil: The mining industry of Cameroon is dominated by crude petroleum. Crude oil production has become one of the most important aspects of Cameroon's economic development. The exploitation of this product started in 1978. Drilling of crude oil is undertaken at the coastal sedimentary basin by several companies which include, Elf Serepca, Shell, Mobil and Gulf Oil of Cameroon. For more than fifteen years Elf Serepca spent enormous sums of money on this coastal sedimentary basin on oil prospection. Finally, discoveries of reasonable reserves were made. Since 1978 production levels have continued to rise. Oil production was estimated to be 4.3 million tons in 1981, 5.2 million tons in 1982 and 6.3 million tons in 1983. However, production was thought to have peaked in 1985. The country's only refinery at Cape Limboh has a capacity of two million tons.

Cameroon has become self-sufficient in some petroleum products, and can hopefully see a higher level of industrialisation and investment if only the oil revenues are well utilised by the nation's authorities.

Bauxite: One of the principal mineral resources of Cameroon is bauxite found mainly at Minim-Martap and Ngaoundal in the Adamawa region. Other deposits of bauxite are at Fongo-Tongo in Dschang. From bauxite we obtain aluminium. The mining and processing of bauxite is usually very complex and would require that the attention of foreign companies be called for. This, of course, will be very costly so that Cameroon for the moment contents herself with imported aluminium. To keep the aluminium industry at Edea working, Cameroon imports most of its aluminium from Fria in the Republic of Guinea.

Natural Gas: Deposits of natural gas have been discovered, one at Logbada near Douala with a reserve of about 350 million cubic metres, another deposit is found at Rio-del-Rey but exploitation has not yet begun. The other deposit is situated some few kilometres off the shore of Kribi. There are high possibilities of the presence of oil and gas in the Chad Basin. The production of Cameroon's natural gas has not yet begun owing to lack of demand on the world market. Plans for a liquified natural gas (LNG) plant near Kribi have been shelved, due to the lack of interest in the private sector for building such an expensive plant unless the long-term demand situation for LNG improves.

Tin Ore (Cassiterite): This mineral is worked at Mayo-Darle. The deposits were almost getting exhausted and the closure of the mines was already asked for. In 1969 SOFIMEC which exploits this mineral could only produce 41 tons, but by 1978 production rose to 11,200 tons.

Gold: The extraction of gold in the East Province at Betare-Oya is still on a low scale. Exploitation is by traditional methods. However, prospection continues for richer reserves.

Copper and Uranium: Studies concerning the presence and studies of these minerals are still continuing in the Maroua and Tchollire areas. Their deposits are already found in Poli, south of Garoua.

Wolfram: This is a mineral that is very precious; it is usually used for the fabrication of the filaments of electric bulbs. Some traces of this mineral have been found in north Cameroon.

Asbestos: Asbestos is a mineral that does not burn easily and is found in the region of Meiganga.

Molybdenum: Molybdenum which is whitish in colour is found in the Mungo region but not yet well located.

Iron Ore: Its deposits are at Kribi. The reserves of this mineral are estimated at 120 million tons of ore with an average iron content of 40 percent.

Diamond: At present it has been discovered at Batouri in the East Province. Exploitation has not yet started.

Building Materials:

a) Limestone: The reserves of Limestone at Figuil in north Cameroon are estimated at 600,000 tons with a 90 percent CaCo3 content.

b) Cipolin and Marble: Their reserves at Bidzar in North Cameroon are estimated at 2,500,000 tons with a 95 percent CaCo3 content.

b) Sand: Sand used for the manufacture of glass is extracted from the Manoka island near Douala. Its beautiful and very clean reserves in this zone are estimated at 115,000 cubic metres. Elsewhere in the country large amounts of sand are being removed from rivers and used locally for building. Rocks of various types are also being quarried in different areas and used for the same purpose. An

MINERALS AND POWER
MINERALS
Bauxite
Limestone
Gold
Gas
Iron
Crude Oil
Tin Ore (cassiterite)
POWER
Hydroelectric stations (H.E.P.)
Functional
Projected
H.E.P. lines
Thermal station
Hydroelectric dam
Maroua
Mokolo
Logone R.
Kaelé
Figuil
Garoua
Benue R.
Lagdo Dam
Faro R.
Ngaoundéré
Minim-Martap
Mayo Darlé
Nkambe
Wum
Mbengwi
B'da
Mbam R.
Mbakaou Dam
Ngaoundal
Bamendjim Dam
Cross R.
Mamfe
Foumbot
Fongo-Tongo
Betaré-Oya
Bafoussam
Sanaga R.
Kumba
Nkongsamba
Nachtigal
Bertoua
Buea
D'la
Song-loulou
Limbe
Edea
Yaounde
Nyong R.
Njock-Mpoumé
Boumba R.
Dja R.
Kribi
Ebolowa
Sangmelima
Dja
Ngoko R.

example is the gneiss of Yaounde which is largely quarried and used for building.

SALT: Salt waters in Mamfe have sometimes been boiled to obtain salt in times of salt shortage.

B. POWER

We need power to run industries, for household and commercial use. Power in Cameroon is obtained mainly from water and oil.

Power obtained from water is what we call Hydroelectric power (H.E.P.) and power obtained from oil is known as Thermal electricity.

HYDROELECTRIC POWER: (H.E.P.)

The necessary conditions for the generation of hydroelectric power are:

1) A regular flow of water.

2) Impermeable rocks to allow for enough surface flow.

3) Very steep escarpment to create a waterfall.

4) A narrow river valley to facilitate the construction of a dam.

Hydroelectricity is generated from waterfalls and Cameroon is proud of a high potential hydroelectric power because most of its river courses are imbued with powerful falls and rapids. Hydroelectricity provides around 95 percent of the country's power.

Presently, the largest power stations are on the Sanaga river at Edea and Song-Loulou supplying power to almost all the major towns of the western and southern parts of the country. It is believed that the Song-Loulou and Edea dams are the third and fourth largest dams respectively in West Africa after the Kainji dam on the Niger River in Nigeria and the Akosombo dam on the Volta river in Ghana.

Reorganisation of the electric power sector in 1974 led to the creation of the National Electricity Corporation (SONEL—SOCIETE NATIONALE D'ELECTRICITE), which had taken over the generation, transmission and distribution of electricity throughout the whole country. These activities were hitherto undertaken by small companies such as POWERCAM, which functioned in the former West Cameroon state and Cameroon Electricity (Electricite du Cameroun) in the former East Cameroon state.

1) The Edea H.E.P. Station:

Due to the growth of towns and the increased demand for the use of electricity in domestic and industrial sectors, the government has had to reinforce the Edea power station through the construction of the Bamendjim and Mbakaou Dams.

The Mbakaou Dam: The Mbakaou Dam on the Djerem, a tributary of the Sanaga, was constructed in 1971 and it has a capacity of 2,900 million m^3. This dam did not meet the increased demand for power and therefore led to the construction of another dam, the Bamendjim Dam on the Noun River, a tributary of the Mbam River which flows into the Sanaga.

The Bamendjim Dam: The Bamendjim Dam measures 235 m long by 17 m high. The reservoir behind this dam has an area of at least 344 km^2 and a capacity of 1,800 million cubic metres. These dams were built to maintain the debit of the Sanaga River in the dry season when the level begins to fall. The construction of these two dams to reinforce the flow at Edea was also followed by the installation of three other extra 20.5 mw turbo-alternator sets at the Edea power station.

By 1974-1975 the Edea station was capable of producing only 1.3 billion Kwh but after the reinforcement the output went up to 2.3 billion Kwh in 1976.

Hydro-electric power from the Edea station serves the towns of Edea, Douala, Yaounde, Tiko, Limbe, Kumba, Nkongsamba, Bafoussam, Bamenda, Mbengwi, and also the numerous smaller towns which are situated near these big towns. Extension work is currently underway to many smaller towns and villages.

2) The Song-Loulou H.E.P. Station: The Song-Loulou hydroelectric plant built on the Sanaga river, some 46 km in the north-west of the town of Edea on the Song-Loulou falls, is by far the largest hydroelectric power realisation in the country.

Started in 1975, the building of this dam was followed up very vigorously and the first phase was completed and inaugurated by the then Head of State, Ahmadou Ahidjo on the 14th of November 1981. This first phase comprised the installation of four generators which produce each 48 megawatts. With the installation of four other generators (expected to be carried out between 1985 and 1990) this second phase will increase the total power output to 384 MW.

The Song-Loulou plant does not only increase the output of power in the country but also holds more water upstream at the Edea station which goes to provide more electricity to smelt the aluminium used by the Alucam industry.

The Lagdo H.E.P. Station: The Lagdo Dam built on the Benue river just 50 km south-east of Garoua is estimated to hold 4 billion cubic metres of water and produces some 84 MW which serve the towns of Garoua, Maroua, Mokolo and other smaller towns in the north which were formerly served by thermal electricity. The construction of this power station was financed with aid from the People's Republic of China.

EDEA H.E.P. Station: In the background is the aluminium smelting factory.

Another dam is under study for the Njock-Poume Falls on the River Nyong and it is expected that when this one is achieved, it will supply electricity to the towns of the south such as Sangmelima, Kribi, etc. Cameroon still has a high potential hydroelectricity considering that other falls have not yet been harnessed; for example, the Menchum Falls in the North-West Province which are believed to be the most powerful in the country.

THERMAL ELECTRICITY:

Electricity in Cameroon is also obtained from oil-worked engines known as thermal electricity. This thermal electricity is generated by diesel engines and they are found in areas which are not yet served with electricity from hydroelectric stations.

Towns formerly served with thermal electricity had to surrender their diesel engines to other towns which are still awaiting the arrival of hydroelectricity.

For the moment, electricity is the main source of power in the country. There is no coal, the exploitation of natural gas discovered near Douala and Kribi has not yet started. Petroleum refining is done at Cape Limbo, west of Limbe in the South-West Province.

CHAPTER 14
INDUSTRIES OF CAMEROON

The industries of Cameroon are diverse and can be classified into various branches ranging from very small-scale to large-scale industries.

An attempt to group these industries will give us the following classes:

a) The local crafts industries, and

b) The manufacturing industries.

a) LOCAL CRAFTS INDUSTRIES:

The peoples of the various traditional societies in Cameroon are very industrious. Before the coming of manufactured goods they produced objects which they needed for daily use. It happened that certain groups became famous for particular crafts in which they were very developed. Some of these crafts were practised almost all over the territory. Important crafts practised by Cameroonians are:

Weaving:

This craft is practised using different raw materials. Basket and bag weaving are done using materials collected from rafia palms. Baskets are woven almost all over in the North-West and West Provinces. Rafia bags are produced in the North-West Province mainly by the Meta and the Bafut people. Cane weaving is almost widespread in the southern towns of Cameroon because of the availability of cane from the thick forest. The craftsmen produce cane objects such as cupboards, chairs, and tables, which have a wide market in the big towns, especially among those who cannot afford the relatively expensive iron and wooden articles. Mat weaving is common all over the savanna zones and they are purchased mostly by moslem populations.

Carving:

Here, we distinguish between those who carve on wood and those who carve on ivory.

Wood Carving is very common among the western and north-western populations and they produce masks, stools, door posts, etc. These objects can be found in their numbers when one pays a visit to a traditional ruler's palace.

Ivory Carving is practised mostly in the towns of the south were one easily has a supply of ivory from such forest animals as elephants and hippopotami. Carved ivory products are found in the markets of the south at Yaounde, Ebolowa, Douala, Lomie, Sangmelima, etc. A few ivory products are also found in north Cameroon.

Craftsmen at work in Bamenda.

Smithing and Casting:

Blacksmiths are dotted all over the territory and mainly in the villages. The craftsmen smelt iron ore to produce cutlasses, knives, swords, traps, spears and hoes. In the big towns there are a lot of gold and silversmiths who produce rings, bangles, and bracelets out of gold and silver.

In the Foumban area, the Bamum in particular, are famous for the casting of imported brass and copper into objects of different shapes.

Pottery:

This craft is fast dying out because of the availability of many modern aluminium and silver pots. The ethnic groups which are good at this craft include the Bamessing and the Babungo in the North-West Province and the Barombi Mbo in the South-West Province. They use clay for the production of pots, bowls, pipes, etc.

Embroidery:

This craft is very important in the northern provinces and among the Bamum in the West Province. It is also practised by the Bafut, Kom, Nso, Baligham, Balikumbat and Bamileke.

They import cloth and sew it into large garments, which are embroidered into different designs.

Leather Work:

This craft thrives in the north of Cameroon where hides and skins are tanned and dyed with local dye. Maroua in the north is the principal centre for this craft. The leather is then used to produce bags, shoes, cushions for chairs, sheaths, etc.

The products produced by local craftsmen are usually a vital source of money for the country because foreign tourists purchase and carry them to their countries. Tourists are particularly interested in ivory, leather and wooden articles.

B) MANUFACTURING INDUSTRIES

1) FOOD PROCESSING INDUSTRIES

a) The Cocoa Industry (Chococam) at Douala produces cocoa powder for the manufacture of chocolate, biscuits, butter, etc.

b) Coffee Industry: It produces consumable coffee from coffee beans at Douala and Dschang.

c) Tea Industry: The processing of tea for local consumption and export is undertaken by the C.D.C. in factories at Tole, near Buea and at Ndu in the North-West Province.

d) Flour Industry: The Cameroon Flour Corporation produces flour from imported wheat at Douala. It is hoped that wheat which is presently being grown at Wassande will eventually support this industry. All the flour produced by this industry is consumed locally.

e) Bakeries: This sector has undergone considerable expansion with modern equipment owned largely by Cameroonians in the large towns of the country.

f) Sugar Industry: The CAMSUCO and SOSUCAM industries at Mbandjock produce sugar from sugarcane cultivated in vast plantations.

g) Rice Industry: Local rice is hulled at the mills at Yagoua (SEMRY); Mbo plain (MIDERIM); Ndop plain (UNVDA); Nanga-Eboko and Tonga. There are also numerous hulling factories in most rice producing areas.

h) Oil Industries

Groundnut oil Industry: The groundnut oil mill set up at Pitoa in the colonial days, is now taken care of by SODECOTON, which produces and labels groundnut oil as "huile d'abbia". At present, a very large mill has been built for groundnut oil production. Other groundnut oil mills are found in Bertoua and Douala.

Cotton oil Industry: The Pitoa and Maroua mills which are engaged in the production of groundnut oil also extract oil from cotton seed and label it as "huile de Diamoar".

i) Palm oil Industry: Most of the para-public corporations engaged in oil palm farming have also set up factory-mills for the processing of their oil. The major palm oil mills are found at Mondoni, Lobe, Edea, Dibombari and Douala. Most of the oil is used locally for the production of soap and for domestic consumption. Only a relatively small quantity is exported.

j) Brewery Industry: Brewing in Cameroon is undertaken by four major companies, namely:

Brasseries du Cameroun, whose main seat is Douala with major branches at Yaounde, Garoua and Bafoussam for the production of beer and soft drinks. Smaller branches are dotted all over the territory for the brewing of soft drinks. *Union Camerounaise des Brasseries,* is based mainly at Douala and produces both beer and soft drinks. *Guiness Cameroun,* is also based in Douala and engages in the production of Guiness beer and Gold Harp beer. *Nouvelles Brasseries Africaines* (NOBRA) in Douala, produces the STAR beer and other favourite drinks.

j) Salt Industry (SOCARSEL): The Cameroon Salt Refinery is located at Bonaberi and is involved in the refining and packaging of salt from two sources. First, it uses imported,unrefined salt and, second, it uses salt produced locally from salt marshes or extracted from rock salt deposits.

2) CHEMICAL INDUSTRIES:

This branch concerns chemicals of many types and they are concentrated essentially at Douala. They include:

a) Paint and Varnish Industry: This industry is in the hands of the C.E.P. (Compagnie Equatoriale des Peintures) and is based in Douala.

b) Soap Industry: This industry depends on locally produced copra and palm oil. The soap industries are found at Bassa, Bonaberi, Yaounde and Tiko.

SAPCAM is another industry which produces camel water, insecticides and candles.

c) Matches Industry: The fabrication of matches is undertaken by a company called UNALOR and its industry is found in Douala.

d) Perfume Industry: Two companies in Cameroon are involved in the production of perfume.

Brewery: Inside a Brasseries du Cameroun factory.

Industries

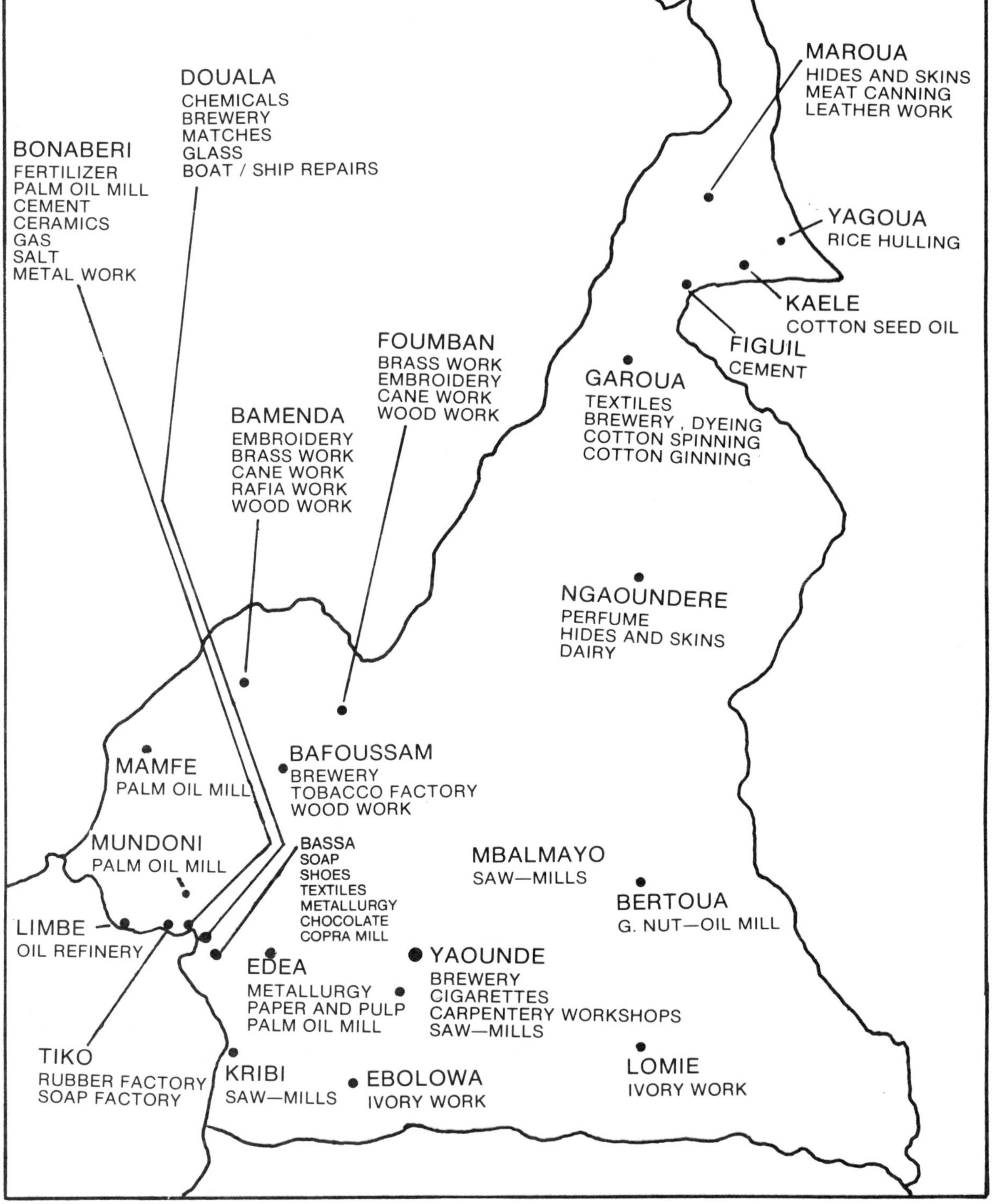
DOUALA
CHEMICALS
BREWERY
MATCHES
GLASS
BOAT / SHIP REPAIRS
BONABERI
FERTILIZER
PALM OIL MILL
CEMENT
CERAMICS
GAS
SALT
METAL WORK
MAROUA
HIDES AND SKINS
MEAT CANNING
LEATHER WORK
YAGOUA
RICE HULLING
KAELE
COTTON SEED OIL
FIGUIL
CEMENT
FOUMBAN
BRASS WORK
EMBROIDERY
CANE WORK
WOOD WORK
GAROUA
TEXTILES
BREWERY, DYEING
COTTON SPINNING
COTTON GINNING
BAMENDA
EMBROIDERY
BRASS WORK
CANE WORK
RAFIA WORK
WOOD WORK
NGAOUNDERE
PERFUME
HIDES AND SKINS
DAIRY
MAMFE
PALM OIL MILL
BAFOUSSAM
BREWERY
TOBACCO FACTORY
WOOD WORK
MUNDONI
PALM OIL MILL
BASSA
SOAP
SHOES
TEXTILES
METALLURGY
CHOCOLATE
COPRA MILL
MBALMAYO
SAW—MILLS
BERTOUA
G. NUT—OIL MILL
LIMBE
OIL REFINERY
EDEA
METALLURGY
PAPER AND PULP
PALM OIL MILL
YAOUNDE
BREWERY
CIGARETTES
CARPENTERY WORKSHOPS
SAW—MILLS
TIKO
RUBBER FACTORY
SOAP FACTORY
KRIBI
SAW—MILLS
EBOLOWA
IVORY WORK
LOMIE
IVORY WORK

They are SOPARCA and SIPCA which are both based in Douala.

e) Fertilizer Industry: The SOCAME company (Societe Camerounaise des Engrais) has a fertilizer industry at Bonaberi which produces many tons of fertilizer daily. Most of its raw materials are imported. The fertilizer produced by this industry serves the agriculturalists of Cameroon to improve their farm yields.

f) Compressed Gas Industry: The CAMOA company (Societe Camerounaise d'Oxygene et Acetylene) has a gas factory at Douala which compresses gas into bottles for household use.

g) Cement Industry: CIMENCAM — the Cameroon Cement Corporation, manufactures cement at Bonaberi in Douala from imported clinkers and at Figuil in north Cameroon from locally quarried limestone. The industry produces over 350,000 tons of cement a year. In 1978, for example, its total production reached 413,976 tons of which 59,200 tons were produced at Figuil and 354,766 tons were produced at Bonaberi.

3) METALLURGICAL INDUSTRIES:

a) The ALUCAM company has a big factory at Edea which converts imported bauxite from Guinea into aluminium. Most of this aluminium is used for the production of roofing sheets, household utensils, such as, pots, plates, and basins at Douala by the ALUBASSA company.

b) The SOCATRAL industry in Douala also produces aluminium roofing sheets and household utensils.

c) TROPIC is another company based in Douala which is engaged in the fabrication of agricultural tools such as, hoes, axes, cutlasses, pick-axes, etc.

d) SOLADO (The Laminated Products Industry): The manufacture of concrete reinforcing rods and other laminated products is in the hands of the company called "Societe de laminage de Douala." It is based in Douala.

Matches factory in Douala.

Fertilizer industry at Bonaberi.

4) VEHICLE ASSEMBLY AND REPAIRS INDUSTRIES

a) Assembly of all purpose vehicles Industry: The only production line used to be the Land Rovers. Its output fell from 560 vehicles in 1970 to 300 in 1973 and work finally had to close down for some time.

b) Assembly of Trailers and Skips Industry: The Cameroon Industrial Corporation (SCI) is the largest concern in this field. Two other small companies manufacture skips locally, one being SOCOMECAM which is licensed by Decaville at Douala.

c) Assembly of bicycles and Motorcycles Industry: Bicycle assembly is undertaken by a company called "Maison du Cycle" based at Douala which assembles both the Zephyr and Peugeot cycles. It also assembles handcarts.

d) Industrial Repairs Industry: SOCAPAR and Union Industrielle du Cameroun are two companies at Douala which are concerned with major undertakings of repairs. Besides these two companies there are other numerous small repair units.

5) MECHANICAL AND ELECTRICAL INDUSTRIES

PILCAM industry at Douala manufactures Hellesens batteries; Societe Emaillerie Nouvelle d'Afrique is a company which produces enamel utensils. Equatoriale Electronique is another industry that is concerned with electrical and electronic equipment. It assembles radios and other electronic apparatus.

6) TEXTILE INDUSTRY

There are two textile complexes in Cameroon owned by the Cameroon Industrial Cotton Corporation (CICAM). One is at Douala and the other is at Garoua. The Garoua industry has spinning and weaving machines and the finishing touches of this industry are done at Douala.

7) FOOTWEAR AND LEATHER INDUSTRY

The BATA factory in Douala produces a variety of shoes. At Ngaoundere the S.T.P.C. (Societe des Tanneries et Peausseries du Cameroun), a tannery industry, makes use of hides and skins from animals in Cameroon. This factory (S.T.P.C.) has a large collecting centre for hides and skins at Bonaberi in Douala. The leather which is produced is used for the fabrication of several articles, such as shoes, bags and other leather products.

8) RUBBER AND PLASTIC INDUSTRY

Latex is processed at Tiko and Dizangue in small factories before being exported. Cameroon in turn imports the well processed latex which it uses in some of its industries. CIAC company has already started to produce tyres and inner tubes for bicycles and handcarts in Douala. PLASTICAM produces plastic objects, such as, buckets, basins, mugs, polythene bags, and toys from imported by-products of oil.

9) CIGARETTE INDUSTRY

The BASTOS company at Yaounde produces cigarettes from the locally produced tobacco from the regions of Mbouda and Batchenga. Part of their output is exported and the rest is consumed locally.

10) CERAMICS INDUSTRY

The ceramics industry, CERICAM, (Ceramique Industrielle du Cameroun) was recently established in Douala to produce tiles and sanitary ware.

11) PAPER AND PULP INDUSTRY

The CELLUCAM is one of the most important industrial installations in Cameroon. It is established at Edea to make use of the vast timber reserves of that area. For now it produces mostly paper pulp but there are plans to produce paper in the near future.

12) OIL REFINERY

(See industries of the Coastal Lowland Region under REGIONAL STUDIES)

INDUSTRIAL REGIONS OF CAMEROON

It is important to remark that more than half the number of industries in Cameroon are concentrated in Douala and less than half the number are found in the other towns.

It is also important to note that apart from the local crafts industries which are spread all over the territory, some parts of Cameroon are completely bare of any factory of reasonable size to be termed an industry. Such is the case with the North-West Province and the forested vast zone of the East Province.

Other parts, on the contrary, are jammed with industries due to reasons which range from economic to political.

However, our concern here is not to investigate the factors underlying the location of an industry in a particular area because they are really numerous, but to present in an easy and superficial way the main industrial zones of Cameroon. It is then left to the individual student to examine each industry and see exactly why it is set up in that area. For example, the ALUCAM industry at Edea needs much power for smelting its aluminium and that is why it is located near the source of power.

In Cameroon, therefore, we can group the industries into five main regions, though some very important industries do not fall under any of these regions because they are situated in very isolated places.

The five main industrial regions are:

a) The Douala Industrial Region
b) The Tiko-Limbe Industrial Region
c) The Edea Industrial Region
d) The Yaounde Industrial Region
e) The Garoua Industrial Region.

A) THE DOUALA INDUSTRIAL REGION:

As earlier indicated, this is the main centre of industries in Cameroon with both heavy and light industries.

The Douala industrial region has industries spread in four main centres.

1) Douala Centre: This refers to the central portion of the town and its immediate surroundings. Here, the population of the town is concentrated to provide for a big market to buy the products. The industries are engaged in the production of necessities which are immediately consumed. This is, therefore, the centre for food processing industries, e.g. bakeries, breweries.

2) Bonaberi Centre: On the right side of the Wouri river and facing Douala town is Bonaberi. Behind the port is the vast industrial zone of Bonaberi which has mainly heavy industries producing equipment which is distributed to the whole country and exported to neighbouring countries. It is here that we find the cement industry, salt industry, soap industry, compressed gas industry, fertilizer industry and many factories which are still under construction. This centre is presently being extended to accomodate more heavy industries further away from the coast.

3) Bassa Centre: The Bassa centre is situated on the Douala — Edea — Yaounde road. This zone has long lodged many industries though the prospects for expansion are not as high as in the case of the Bonaberi centre. At the Bassa centre the main industries include, the CICAM Textile Industry, the Bata Shoe industry, the ALUBASSA industry

THE DOUALA INDUSTRIAL REGION

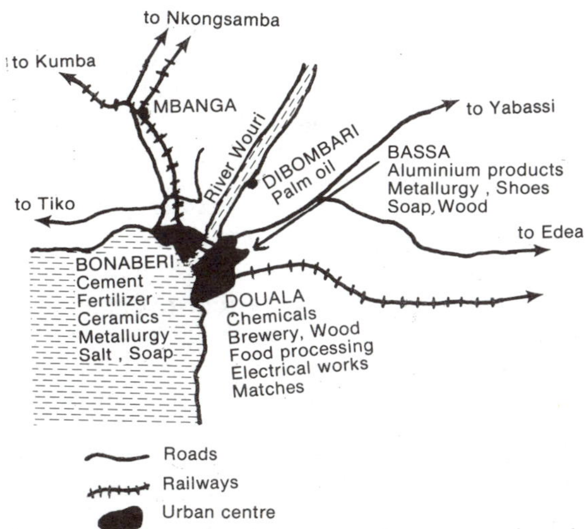

which produces aluminium products, wood and soap industries.

4) Dibombari Centre: Of the four centres, this one is furthest away from the town of Douala and is situated on the right side, just some few kilometres from the Douala-Nkongsamba road. Its industries are few in number. The palm oil mill is the main industry here because it is situated in the oil palm zone.

CIMENCAM: Cement factory at Bonaberi.

B) THE TIKO-LIMBE INDUSTRIAL REGION:

The industries of this region are far from being concentrated in one place. They are separated by distances of many kilometres. Almost all of them are located at the source of their raw materials. The main ones are:

a) *The Mondoni oil mill,* which is one of the very large and modern oil mills in the country, is situated in the centre of the C.D.C. oil palm plantations.

b) *The Tole tea factory:* It is also situated in the centre of the C.D.C. tea farm at Tole, just a few kilometres south of Buea.

THE TIKO—LIMBE INDUSTRIAL REGION

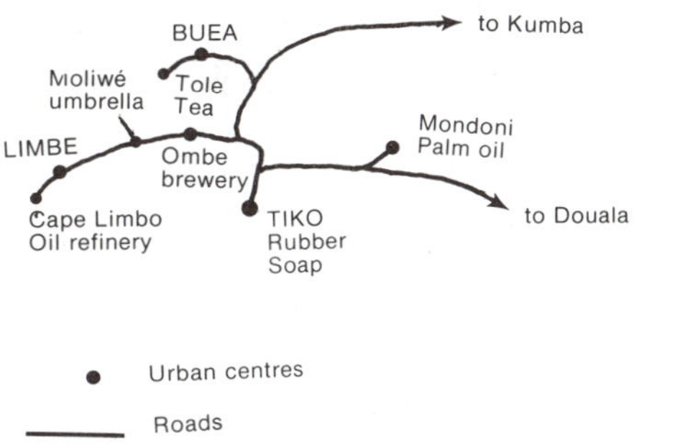

c) *The Tiko Rubber and soap industries:* They get their raw materials from the rubber and oil palm plantations respectively.

d) *The Ombe Brewery:* At Ombe there is a brewery for soft drinks operated by Brassseries du Cameroun.

e) *The Moliwe factory* produces Umbrellas. Like the Ombe brewery, its location is not determined by the presence of raw materials.

f) *The Oil refinery at Cape Limbo* — At Cape Limbo, just a few kilometres from the town of Limbe is the country's petroleum refinery, run by the National Refining Corporation, SONARA.

C) THE EDEA INDUSTRIAL REGION:

Besides the large hydro-electric power station at Edea on the Sanaga, this region has two other very important industries. The ALUCAM industry smelts aluminium which is imported from Guinea for the production of aluminium products.

THE EDEA INDUSTRIAL REGION

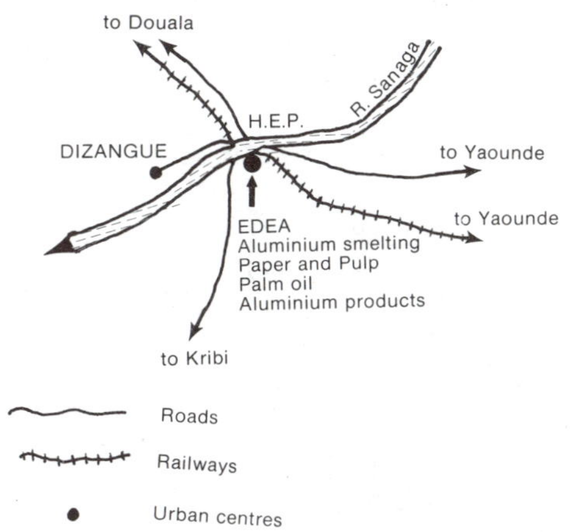

The CELLUCAM industry is a paper and pulp industry located at Edea. Further away from the town at Dizangue is a rubber factory operated by SAFACAM.

D) THE YAOUNDE INDUSTRIAL REGION:

Here, we are also considering the industries in Yaounde and those located in the neighbouring towns. In Yaounde there is the cigarette factory run by the BASTOS Company; a big brewery owned by Brasseries du Cameroun; saw mills at Coron and many carpentry and goldsmith workshops.

THE YAOUNDE INDUSTRIAL REGION

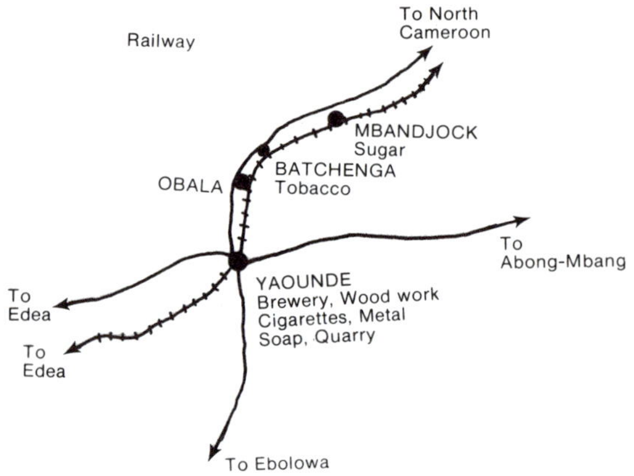

At Mbandjock, just over 90 km from Yaounde, are the two sugar factories operated by CAMSUCO and SOSUCAM.

At Mbalmayo there are the plywood and veneer factories. There are relatively few industries in this region because Yaounde is mostly an administrative and an educational rather than an economic centre.

E) THE GAROUA INDUSTRIAL REGION:

Here, one is counting the industries established in Garoua and those found in smaller towns which surround the town of Garoua. In Garoua, there is the textile industry, which is engaged in spinning and weaving of cloth. Garoua also has a branch of the Brasseries du Cameroun brewery.

THE GAROUA INDUSTRIAL REGION

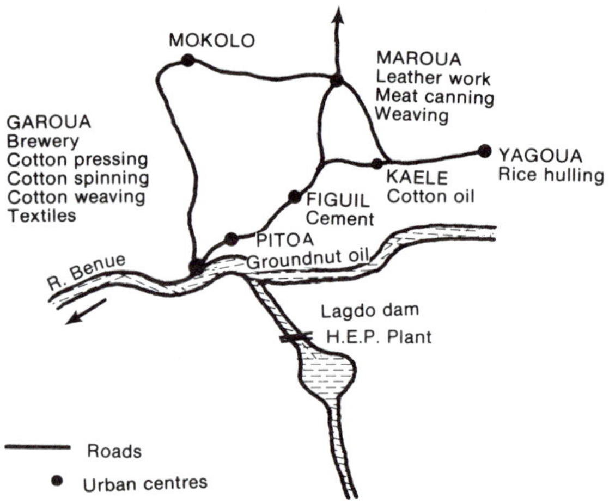

At Kaele there is a cotton seed mill and at Pitoa one finds a groundnut oil mill. A cement factory is located at Figuil. Further away at Maroua there is a meat conservation industry and it is the centre of local crafts industry. In the middle of the rice belt, the town of Yagoua has many rice hulling factories.

PROBLEMS OF INDUSTRIALIZATION INCAMEROON

The rate of industrialization in Cameroon is rather low, especially when we examine the kinds of industries involved.

A look, first of all, at the food industries in Cameroon reveals that most of them have not had that large market of consumers as is the case with the breweries which bring little benefits to the population. The population's high rate of drinking accounts for the fast expansion of breweries. The rate of expansion of breweries far supercedes that of any other vital industry in Cameroon. Moreover, there are no big iron and steel industries which could form a strong industrial base for the country.

In order to encourage the necessary food industries, the government has therefore set up an *investment code* and the *National Investment Corporation* (S.N.I.) for the more important projects. By the intermediary of the S.N.I. (Societe Nationale d'Investissment) the government provides part of the capital necessary for financing some major investment projects. The *Cameroon Development Bank* (C.D.B.) is another intermediary by which the government assists financially in investment.

b) Insufficiency of Market:

Another problem which hinders the industrial expansion of Cameroon is the low purchasing power of Cameroonians on the one hand and the relatively high prices of industrial products on the other hand.

Given that the average citizen has only a modest income and given the present rate of inflation in the world, the Cameroonian citizen barely has enough for his necessities and almost nothing left over for industrial goods which are also produced at high costs.

Psychologically too, the general misconception that imported goods are superior to the same goods produced in the country, greatly reduces the population's buying incentive of the locally manufactured goods. The solutions to these problems have to come with time following the general improvement in the standards of living and change of mentality.

The market opportunities for industrial products, though automatically widening as economic development and population growth take place, can be widened still further by improving internal lines of communication because that will increase the extent and intensity of market contacts within the territory. With the completion of the trans-African roads, especially those to other UDEAC countries, Cameroon's industrial products will have a much broader market.

C) Uneven Distribution of Industries:

Another problem which is worth noting is that of decentralizing the industries which are congested only in Douala. Over seventy-five per cent of Cameroon's industries are located in Douala and Bonaberi.

With such congestion one is bound to have higher prices for the same products in the localities situated very far away from Douala, and moreover, there is going to be rural exodus and overcrowding of people in the areas equipped with industries.

Industries follow such facilities as, presence of electricity, telephone, good roads, railways, airports and seaports for the transaction of their businesses. To effectively decentralize industries, other zones should be equipped with such facilities.

The government policy of *Balanced Development,* if well instituted, could attract industries to other towns.

d) Lack of Capital:

Capital accumulation and industrialization are, in practice, virtually identical processes. For an increase in industrial production to cope with a country's growing needs, capital formation on a large scale is needed. The meagre incomes of many citizens allow limited room for any savings or sacrifices for the amount of capital required for industrial expansion.

e) Lack of High Quality Manpower:

There can be no doubt that Cameroon has an abundance of labour force, but a good portion is unskilled. A person's capacity for mental and physical energy and even his physical comfort are affected by a number of factors — nutrition, cultural habits, training, religion, standards of living, and even psychological factors. All these factors affect the quality of labour force in one way or another, and the Cameroonian population is only beginning to find solutions to them. There is therefore a shortage, at present, of high quality labour force. Even the limited skilled labour is underemployed or placed in situations where they do not perform useful functions.

f) Entrepreneurial Shortage:

Little attention has been given to this problem by many writers even though it is an important handicap to industrialization. Capital cannot by itself produce anything — it must be harnessed for producing capital goods and the capital goods in turn harnessed for producing consumer goods. Such investment requires the services of some agent or entrepreneur who initiates, organizes, makes decisions, takes risks, innovates, and sometimes also manages.

Many Cameroonians are content with the idea of belonging to the civil service and a large portion of the business sector is controlled by non-Cameroonians typically, Nigerians (Ibos), French, Greeks and Lebanese. The few available Cameroonian businessmen lack the real entrepreneurial skills; they delight essentially in acting as middlemen or sales representatives of some foreign produced goods and do not engage in the actual production of tangible goods. Others are unable to believe that developing creative and enterprising attitudes are the key factors for success; instead, they depend on magic or superstitious beliefs to be successful. Others who venture into business and fail, attribute their failure to supernatural or mystical causes.

g) Inadequate and Inappropriate Government Action:

Every nation and every group of people in the world are capable of developing a strong industrial economy if the right policies and programmes are implemented. To have a good minister of Economy and Planning and a competent team of economic advisers capable of implementing a good policy is certainly a rare historical accident, not only for Cameroon but for most of the less developed countries of the world.

Many industrial undertakings in Cameroon have failed because they were either poorly conceived, or badly managed by incompetent and selfish authorities. This practice continues because there is little accountability on the part of many public officials. Many western firms and businessmen invest millions of francs and dollars, often under excellent conditions which only exploit the wealth to enrich the foreigner and a few local collaborators at the expense of the whole nation. Moreover, due to inappropriate public policy a good portion of Cameroonian engineers find themselves sitting idle in offices instead of exercising their knowledge in industries while expatriate technicians are imported and paid heavily for minor jobs which could have been performed by Cameroonians.

The government has, as a duty, to encourage a general attitude in the population about amount and quality of work, level of savings and consumption, and favourable conditions for enterprise. Government action in this line, so far, has been limited, and in some cases helped to destroy local industries to the advantage of foreign firms and businessmen. Only the Bamileke have been reputed for being hardworking and enterprising. There is little public action to activate many ethnic groups that are noted for their laziness and extravagance. As a matter of fact, many local businesses are run by foreigners, notably, the Ibos from Nigeria, the French, Lebanese and Greeks. The increasing involvement of American businessmen is also to be noted.

h) Lack of Management and Technical Staff

Most of the high-level management and skilled labour force in the industries are expatriates. Whereas the management staff is gradually being Cameroonized, skilled technical labour supply remains an acute problem. The shortage in skilled labour results essentially from inappropriate educational structures, curriculae, school system and lack of facilities in technical schools. The schools then turn out graduates who only know scientific facts in theory with little ability to apply them to practical life.

Another reason for the lack of skilled technical staff is the "international brain-drain" which affects Cameroon. By this we mean, the emigration of professional and skilled personnel educated in the country but living and working in the more industrialized nations. Typical among such professionals are economists, engineers and doctors, some of whom were sponsored abroad but decided to settle there after their studies for higher salaries.

i) The Legacy of Colonial Rule:

Much of Africa was colonized by European countries — Britain, France, Belgium, Germany and Portugal — which had gone through the industrial revolution and were important producers of manufactured goods in the late nineteenth century and the twentieth century. Cameroon experienced colonialism under three of these major colonizing powers — Germany, Britain and France. These European countries saw their colonies simply as sources of tropical crops and raw materials for their home industries, and as markets for the export of manufactured goods. Economic activity in these colonies meant concentrating on the production of raw materials for European industries and this process has continued even after independence. The tendency in Cameroon is to concentrate on agricultural activities while waiting for the European nations to produce industrial goods which will then be purchased with money obtained from the sales of agricultural products. With easily available manufactured goods in the market, there is little incentive to industrialize.

j) Lack of Savings for Capital:

To provide factories and good infrastructure the government and entrepreneurs need to set aside adequate sums of money as savings. Such savings constitute what economists term, *capital,* but Cameroon, like most African nations, has little surplus for saving after spending on education, wages, defense and other social needs. Lack of capital therefore, is the result of failure to create savings, and not the cause of the low level of industrialization in less developed countries. Some countries have tried to get around this problem by rushing to raise capital from international financial institutions (e.g. The World Bank, IMF) but such institutions have become so politicised by the capitalist nations which control them, and lend money out at very high interest rates that borrowing from them only creates problems of indebtedness for most of the less developed nations.

CONCLUSION:

In Cameroon the industries of the secondary sector are very diverse, comprising industries of small, medium and large scale, whose capital is both national and foreign. The secondary sector is still in the process of being Cameroonised and also being extended to all the branches which range from minor local crafts and food processing factories to the heavy manufacturing industries.

These industries have bright prospects because some of them have available forest and agricultural raw materials, but are handicapped by limited markets, inadequate means of commercialization, competition from foreign industries and employment of imported qualified labour and managerial personnel. Because of these handicaps Cameroon's industries are still far from solving the unemployment problem in the country.

In order to Cameroonise the capital, the S.N.I. (already mentioned above) was created. To provide Cameroonian technicians, the government has set up the Polytechnic Higher School and other numerous technical colleges in the country. To also provide qualified management staff, the Higher School of Business was created in Douala. However, the problems still persist because the graduates from these institutions are yet very insufficent in number and still expected to prove their competence.

CHAPTER 15
TOURISM IN CAMEROON

A few decades ago tourism in Cameroon was virtually nonexistent but in recent years there has been a rapid growth of this industry because the outside world is becoming more and more aware of the touristic opportunities available in the country. The striking physical and human diversities available in Cameroon provide a panoramic view of Africa within a small geographical area. That is why many people refer to Cameroon as "Africa in miniature."

Starting from the desert fringes near Lake Chad in the north, Cameroon extends down to the dense equatorial forest in the south, thus offering many touristic sites to a tourist. Because of these numerous opportunities found within the small 'triangle' of Cameroon it could rightly be asserted that, when a person is bored with Cameroon, he or she is indirectly bored with the whole of Africa, for there is in Cameroon almost all that can be attractive to the eye in Africa.

Factors favouring the Development of tourism in Cameroon:

a) Government Effort.

The government of Cameroon, in recent years, has made some efforts to exploit the potential wealth involved in tourism. On December 3, 1974, the President of Cameroon passed an order granting special status to the tourism industry, and on June 28, 1975, the *General Delegation for tourism* was created. This delegation coordinates and promotes tourism. It creates the necessary conditions so that hotels, travel agencies and carriers can function smoothly.

Other government orders have been issued to preserve fauna by reducing the killing of wildlife and creating game reserve areas. Considerable efforts are also being made to improve the transport system to touristic sites. Roads, railway lines, water and air transport facilities are being created to facilitate the movement of tourists in the country.

b) Private Effort.

Private effort is also increasing in that individual businessmen and companies are becoming more and more involved in hotel building, management, etc.

c) Wide Variety of Natural Resources.

Cameroon is rich in animals and birds of both the forest and grassland areas. There are such rare species as lions, gorillas, cheetahs, giraffes, rhinoceroes, and buffalos. Cameroon is captivating for its contrasting and picturesque scenery. The country has a variety of relief features which range from fascinating volcanic landscape to high mountains, depressions and plains. There are fine sandy beaches bordered with coconut palms and mangroves at the edge of a luxuriant forest.

d) Cultural Diversity

The nearly 200 ethnic groups in Cameroon are engaged in diverse cultural practices. Distinct cultural and social organizational patterns have developed in the different natural environments of Cameroon. The forest and coastal groups have egalitarian social structures and village organizations while the 'grassfielders' have centralized political institutions marked by chieftaincies, regulatory societies and judicial organizations. Much further in the north, Islam has become the main cultural pattern as the Islamised Fulani outnumbered the non-Muslim kirdi groups. The different groups of people in Cameroon also produce a variety of local crafts.

Above all, Cameroonians have always been regarded as a pleasant and hospitable people, though quite sensitive to all forms of outside domination or exploitation. As the foreign traveller would realize, they are a very generous people. When they offer you food or drinks they expect you to, at least, taste them even if you are not hungry or thirsty. Outright rejection is seen as a manifestation of a superiority complex.

It is advisable for the foreign tourist to note that Cameroonians generally detest anyone taking a picture of them without their permission. Like members of other less developed nations, they are very sensitive to picture-taking, as foreigners tend to take pictures of only the bizzare and worse aspects and generalize about the whole country. Moreover, many governments have also become concerned about spying and mercenary activities by some of those who travel under the guise of touring. In Cameroon, for example, a tourist should not be surprised that he or she cannot freely take pictures of the Presidential Palace, airports, military and police installations, government officials, policemen, soldiers, bridges and markets. Before taking the picture of someone, the visitor should ask for permission and if possible, explain what the picture is intended to be used for.

Foriegn visitors should also be careful with the use of words which seek to portray a culture that is different from theirs as inferior. For example, huts (instead of mud and wood houses), primitive (instead of traditional), and paganism or animism (instead of traditional religion).

TOURISTIC SITES

The Coastal Area:

Arriving at the beautiful city of Douala either by air, land or sea the tourist embraces the numerous industries, the 1.8 Kilometre-long bridge on the Wouri River and many colonial relics. Some over 80 Kilometres from Douala is the much smaller town of Edea with the Sanaga falls and the hydro-electric power station including the Alucam and Cellucam industries. To the south is Kribi town, with its beautiful sandy beaches bordered with coconut trees. This town served as a port right from German times and today it still handles timber export.

North of Douala are the creeks of Tiko, the beach of Limbe with solidified lava flows which extended down from the Cameroon Mountain during past volcanic eruptions. Beyond this coastal zone stands the impressive Cameroon Mountain, which is one of the most active volcanoes in Africa and the highest in West Africa.

Also bordering the coast are the vast and modern C.D.C. plantations of rubber, palms, tea, bananas, and pepper. Further inland at the outskirts of Kumba town one finds the magnificent Barombi Lake. These and many others are captivating sites for tourists.

Yaounde and its Environs:

The beautiful city of Yaounde performs essentially educational and administrative functions. It has many aspects and sites which appeal to tourists. Its decent buildings and numerous lakes give it a fantastic look. Perched on the slopes of Mt. Febe is one of the most beautiful hotels in the country — the Mt. Febe Hotel.

Downtown, the Unity Palace, the Municipal Stadium, the Reunification Monument, and many other structures are great edifices demonstrating modern art and architecture. The local showroom in Yaounde, *Office National Camerounais,* contains works of art such as paintings, statues etc.

The Western Region:

On the Western Highland region, the tourist is at the threshold of a really magnificent country, a vast area of mountains, many rising over 2,000 metres, fantastic rock massifs, innumerable waterfalls, clear streams and lakes and many deep valleys. Impressive lakes here mainly of volcanic origin include the 'male' and 'female' Manenguba

The Wouri bridge linking Douala with Bonaberi (1.8 km long).

lakes, Baleng, Awing, Nyos, Wum etc. The scenery is superb. At the beginning and end of the wet season, when the atmosphere is crystal clear, it is possible to see for over 80 km and at some vantage point for over 200 km. The rolling hills are covered with grass, and the valleys with raffia palms. Eucalyptus trees grow very rapidly in this region.

The air is dry and cool and people coming from the damp heat in the south find the climate very stimulating. It is seldom very hot, the nights are quite cold and many people in the evening time are to be found sitting around fireplaces keeping warm with the heat from blazing Eucalyptus tree logs.

This region is renowned for its traditional crafts, cultural dances and music. The greatest production of art in Cameroon is found in this part of the country with its chiefdoms, palaces and royal treasuries. The art of royalty, wealth, and status, affirm the values of the social system in which the kings — generally referred to as *fons,* and the regulatory societies promote and protect the well-being of their people. Funerals and other traditional ceremonies are important occasions to observe the personal prestige items of the *fons* and other high-ranking officials of each tribe. Some of these prestige articles include embroidered caps and garments, jewelry, drinking horns, pipes, gongs, drums and elephant tusks.

In Foumban the tourist is impressed by the palace of Sultan Njoya, constructed some 50 years ago, the bronze cavaliers, and numerous sculptures produced by the Bamum. In the Bamileke and the rest of the *grassfields* area the people are famous for traditional crafts which include local pipes, figurines, ornamented objects, masks, etc. The *Centre Climatique* at Dschang and the *Centre Touristique* at Batie are good resort spots for tourists.

In Bamenda there are numerous handicarft centres with many showrooms and sales rooms. The Bamenda Museum, and the traditional palaces, owned by *fons* and chiefs in the neighbouring villages provide striking views to tourists.

The North:

This is the heart of the tourist industry in Cameroon. The north is linked to the south by air, which is the fastest means of transport for tourists who want to journey from the south; there is also road transport, and the railway from the south extends only up to Ngaoundere.

In the north, there exists a variety of touristic aspects, paramount is the wildlife. North Cameroon is the paradise for wildlife lovers. It is incredibly rich in all types of wildlife, which inlcude mainly giraffes, hippos, deers, rhinoceroses, lions elephants, birds etc. The government has created national wildlife reserves at Waza, Bouba-Ndjida and in the Benue area.

Partial view of a traditional ruler's palace — the impressive Palace of the Fon of Bafut.

The people of this region are quite different in many ways from those of the south. This is the home of the Sudanese, Fulbe and Shuwa Arabs. These peoples are engaged in many activities. Their significant forms of art include body ornamentation, jewelry, and tera-cotta.

Maroua is the chief crafts town and its museums are very attractive. Tourists enrich themselves here with ivory, metal and wooden objects. Other articles are made of leather, snake and corcodile skins. The dry season is the best time to make a journey to the north.

In the town of Rey-Bouba one comes in contact

Musicians of the Lamido at Banyo.

A scene at a folk dance in Sangmelima, South Province.

with well preserved medieval traits. It is the base of an important sultanate which still has many of its traditional customs.

The dances and music of the northern populations are very unlike those of the south. Northerners are easily dragged into ecstasy by the music produced by the flute and drums, while southerners rely essentially on wooden gongs and xylophones.

NATIONAL WILDLIFE AND FOREST RESERVES

Cameroon counts three main and two minor national wildlife reserves covering an area of over 575,000 hectares. Wild animals are well protected by government regulations in these reserves. The wildlife reserves are equipped with comfortable camps to lodge tourists. There are eight forest reserves which also contain a good variety of animals.

The national wildlife reserves are as follows:

a) The Waza reserve
b) The Benue reserve
c) The Bouba-Ndjida reserve
d) The Kalamaloue reserve
e) The Mozogo-Gokoro reserve

It is worth noting that all these wildlife reserves are concentrated in the northern part of the country.

a) THE WAZA WILDLIFE RESERVE

The Waza Reserve was created in 1968, and covers about 170,000 hectares. It is situated 120 kilometres from Maroua town and 135 kilometres from Kousseri. It is the best known and most spectacular wildlife reserve in Cameroon and in all of French-speaking Africa. The principal species of animals found in this reserve are, the lion, cheetah, elephant, hippotragus, waterbuck, hartebeest cob, giraffe, warthog, panther, etc. Apart from animals, there is also a wide variety of such birds as, the ostrich, heron, goose, pelican, egret, vulture etc.

From mid-November to mid-June, this reserve is accessible by DC-4 planes, which may land on a 1,800-metre airstrip along the Maroua-Waza road. The western part of this reserve is more forested than the eastern part. The eastern part is covered by vast expanses of grass. It is advisable to visit the forested part between November and June, and the grass plains *(the Yaeres)* between

March and June. These two periods are the most exciting for tourists wishing to admire the herds of antelopes, elephants, giraffes and lions moving gracefully and feeling very undisturbed.

Accommodation:

There is a lodging camp for tourists perched on the southern slopes of the hills overlooking the town of Waza. The camp has many beds and tourists can safely pass their nights there, paying a moderate fee.

b) THE BENUE WILDLIFE RESERVE.

The reserve which covers 180,000 hectares is located along the Ngaoundere-Garoua highway. The landscape here is hilly and culminates at Mt. Garoua, which has a height of 1,100 metres. Unlike the Waza Reserve, the Benue is open to tourists at all seasons. The Benue Reserve is watered by the waters of River Benue.

The hippopotami encountered in the swamps and in the Benue River are the reserve's main inhabitants. In hot weather, one sees several hippopotami and crocodiles lying in the ponds, especially on the shallow bed of the Benue River. Other important animals here include buffaloes, Derby elands, antelopes, lions, and elephants.

Accommodation:

The Benue Reserve offers accommodation in its two main camps, the *Buffle noir,* which has comfortable rooms, showers and toilets, and the *Grande-Capitaine,* which is less equipped when compared to Buffle-Noir. The Grande-Capitaine is located on the banks of River Benue, along the Guidjaba-Tchollire road.

c) THE BOUBA-NDJIDA WILDLIFE RESERVE.

This reserve is located between latitudes 8° and 9° N, and lies in the Cameroon-Chad frontier. It occupies an area of over 220,000 hectares and is the home for rhinoceros, giant eland and antelopes. Lions and buffaloes are also found here.

The other less important reserves in the north include Kalamaloue, 4,500 hectares, which is found near the border town of Kousseri, and the Mozogo-Gokoro reserve with only 1,400 hectares. It is found near the Koza pass.

REGULATIONS GOVERNING VISITS TO NATIONAL WILDLIFE RESERVES.

The regulations imposed on visits to wildlife reserves in Cameroon are not intended to interfere with the pleasure that has to be derived from visits to the reserves, but are aimed at protecting the tourists from danger. The regulations are also aimed at preserving the species of the various animals.

1) A small fee is usually charged for the individual

Giraffes in the Waza Park.

tourists entering a wildlife reserve. This fee is much reduced when it concerns a group of tourists conducted by a travel agent. The permit issued after paying the fee is valid for a specified period and it is not transferable to another person. It cannot be lent or sold to another person.

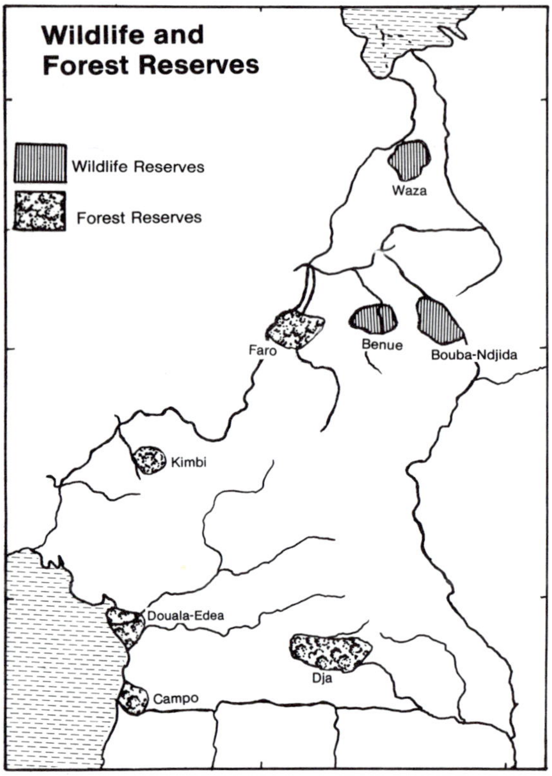

2) It is forbidden to organize picnics, reside or camp in a wildlife reserve. All visitors have to leave the reserve before sunset.
3) There shall be no driving on paths other than those cleared up for that purpose.
4) Visitors are advised not to get out of their cars to run after animals or go too close to them. Visitors shall keep a reasonable distance from elephants and lions in particular.
5) Riding motorcycles or driving lorries is strictly prohibited in the wildlife reserves.
6) Visitors are urged not to scare the animals by either making too much noise or wearing clothing with very bright colours.
7) Hunting within the reserves is strictly prohibited and defaultes shall be liable to severe penalties. Tourists living in a camp and who have to go across the reserves must keep their guns in their sheaths, unloaded and inside their cars. Anyone caught with a gun in hand or a loaded gun shall be deemed a hunter and shall be prosecuted.
8) All arms and other trophies found in a wildlife reserve shall be handed to the game warden.
9) All visitors to a wildlife reserve must be accompanied by a licensed guide or a game warden. There is usually a very small fee for providing guides.

Restrictions on Killing:

Animals and birds in the reserves are of two classes — 1) those that are fully protected and whose killing is completely prohibited; 2) those that are only partially protected and whose killing carries some restrictions. The partially protected animals carry the following restrictions:
a) It is not allowed to kill more than two animals of the same species or more than four different species on the same day.
b) It is unlawful to kill, in the course of one week (seven consecutive days) more than fifteen animals of different species.

FOREST RESERVES.

The main difference between wildlife and forest reserves is that, whereas the wildlife reserves are well maintained and equipped with route and lodging infrastructure to facilitate the visits of tourists, the forest reserves do not have such facilities. The forest reserves do not receive tourists except the specialists who work in them. It is in these forest reserves that one comes in contact with the natural luxuriant forest having both exuberant trees and grass.

The forest reserves are similar to the wildlife reserves in that they also have enough wildlife. Cameroon has at least eight main forest reserves, namely:
1) The Kalfou Reserve — 4,000 hectares
2) Faro Reserve — 335,000 hectares
3) Pangar and Djerem Reserve — 300,000 hectares
4) Kimbi Reserve — 5,000 hectares
5) Dja Reserve — 520,000 hectares
6) Douala-Edea Reserve — 160,000 hectares
7) Campo Reserve — 300,000 hectares
8) Lake Ossa Reserve — 4,000 hectares.

Apart from the first three reserves whose animals are the same as those of the wildlife reserves because they are found in the savanna region, the rest have principally forest animals, such as, chimpanzees, gorillas, monkeys, forest elephants and many kinds of multicoloured birds.

CRITICAL EVALUATION OF THE ROLE OF TOURISM

Tourism has always been regarded as a positive and useful venture. Those who benefit from this essentially international industry, have always

Jubilant Villagers with a dead elephant after a fruitful hunt in a forest reserve in the southern part of cameroon.

extolled the virtues of it — usually outlining the contributions tourism can make to a nation's economy and the pleasures individuals can derive from visits to particular sites.

There is a growing tendency to re-evaluate the role of tourism, especially in less developed nations. Tourism, in itself, is not bad. However, less developed nations have become increasingly critical of the industry because of the way it operates between more developed and less developed nations.

In most traditional societies, it is customary to be hospitable to visitors — treating them with respect, being courteous to them and sharing with them what the society produces. But there are many puzzling issues which are never understood by local inhabitants about this new kind of visitors (or tourists). For example, who are these visitors? Where do they come from and what do they want? Why do these uninvited guests keep coming, especially to a less developed society, which supposedly, has little or nothing to offer to people from the "developed world?" Criticisms about the operation of the tourist industry in recent years have challenged the traditionally held views that tourism contributes to national development.

We are going to re-evaluate some of the "advantages" of tourism in the light of what actually happens in most less developed nations.

It is often asserted that tourism enables a country to improve its earnings from the payments made by foreign tourists and that new jobs are created for local inhabitants. Our contention here is that many of the claims made for real development are often exaggerated, and that the economic benefits to a less developed nation are not nearly as large as the promoters of tourism would like us to believe.

a) Source of Foreign Exchange?

It is often held that tourism increases a nation's revenue. This can be the case, but the manner in which this industry operates makes it difficult for less developed nations to increase their foreign exchange earnings. Gross tourism earnings must be balanced against tourism expenditures. It is important to look at the net profits after all the invisible items have been deducted, but it is difficult to discover figures on or assess such transactions as smuggled goods and private agreements. A good portion of the money spent by foreign tourists finds its way back to the tourists' home country in a number of ways.

Many tourist-generating nations operate chains of hotels, which are dotted all over the world to lodge members of their country whenever they travel abroad. Examples of such hotel chains are: Holiday Inn, Hilton, Sheraton, Meridien, and Novotel. Many foreign tourists like to lodge and spend their money in hotels in which their countries own shares. In this way, the host country barely benefits from only a small share of what tourists spend in the country.

Since the tourists often prefer their own food, they will eat in hotels where imported food items and drinks are served. Few or no local foodstuffs are served in these hotels. To meet tourists' tastes, bars and night-clubs have to stock a variety of foreign drinks: Scotch Whisky, Gin, Campari, Martini etc. When a Cameroonian travels abroad, he adapts completely to new dishes and drinks and lodges in hotels that are not Cameroonian owned.

Also related to this point is the fact that the interests of a poor country are not well served if the increase in foreign reserves is squandered by the elites of the country on their own overseas travel. In the case of Cameroon, for example, the *1986 Britannica World Data Annual* (p. 646) maintains that tourism (1983) receipts from visitors amounted to 29 million (U.S.) dollars whereas expenditures by nationals abroad (1982) amounted to 70 million (U.S.) dollars.

Shopping presents another contradiction. It is often said that the purchases made by tourists in the country encourage the growth of local small-scale cottage industry. This point cannot go unchallenged. Tourism souvenirs which show little artistic merit, are increasingly being produced in bulk at factories in industrialized nations. In this way, local small scale industries are facing stiff international competition. Even photographs of geographical features on Cameroon are produced by French photographers and mass printed in France, and Cameroonians barely have to buy them. The money is then repatriated to France.

b) Employment for the People?

It is also asserted that tourism creates employment for ordinary people. Any effort to create a means to earn a living needs to be praised if the people find dignity in what they do. The problem with tourism in less developed nations is that the kind of work it offers to the local population is unskilled, low-paid and servile. The tourist industry, in most cases, employs people as labourers, gardeners, room-maids, baggage carriers, drivers, waitresses, and in related occupations. These are jobs, no doubt, but management and those they serve regard such workers as undignifying and treat them accordingly. We see here, how local inhabitants abandon their fishing nets and farms in the rural areas to take up temporary and less dignifying jobs in cities.

The skilled jobs, and the top-paid employment in the industry, often go to expatriates. Professional workers, consultants and managerial staff are usually imported and their allegiance is to their hotel owners and not to the host country. Even at the level of intermediate skilled labour, as in the installation of electric and electronic facilities, labour is often imported, with the excuse that there are no qualified Cameroonians to handle such jobs.

c) The Destruction of a Living Culture.

Every living culture is dynamic. That means, a culture is always in the process of changing. The process of cultural change, in itself, is not what matters. What matters is the direction of change, the way the change takes place, and why the change occurs.

There is the type of change resulting from the dynamic nature of the people of a particular society, or from decisions of that society. There is also the kind of change imposed on a culture by external forces which are not always in line with the desires of the society. For example, the artist is recognized and respected in his society because of the kind of art he produces which is understood by the members of that society. He expresses himself and communicates with his society through his work. Almost every traditional society in Cameroon has a particular form of art. Let us examine what happens to this piece of art when the tourist arrives. The visitor (or tourist) might not like the shape or might want to take the piece of art along with him but it might be too bulky, and this is the way it was meant to be. The visitor wants the size to be reduced so that he can easily carry it along in the plane. The artist then reduces the size and shape in a dramatic way to suit the taste of the visitor. The visitor who is ignorant of the meaning and the significance of the colours and form of the art, influences the artist to alter everything to suit his taste. In this way, the art changes in the direction of the market.

Equally repugnant is the way cultures are being

commercialized and perverted to please the tourist. Important rituals reserved for special occasions because of their meaning, are performed at random to please any person who comes as a tourist and is able to throw away a few coins at the natives. In the North-West Province, for example, venerated chiefs (or *fons),* by tradition, are not supposed to shake hands with ordinary people. This is a way of life for the people, yet, some of the chiefs immediately break the tradition with anyone believed to be a tourist.

d) The Loss of a Heritage.

The uniqueness of a people's culture is expressed in its music and art, in its dance and architecture. It is these which draw the tourists to a new country in the first place. It is a sad irony that the culture which attracts the tourist is the same their presence will destroy.

Western nations recognize the riches of the cultures of less developed nations. Museums, private collections and rich individuals in more developed nations have come to covet the irreplaceable historic treasures of some ethnic groups. Some pieces of art are of paramount significance to the history and tradition of an ethnic group and are not available for purchase to the tourist. The concern for some tourists is how to secure such articles for display in their home museums. Some tourists then embark on negotiations with some poor natives to steal and sell a significant piece of art valued by a whole tribe or ethnic group. The tourist, in turn, smuggles it out of the country. Such losses cannot be assessed in monetary terms.

In the summer of 1966, for example, *Afo-A-Kom,* the century-old beaded statue, regarded with reverence by the people of Kom (North-West Province) disappeared, reportedly stolen and sold. It made its way out of Cameroon and out of Africa. For over seven years, its whereabout was not known. It was later displayed for sale in a New York art gallery.

When information about the whereabout of Afo-A-Kom was obtained, diplomatic moves were initiated by the Cameroon government to have the sacred wood carving returned. Finally, on December 4, 1973, the Afo-A-Kom returned to African soil, landing first in Douala, then going on to Yaounde before being escorted to the hilltop settlement of the Kom by Cameroon's Minister of Information and Culture.[1]

e) Change of Life-Style and Consumer Pattern.

The way tourists dress and eat has an impact on the local population, given the misconception that what people of the more developed world do, is always good and "modern". Local teenagers begin to reject the more sensible attires of their ancestors, and take to the jeans, hats and shoes of the visitors. Others wear T-shirts bearing flags and propaganda inscriptions of their former colonizers or of other imperialist nations. Feeding habits also change. Some natives begin to switch over from their fresh, nutritive food coming directly from the farms to imported tinned food containing a lot of additives and preservatives. The tourists also bring more expensive images: a complicated watch, radio or camera, which become valued status symbols, and as the demand for such articles grows, the problems of the poor increase.

Many young girls come from the countryside and from the provinces to the big cities, mainly Douala and Yaounde, lured by the promise of work and wealth. A few of them succeed but a majority end up as prostitutes who cling to tourists because they give them much money. Public and private morality declines, particularly in urban centres. This, in no way, suggests that tourists are solely responsible for such decline, but they have been a significant influence on the lives of many local girls. There have been numerous cases where Cameroonian girls and women are attracted to high-status hotels and abused by foreign male sailors and tourists. In recent years, Cameroonians, especially those in big towns, have become increasingly concerned about contracting the deadly disease — AIDS[2], which is common in Europe and North America.

f) Reinforcement of Stereotypes.

Stereotyped images and beliefs are reinforced on both sides. The tourists on the one hand and their hosts on the other hand do not have enough time to interact with each other to learn something new about other cultures. For the tourists coming from the so called developed country to a developing country, they drive through towns and villages in air-conditioned buses as spectators. They never have the opportunity of developing a meaningful relationship with their host country. Insulated in their coaches, they photograph poverty from afar using their zoom lenses — the poor people, ramshackled buildings and many things that contrast with what they have in their home countries. They seek essentially those things that will correspond to the ridiculous images shown on their home TV sets about Africa.

The tourist may become upset by the persistence of the beggars, but would never be able to understand why the host country is the way it is. The poor people are those who leave lasting

---

[1] *New York Times,* Oct. 25, Dec. 1, Dec. 6, Dec. 14, 1973.

[2] AIDS — Acquired Immune Deficiency Syndrome. For the concern of citizens, see *Cameroon Tribune* Nov. 20, 1985, p. 4

impressions in the minds of the tourists. It happens that such tourists who left their home country with the assumption that some groups of people are really poor and wretched finally return home with those views unchallenged and unchanged. This is so because the tourists' contacts are often with shopkeepers and street people who only seek to make gains from the "rich foreign tourists". Many tourists, especially those engaged in group tours, hardly have the opportunity of confronting intellectuals, labour leaders, religious leaders, community workers, radicalised students who would challenge and enable them to revise their beliefs and impressions.

On the part of the host country, the poor citizens and hotel servants are overwhelmed with joy at the small tips handed out to them by the tourists. There is a reinforcement of the stereotyped image of the tourist as a good-natured, rich foreigner who comes into the country with a camera ready to take a picture at any time and riding in air-conditioned buses. These servants tend to be submissive and even servile; they never have the opportunity of learning or knowing about the poor and miserable conditions that exist in the tourists' home country. Not all tourists are rich people in their home country, but when they save up money and travel overseas, they suddenly become rich and important. For the first time in their life, some of them have others to serve them, make their beds, and drive them around. The new extravagant lifestyles and importance become possible only because of the economic disparity and more especially because of the difference in costs of goods and services between the two countries. So, while the tourists often live very humbly in their home country they become the *new rich* in the host country.

In conclusion, there is no doubt that tourism has always been considered an important source of foreign exchange, especially among the 'developing' countries. But it would be unwise to invest enormously in this industry without caution as to who will benefit from it and what may be the underlying negative aspects. To fully benefit from tourism, a nation needs to encourage the construction of hotels by nationals; the control of the industry should be in national hands and the citizens themselves should be concerned about safeguarding their culture from destruction by outside cultures. Only the international aspect of tourism is given elaborate discussion in many books, but internal tourism can also have many advantages to the nation as a whole.

CHAPTER 16
FISHING IN CAMEROON

We recall from Chapter One that Cameroon derived its name from the numerous prawns that were seen swarming in the mouth of the Wouri River and the Portuguese at the time called the river "Rio dos Camaroes" meaning "River of Prawns." This, in itself explains the position of fish and shrimps in Cameroon but it does not, by any means, complete the story. See how important fish is to the diet of coastal populations, particularly the Douala people and those who live near rivers and lakes.

The role of fish is important not only to the diet of the citizens but also to the economy of the country. It is estimated that over 20,000 subsistence fishermen depend on fishing along the coast while the organized fishing industry provides jobs for about 5,000 people.

Cameroon produces about 100,000 tons of fish a year from its coastal and inland waters. Fishing in the coastal waters, or what we term *Maritime Fishing,* is of two types, namely:

a) Subsistence Fishing
b) Industrial Fishing

A. SUBSISTENCE FISHING

It has already been mentioned that over 20,000 fishermen are engaged in this activity along the coast and in the estuaries of the rivers which drain into the sea. These fishermen use over 10,000 dug-out canoes with only about 10% of them equipped with engines. The rest are moved by hand paddles.

Production:

The annual catch of this type of fishing is estimated at 35,000 tons, which comprise 15,000 tons of shrimps *(njanga)* and 20,000 tons of fish, especially ethmalosa, commonly called *Bonga.* These 35,000 tons represent about 60% of the total maritime catch for the whole country.

The average output of these small fishing concerns is low, perhaps 2-3 tons per year per ordinary canoe, and 10-20 tons per canoe equipped with engines.

Only a small fraction of the subsistence catch is sold in the fresh state. Most of it is sold after drying and smoking over a wood fire and sold to Nigeria and the inland towns of Cameroon. A large percentage of the subsistence fishing is done by Nigerians who are settled along the coast and send their catch back to Nigeria to satisfy their country's considerable needs. Subsistence fishing is concentrated in the 'West Coast', between Rio-del-Rey and Betika, near the border with Nigeria. Elsewhere in the other provinces, which share the coast of Cameroon, there are some small fishing camps such as Campo Beach, Ebodje, Kribi, Londji, and Lokoundje in the Ocean Division; Dehane, Mouanko, Badangue, Mbiako and Yoyo in the Sanaga Maritime Division; Manoka, Kombo Moukolo, Cameroon Cape, Bessoukouduou, Toube, Toupwe (Douala) in the Wouri Division and M'Bomo, Mabeta, Limbe, Enyengue in the Fako Division. One of the greatest markets for subsistence catch is at Bonaberi where fish is sold in big bales and sacks.

The fishing methods are many and varied. Some of the fishermen use motorized canoes from which they throw their nets into the sea and rivers, while others simply use hooks and nets without canoes.

Perspectives:

Looking at the situation of the Subsistence Maritime Fishing, we realize that left on its own the prospects for development are low, despite the significant position it occupies in Cameroon's fishing industry. It is for this reason that the government, with the assistance of some countries, (Canada and the United States of America), created in 1977 *The Mission for the Development of Subsistence Coastal Fisheries* (MIDEPECAM) which aims at improving the fisheries and well-being of fishermen and fish farmers. It is presently undertaking projects in the East, West and North-West Provinces geared at modernising fish culture.

B. INDUSTRIAL FISHING

Industrial fishing in Cameroon accounts for about 20,000 tons of fish and shrimps every year.

Fishing at the coast of Limbe.

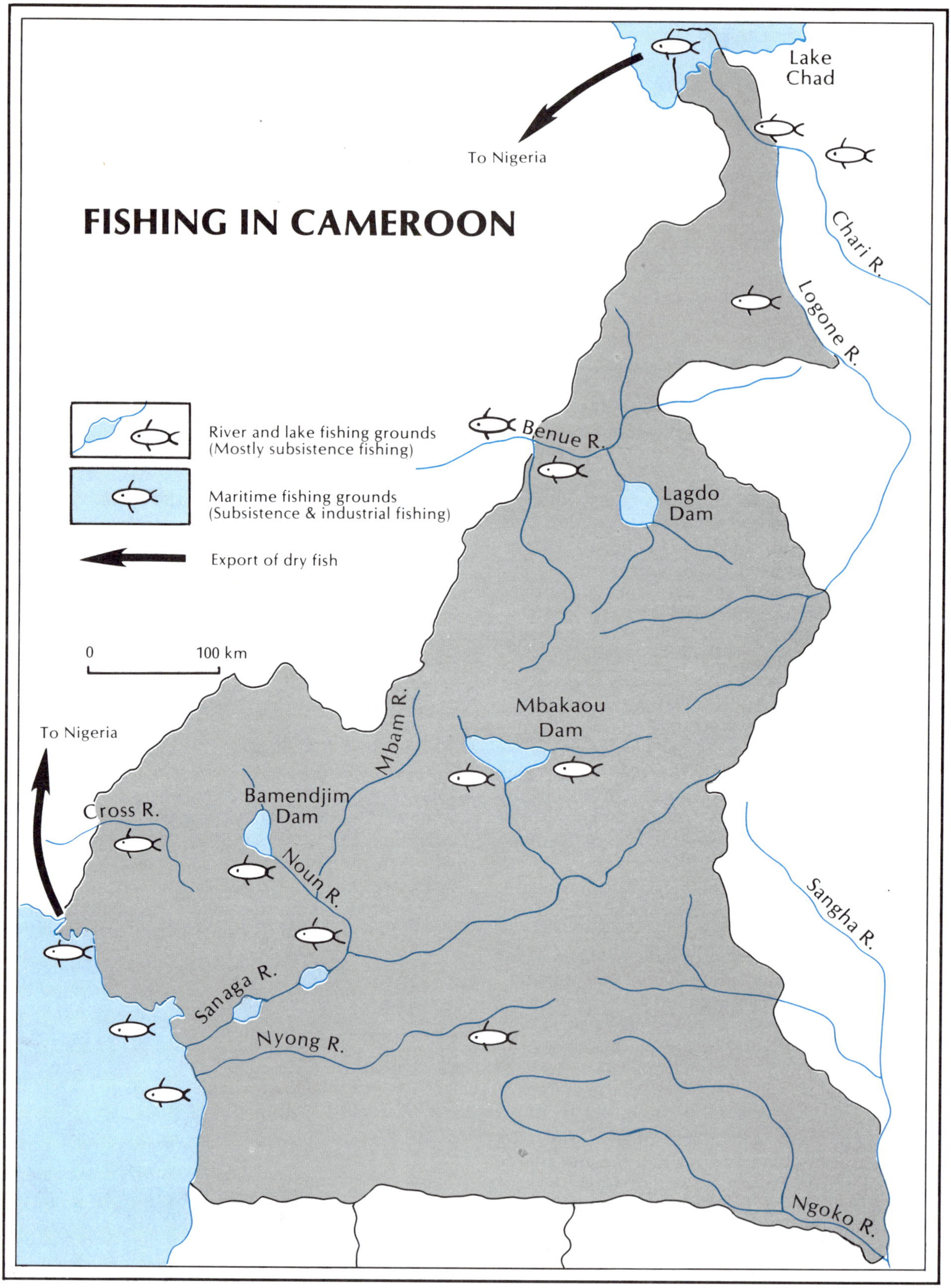
FISHING IN CAMEROON
River and lake fishing grounds
(Mostly subsistence fishing)
Maritime fishing grounds
(Subsistence & industrial fishing)
Export of dry fish
0
100 km
To Nigeria
Lake
Chad
Chari R.
Logone R.
Benue R.
Lagdo
Dam
Mbakaou
Dam
Mbam R.
To Nigeria
Cross R.
Bamendjim
Dam
Noun R.
Sanaga R.
Nyong R.
Sangha R.
Ngoko R.

This is by far lower than the output of subsistence fishing.

The authorities concerned with the development of industrial fishing have enforced the modernization of the Douala fishing fleet facilities by equipping the harbour with an ice factory (150 tons of ice per day) and by providing a series of sheds with refrigeration facilities.

Presently, the industrial fishing fleet is made up of seven fishing companies which altogether own 42 fishing vessels: 24 trawlers and 18 shrimpers. The companies and their vessels are as follows:

and secondly, because the means of quick transportation are lacking. It is for these reasons that towns in the interior of the country only receive imported fresh fish whose conservation methods are good enough to keep the fish fresh. Most of the fresh fish sold in the interior is imported from Morocco and Russia.

As regards shrimps, they are destined essentially for export to the United States of America and other foreign markets where they have become very popular.

COMPANY	NO. OF TRAWLERS	NO. OF SHRIMPERS	TOTAL
COTONNEC	6	—	6
PECH-CAM	6	—	6
SOPAC	—	5	5
COPEMAR	4	—	4
CHALUTCAM	2	—	2
CRECAM	6	12	18
CAFISTRO	—	1	1
TOTAL	24	18	42

Six of the seven companies are financed entirely by private capital. Only CRECAM (Crevettes du Cameroun) has part of its capital supplied by the National Investment Corporation (SNI).

Areas of Operation:

The industrial fishing is practised mainly in the maritime zone stretching from the mouth of the Lokoundje river right up to the borders of Cameroon with Nigeria. Important zones of operation here include Rio-del-Rey, Bibundi, South Limbe, Bimbia, Wouri, and sometimes fishermen move far upstream on the Sanaga and Nyong rivers.

The depth into the ocean depends mainly on what type of catch is needed. To get fish, the trawlers reach a depth of 8 to 25 metres, but to get shrimps (njanga or melolo) the shrimpers need to plunge to a depth of about 20 to 50 metres into the ocean.

Conservation and distribution of the catch:

The commercialization of the catch is done partly by the companies themselves and partly by other private individual traders who are not involved in the fishing process. An important fact to note is that, most of the catch does not go far from the fishing grounds. It is only sold for immediate consumption to the neighbouring towns. Those towns situated far away from Douala do not get the fresh fish, primarily because the conservation methods are unsatisfactory and limited,

INLAND FISHING:

Besides the fishing done in maritime areas, we also have local fishing methods practised inland in the fresh waters by the peasant populations who live in river and lake zones, such as, the Barombi people on the Barombi Lake region; the Bambalang people near the Bamendjim dam; the Maka and Sso people of the Sanaga and Nyong river zones; and the Kotoko along the Logone and Chari rivers. In most of these areas individual fishing dominates. Inland fishing in Cameroon accounts for about half of the total catch for the whole country. It is in the north of the country, just beyond the Adamawa Plateau, that fishing in rivers and lakes is of great significance because the rivers of this zone are extremely rich in fish of all types.

In the dry season in particular, fishing is done on a large scale in Lake Chad, the Benue, the Logone and Chari rivers. The Kotoko in the north depend almost entirely on fishing for their livelihood.

A good quantity of fish in the north is smoked *(Salanga)* or salted and the surplus is sold to the southern part of the country and to Nigeria. Fishing in the north is done in cooperative societies which are called *SOCOPELI.*

Another important aspect of inland fishing in Cameroon is the prevalence of fish culture in ponds, mostly in the West and North-West Provinces. Tilapia, Carp, Catfish and Heterolis are the major varieties of fish reared in the over 5,000 ponds owned by fish farmers in the country.

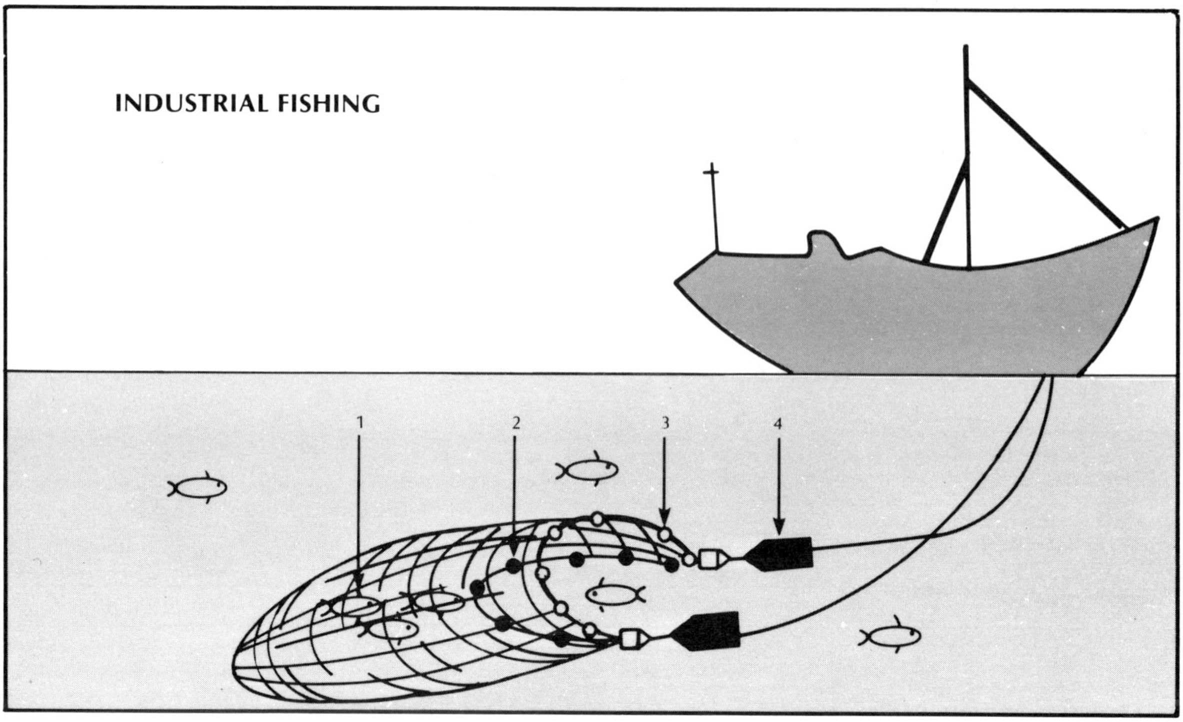

1. The net traps fish in the sea as it is pulled along.
2. Floor line with weights to hold the line on the ground.
3. Head line with floats to keep the line above the ground.
4. Boards to lift net over rough areas of the sea-floor.

CHAPTER 17:

FORESTS AND TIMBER EXPLOITATION

The forest in Cameroon covers a total area of 20 million hectares, which is about 42% of the total national area.

The forest which extends from latitudes 2° to 5° N with protrusions on the eastern slopes of the western highlands, comprises mainly the dense evergreen rain forest, deciduous and gallery forests. The exploitable forest, which covers 17.5 million hectares, forms almost 37% of the national area.

IMPORTANCE OF THE FOREST

The role played by the forest in Cameroon is significant. Besides its ecological role of providing shelter and fruits to wildlife, it also fulfills a very important role in the economy of Cameroon.

Timber is a valuable commodity of trade, because certain types of hardwood trees such as Mahogany, Iroko, and Teak fetch good money from makers of furniture in Europe, America and also in Cameroon. Timber exploitation also provides employment to many Cameroonian citizens. It provides building materials, medicinal plants and also electric poles to the population. Timber occupies a significant place in the external trade of Cameroon.

Wood occupies the fourth position by value after crude oil, cocoa and coffee, and first position by tonnage. With the establishment of the paper pulp industry (Cellucam) at Edea the position of timber in the country's economy has become even more important than had been before. However, despite efforts to 'Cameroonise' many industries in Cameroon, timber exploitation remains dominated by large foreign firms.

FORESTRY LEGISLATION

Given, therefore, the importance of the forests in the nation's economy, the government has had to outline regulations under which rational exploitation has to be undertaken. These regulations are aimed at maximizing the use of the forests while taking enough precaution not to exhaust this resource.

Forest exploitation in Cameroon cannot be embarked upon except under an agreement with the authorities of the country. The individual intending to undertake the exploitation exercise must, first of all, guarantee that he is capable of financing the activity; that he has adequate knowledge of forest exploitation and, of course,he should be of good moral standing.

Before 1973, the laws regulating forest exploitation in the former East Cameroon were different from those of the then West Cameroon.

The May 1973 Forestry Ordinance with its April 1974 decrees of implementation, now operate in the whole Republic of Cameroon.

From the Ordinance, the following aspects are of paramount importance.

1. Any natural person or corporate body wishing to exploit the forest under licence or sale of standing volume must be approved by the forestry authorities according to a procedure fixed by a decree.

Furthermore, the exploitation licences may be issued only to natural persons resident in Cameroon or to companies whose registered offices are in Cameroon.

2. Forest exploitation licences shall be granted or renewed on the recommendation of a technical Commission, the composition and functioning of which shall be fixed by decree.

— By the President of the Republic in the case of licences covering an area of 10,000 hectares or more, or bringing the total area to be exploited by the same natural person or corporate body to 10,000 hectares or more.

— By the Minister of Agriculture, when the total area to be exploited under licence by the same natural person or corporate body is less than 10,000 hectares. Licences whose total surface areas are less than 10,000 hectares are exclusively reserved for Cameroonians and the companies of Cameroonian nationality.

3) All licences shall be granted for the period of five years which shall be renewable.

4) The sale of exploitation right without the consent of the forestry authorities is forbidden.

5) All forest exploitation licences, sales of standing volume, all permits or authorizations to cut poles, firewood and wood for charcoal confer on their holder throughout the area conceded the exclusive right to collect the produce described in the exploitation licence for the specified period but give no right of ownership over the corresponding land. Consequently, the holder may not prevent the traditional harvesting, gathering of seeds, roots and medicinal plants from that concession.

6) Before a licence is granted or renewed, the prospective forest exploiter must agree to the obligations to construct roads and bridges of

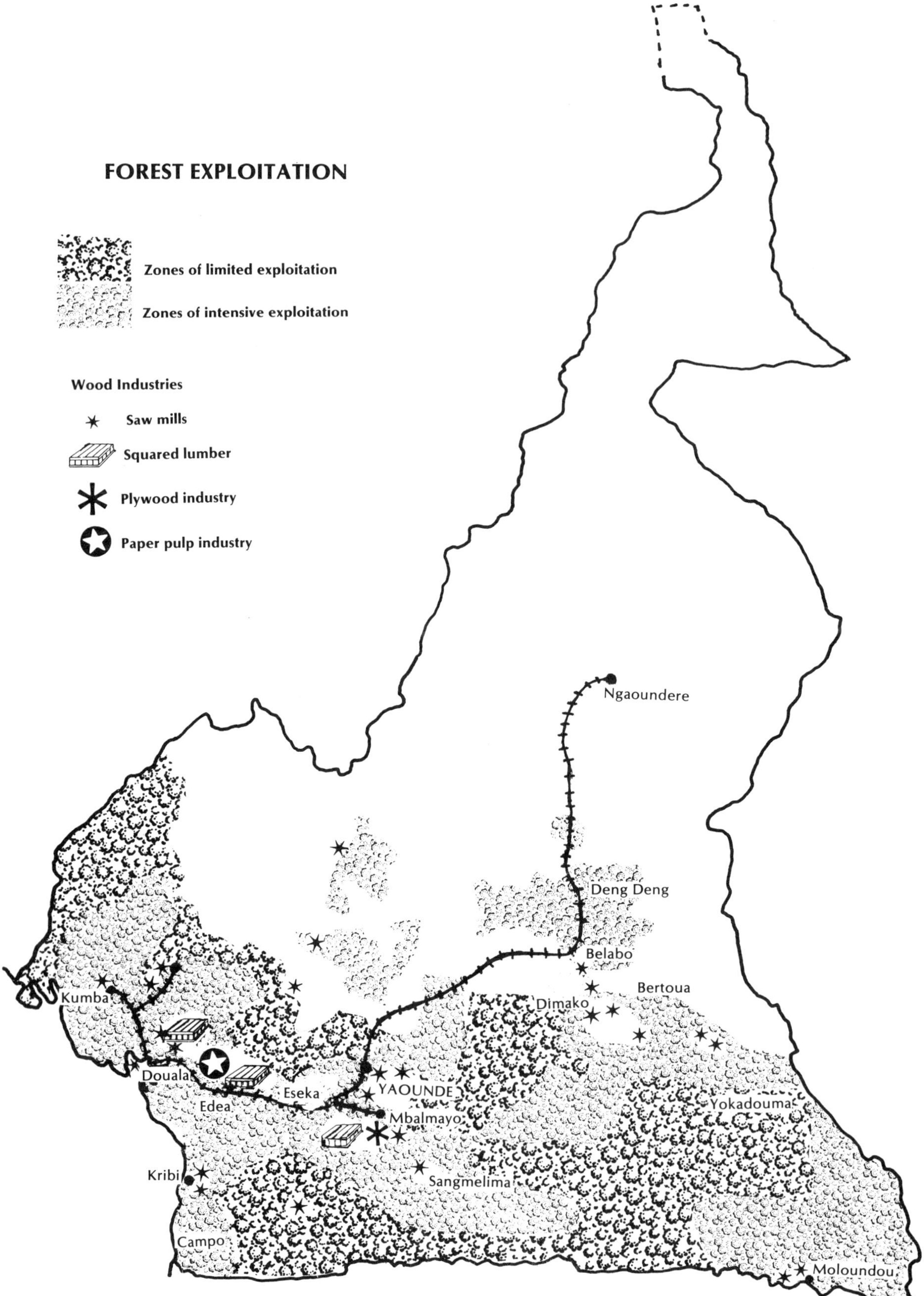
FOREST EXPLOITATION
Zones of limited exploitation
Zones of intensive exploitation
Wood Industries
Saw mills
Squared lumber
Plywood industry
Paper pulp industry
Ngaoundere
Deng Deng
Belabo
Bertoua
Dimako
Kumba
Douala
Edea
Eseka
YAOUNDE
Mbalmayo
Yokadouma
Kribi
Sangmelima
Campo
Moloundou

national utility as follows: One kilometre of road or five metres of bridge per 2,000 hectares granted or renewed.

i) Industrial Installations: The total area of forest which may be granted under forest exploitation licences to one and the same natural person or corporate body shall be in proportion to the industrial installations for wood processing to be set up in Cameroon according to the scale shown below.

ii) The maximum area which may be granted to one and the same natural person or corporate body exploiting the forest shall not exceed 250,000 hectares. The site of industrial installations for processing timber shall be decided by agreement with the local administrative authorities taking into account the effects flowing from its location.

a) The holder of a forest exploitation licence is entitled to open roads for the evacuation of timber.

b) Service roads may only be built across cultivated land on payment of compensation for the destruction of crops, the amount of which shall be fixed by order of the Ministry of Agriculture.

d) No compensation shall be payable by the licensee if it is proved that the crops were planted after the road was opened.

EXPLOITATION OF THE FOREST

Two lumbering methods can be distinguished; the traditional and modern methods.

THE TRADITIONAL METHOD:

This method which is fast dying out can still be seen only with poor exploiters, and their products are usually destined to satisfy the local markets. In this method, those felling the tree use axes and cutlasses, and usually spend a long time to put down the tree.

With large-trunk trees, stands are usually erected round the trunk so that more than one person can stand on them and the chipping would be done on all sides.

The alternative to axes and cutlasses is a long saw which two people handle at both ends. They stand in opposite directions and saw it simultaneously. After the felling, the next step is often to cut the tree into logs of the desired lengths. The logs are then sawn on the spot.

A trench is dug near the log and smaller sticks are put across it before the log is rolled onto them. Two sawers, one in the trench and the other standing on the log then embark on the sawing process. They handle at both ends a single saw and do the sawing at the same time. The logs are sawn into wood of various lengths and sizes.

THE MODERN METHOD:

After granting the exploitation licence and delimiting the areas of operation, the individual or company, first of all, constructs access roads to the area. The trees for felling are then identified and felled using chain-saws. An individual, operating on this instrument takes just a few minutes to fell even trees of big girths.

After felling the tree, the sawer then cuts it into logs. Using a tough rope or chain, a tractor drags the logs out of the bush to a spot where they are then loaded onto special timber trucks or *trailers* and taken to the port of export — either Douala

Total area of forest	Industrial installation to be brought into operation	Minimum % wood to be processed locally
Less than 10,000 hectares	No fixed norms	60%
10,000-75,000 hectares	At least one sawmill comprising two head-saws, one of which is not less than 140 cm in diameter	Not less than 60%
75,000-130,000 hectares	At least one sawmill comprising two head-saws, one of which is not less than 160 cm in diameter	Not less than 60%
130,000-175,000 hectares	At least one sawmill comprising two head-saws one of which is not less than 180 cm in diameter	Not less than 60%
175,000-250,000 hectares	A wood-peeling factory manufacturing plywood or panels, together with a saw-mill comprising two headsaws, one of which is not less than 160 cm in diameter	Not less than 60%

Traditional method of Forest Exploitation: Usually, a difficult and time-consuming task.

or Kribi, depending on the region of exploitation. They may also be transported to sawmills within the country to be sawn before exportation or for local consumption.

There are numerous holders of exploitation permits who are presently involved in the exploitation of nearly eight million hectares of forest. Over sixty of these forest exploiters own sawmills and at least four of them own integrated industrial complexes, producing sawn timber, plywood, squared lumber, and veneer.

Some of these companies include:

COCAM (contreplaque du Cameroun) whose factory is based at Mbalmayo. It is an Italian establishment in which the National Investment Corporation (SNI) has shares.

SFID (Societe Forestiere et Industrielle de la Doume) is based at Dimako in the East Province.

CFGG (Compagnie Forestiere du Golfe de Guinee). The SNI also has shares in it. It is based in Douala.

SOFIBEL (Societe Forestiere et Industrielle de Belabo) operates in the locality of Deng-Deng in the East Province with its industrial complex at Belabo.

Most important of all these, is the only giant paper pulp industry (CELLUCAM) at Edea whose establishment has added more value to the timber exploitation of the country. (Its detailed significance can be seen under regional studies).

Besides these integrated industrial complexes, we also have other sawmills spread all over the forest region. Other important forest exploitation companies operating in the country include the Timber Industries (Cameroon) LTD; Cameroon Industrial Forest at Etam in Kumba; Afri-Stevedoring enterprises; UNALOR which produces matches; SIFFCCA; and R.C. Coron in Yaounde.

Felling of timber using a chain-saw can be fast and easy.

Peeling of the logs.

TRANSPORTATION OF WOOD

Wood transportation is usually one of the most difficult and most important aspects of forest exploitation. It is more burdensome than any of the processes in the forest.

1) Road Transportation of Wood:

The logs are transported on big vehicles called trailers. This is the most common means of wood transportation in Cameroon, though appearing to be the most expensive of all the methods.

2) Rail Transportation of Wood:

This means is used only in areas that are crossed by the railway line. Here, the logs are brought out of the forest and parked at particular spots from where they are loaded onto the rail carriages.

3) Water Transportation of Wood:

Where river courses exist in forest areas, their use for wood floating is usually envisaged, especially if the floating can take place for a long distance in order to reduce transport costs. There are two major ways of timber floating in Cameroon.

a) Single log floating: Single logs are floated for short distances on small rivers.

b) Raft floating: A number of logs are tied together using a rope and then floated down a river.

Wood transportation by water is practised on rivers Ngoko, Sangha (up to Brazzaville) in the East Province; on the Dibamba and on the Wouri, from Yabassi to Douala in the littoral Province; and also on the Nyong in the Centre Province.

SPECIES OF TREES

Most of the trees in Cameroon, unlike those in the temperate regions are hard woods of different colours.

The principal types of trees felled are azobe, doussie, ebony, ilomba or caraboard, ayos or obeche, mahogany, camwood, iroko, utili, sapele and bubinga.

Many other species of valuable importance are also being exploited.

MEASURES ADOPTED FOR FOREST CONSERVATION IN CAMEROON:

Forests in Cameroon, like in other African countries, have suffered a lot of destruction for decades because of the practise of shifting-field cultivation, establishment of new towns, and removal by colonial powers with little thought of future needs. A policy of forest conservation is necessary for several reasons: to provide an adequate reserve of suitable land for farming under the system of shifting-field cultivation; to safeguard climate and vegetation; to maintain adequate water supplies, and to assist in the preventing of soil erosion.

Transportation of timber on vehicles.

Considering the importance of forests and also the growing population of Cameroon that will eventually lead to a greater demand for timber, the government has, consequently, laid down rules which will reduce the indiscriminate removal of trees, and also encourage the program of afforestation.

The measures adopted are as follows:

a) At least 20% of Forest Reserve:

The government has legislated that at any particular time, there should be at least 20% of the country's forest set aside as Forest Reserves for timber provision, and that, these reserves shall be fairly distributed all over the country so that all the regions will easily have available supplies of timber in future.

Transportation of timber by water: Raft-floating of logs on the Dibamba river.

b) Minimum felling girths:

In order to ensure a rational felling of trees, the minimum girths of the various tree species have been prescribed, so that the young ones are protected and left to mature because different species of trees mature at different sizes. These sizes are prescribed in the Forestry laws.

c) Granting of exploitation permits:

Not anyone at any time can fell a tree. Licences, permits and other forms of authorization are necessarily granted before a tree may be felled for any purpose, and whosoever fells one, must put it into the best possible use. It is for this reason that any intending exploiter has to show proof of his ability to finance the activity.

d) Contribution towards forest regeneration programs:

Individuals and companies involved in forest exploitation are supposed to replace the trees they remove, but since this is never practised, they are then obliged to contribute financially towards the replacement by paying a sum of money to the Credit of National Funds for Forestry and Piscicultural Development. The rate to be paid is assessed per hectare of the total area held under exploitation.

e) Afforestation of grassland areas:

While regeneration is being pursued in the forest zone, afforestation of grassland areas is practised on the Bamenda and Western highlands, and in the Adamawa and northern parts of the country, to provide for wood needed for local use.

Vast hectares of Eucalyptus, Pines, Cypress and other species of trees are common in these regions planted through private and public initiative.

The "Green Sahel" program for the north of Cameroon is undertaken through government initiative and it is yielding some favourable results. In this program, trees are planted every year in the grassland areas which lie at the fringe of the desert.

PROBLEMS OF TIMBER EXPLOITATION IN CAMEROON

1) Climate:

Rainfall in the forest region varies between 1,500-4,500 mm annually. This hinders exploitation activities especially in the wet season so much so that some companies halt temporarily to wait for the dry season.

2) Relief:

In a country like Cameroon whose relief is accidented, forest exploitation becomes difficult in certain areas.

3) The Thick Nature of the Forest:

Penetration in the forests is usually difficult since they are neither maintained nor prospected because of lack of roads and scanty population.

The trees do not grow in "stands", and the work of finding and then extricating the suitable trees is laborious and expensive.

CHAPTER 18
TRANSPORT AND COMMUNICATIONS

Human communities which are not served with the means of transportation live in a type of closed economy where trade or the exchange of goods and services is absent and production is only aimed at local consumption and never for export. With the disappearance of such communities in Cameroon (except for the few Pygmy settlements) many groups of inhabitants, assisted by government effort, are increasingly seeking ways of linking their localities with major urban centres for the sales of their commodities. It is for this reason that economically important regions of the country are fairly well served with the various means of transportation.

A. ROAD TRANSPORT:

Before the coming of the Germans, Cameroon could only boast of several foot-paths linking one village to the others. There was greater contact between the villages and ethnic groups in the north than was the case in the south. Two factors were responsible for this, namely: (1) the use of animal transport (camels, horses, oxen, donkeys) for transporting people and goods in the north was easier than trekking and head porterage in the south, and (2) the open nature of the grassland area, in the north was easier to cross than the forest in the south.

The greater ease of movement in the north was an important factor which made it possible for rulers to administer vast territories which formed part of their empires. Ancient empires or early civilisations in Cameroon (Kanem-Bornu, Sao, etc.) were established in the northern part of the country. The western grassland areas also saw the formation of smaller, but cohesive geo-political units currently referred to as *fondoms*. In the south where transport was either on foot or dug-out canoes, there were no large geo-political units that could be compared to those of the northern and western grassland areas.

During the thirty years of German rule, Cameroon had a few narrow roads with solidly built bridges across rivers and streams. These roads led from one dense settlement to the other. Even though they served mainly the interest of the colonialists they were constructed at the expense of indigenous hard labour and high human toll.

The period which saw the presence of the French and British in Cameroon was marked by the construction of roads which concentrated in the south, where the sea ports are located. Settlements in the south during the colonial period became more accessible than those in the north, particularly because of the large scale expansion of tree-crop farming. Thus the situation in the north and south was reversed since the primary products needed in Europe had to be exported through the ports located in the south, and through which imported goods also had to get to the north. Many roads were constructed in the south at this time.

Since independence more attention has been given to the improvement of this means of transport for the simple fact that the evacuation of the country's crops from their places of cultivation depends essentially on the available roads.

The roads of Cameroon can be classified into two categories, namely; *Trunk A* and *Trunk B* roads. The Trunk A roads link the major towns of the country. Considering a north-south orientation of the Trunk A network we distinguish four main lines:

a) *The first line* is the Limbe-Nkambe line. It starts from Limbe through Kumba, Mamfe, Bamenda, Kumbo to Nkambe. From Bamenda to Nkambe it continues to Wum and back to Bamenda forming what is commonly known as the *Ring-Road*.

b) *The second line* starts from Douala to Kousseri, passing through Nkongsamba, Bafoussam, Foumban, Banyo, Tibati, Ngaoundere, Garoua and Maroua.

c) *The third line*, otherwise called the *central line*, starts from Douala to Yaounde, then up to Yoko and meets the second line to continue to Kousseri. This line is not very busy, given the sparse population of this region.

d) *The fourth line* starts from Douala through Yaounde, Bertoua, Meiganga and meets the second one at Ngaoundere.

As regards the secondary category called Trunk B roads, they originate from local communities in the interior of the country and link up to the Trunk A roads.

Apart from the roads linking areas within the country, there are others which link the country with the neighbouring states and these are termed, *Transafrican Roads*.

The First Transafrican Line: The Lagos-Mombassa (Kenya) line passes through Cameroon. It enters the country through Ekok and passes through Mamfe, Bamenda, Bafoussam, Tibati, Meidou-

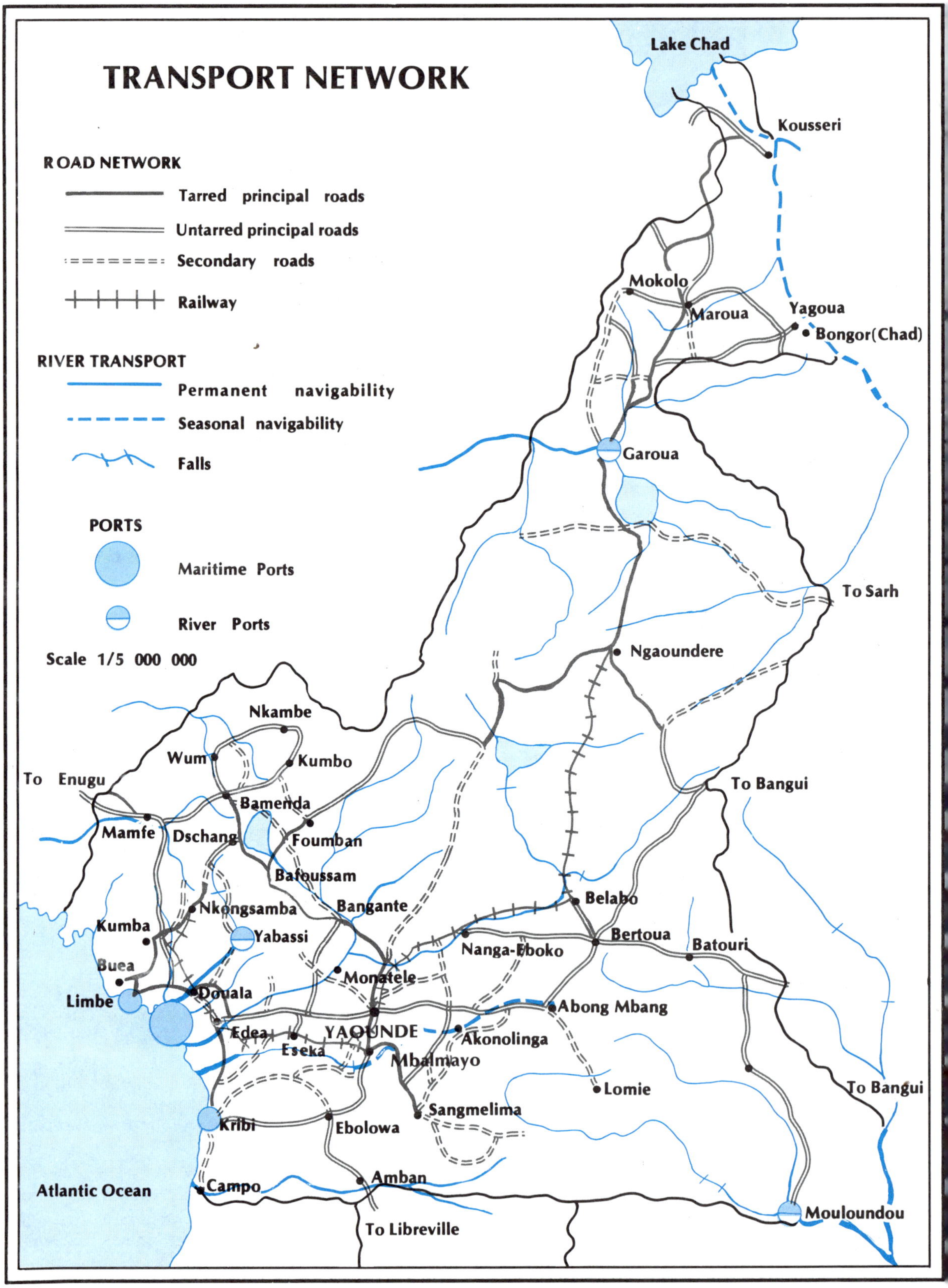
TRANSPORT NETWORK
ROAD NETWORK
Tarred principal roads
Untarred principal roads
Secondary roads
Railway
RIVER TRANSPORT
Permanent navigability
Seasonal navigability
Falls
PORTS
Maritime Ports
River Ports
Scale 1/5 000 000
Lake Chad
Kousseri
Mokolo
Maroua
Yagoua
Bongor(Chad)
Garoua
To Sarh
Ngaoundere
Nkambe
Wum
Kumbo
To Enugu
Bamenda
Mamfe
Dschang
Foumban
Bafoussam
Nkongsamba
Bangante
Belabo
Kumba
Yabassi
Bertoua
Batouri
To Bangui
Nanga-Eboko
Buea
Monatele
Limbe
Douala
Abong Mbang
Edea
YAOUNDE
Eseka
Akonolinga
Mbalmayo
Lomie
To Bangui
Sangmelima
Kribi
Ebolowa
Amban
Atlantic Ocean
Campo
Mouloundou
To Libreville

A section of the Bamenda-Bafoussam Road.

gou to the Central African Republic.

The Second Transafrican Line: This line is termed the 'coastal line', it leaves the first one at Bafoussam and passes through Yaounde to Gabon with another line continuing to the Republic of Congo.

The Third Transafrican Line: It enters the country in the north, coming from Dakar and passing through Maiduguri in Nigeria, and through some towns of north Cameroon to Ndjamena in Chad.

Most of the roads in the country, especially those in rural areas, have gravelled surfaces so that their viability is mainly seasonal. Almost all the roads are pliable in the dry season, but with the arrival of rains some of them become virtually impassable.

In recent years many more roads have been constructed, and old ones improved. There are feasibility studies and many plans are underway for the tarring of the major roads. Given the importance of road transport in Cameroon, the government is paying much attention to the opening and improvement of existing roads.

PROBLEMS OF ROAD TRANSPORT:

a) Contrasts in population distribution: A serious problem associated with road transport is the fact that there exist sharp contrasts in population distribution, and low average population density in the country. This means that roads have to pass for long stretches through uninhabited country and so face large initial operating costs.

b) Accidented relief: The presence of the Adamawa Plateau and the Western Highlands calls for a lot of excavation of hills and filling up of valleys before roads are constructed. Moreover, the rugged relief has always led to the construction of winding roads which are always the cause of many fatal car accidents.

c) Rivers: The presence of many rivers necessitating the construction of bridges, has hindered the passing of roads through certain areas. An example is the Djerem Valley which is avoided by roads because of many rivers.

d) Vegetation: The vast and thick equatorial forest in the south of the country has always presented problems of felling and stumping to the construction of roads. The many trees hold much water in their leaves after periods of heavy rains. The water keeps dripping for many hours thereby rendering the roads wet and slippery for the greater part of the rainy season. Fallen trees across roads often interrupt the free flow of traffic.

e)Soils: Soft soils, coupled with a lot of landslides are serious problems, especially during the wet season. Most untarred roads, both principal and secondary, become almost impassable because of muddy conditions during the wet season.

f) Swamps: The coastal lands have swampy zones which hinder road construction, especially at the Douala and Ndian lowland areas.

B. RAILWAY TRANSPORT:

Railway transport in Cameroon has evolved along the lines of the main historical stages through which the country has passed.

a) The German period (332 km):

The first railway installations in Cameroon were undertaken by the Germans and were oriented towards two directions.

The first line was from Bonaberi to Nkongsamba and was termed the *Western Line.* By the convention of March 6, 1907, the construction of this line was given to a German company. The work started in 1907 and by 1911 it was completed and inaugurated on April 1, 1911. It measured 172 kilometres.

The second line, called the *Central Line* was to start from Douala to Yaounde passing through the Sanaga area. By June 17, 1914, only part of

Ferryboat on River Mbam before the construction of the bridge.

this line was open to traffic and measured 160 km up to the town of Eseka. The work on the rest of the line was interrupted by the outbreak of the First World War.

b) The French Mandate Period (184 km):

The Central line was continued after the war to Yaounde. The work lasted from January 1922 to March 1927 and the line measured 134.7 km. Within this period, an extension stretched from Otele to Mbalmayo measuring 37 km. This extension was later modernised in 1933. After the construction of the Wouri Bridge, Bonaberi was then linked to Douala by rail in 1955.

c) The Period after Independence (651 km):

In order to reinforce the idea of reunification of the two states of Cameroon the "Western Line" was extended from Mbanga to Kumba, measuring some 29 km. Work was completed on this extension in September 1968.

Between October 1964 and February 1974 the *Transcameroon Line,* measuring 622 km, was completed in two phases. The first section extended from Yaounde to Belabo (293 km) and the second from Belabo to Ngaoundere, measuring 329 km. This work was financed by the Cameroon government with loans from France, the U.S.A., Germany, and the European Economic Community.

Presently the main line is known as the *Transcameroon Rail Line* and it starts from Douala through Yaounde to Ngaoundere. The first part stretching from Douala to Yaounde, known as *Transcam I,* has completely been reworked in recent years and the distance reduced from 309 km to 265 km. *Transcam II* extends from Yaounde to Ngaoundere and measures 622 km.

The Douala-Nkongsamba line measures 172 km. It serves in the evacuation of plantation products from the Mungo Region to Douala. The short extension of 29 km from Kumba carries mainly passengers and agricultural produce to Douala.

Railway transport in Cameroon is controlled by the National Railway Company called REGIFERCAM (Regie Nationale des Chemins de Fer du Cameroun). The total railway length of over 1,170 km controlled by the Regifercam is used for the transportation of heavy exports such as timber, bananas, cocoa and coffee from the Mungo region to Douala. From the north the rail carries down groundnuts, cotton, cattle, hides, and skins to the port town of Douala. In the opposite direction it facilitates the transportation and distribution of manufactured products such as cement, fertilisers and heavy building materials to the main towns through which it passes in the

Kumba railway station.

north. Petrol and imported new vehicles are also transported from Douala to Yaounde and other towns in the north by rail.

Besides facilitating easy and quick movement between the north and south, railway transport also serves as a means of cultural exchange between the peoples of north and south Cameroon.

In the South-West Province narrow-gauge railway tracks were constructed by the Germans for easy transportation of agricultural produce from the plantation zones to their assembly spots. These lines are at Ekona, Missellelle, Meanja, Tiko and Limbe. They are controlled by the C.D.C. and not by Regifercam.

Railway lines are concentrated in the most developed regions of the country. The high relief excludes rail transport from such places as the Bamileke and Bamenda highland areas. The rail likes the low-lying lands in the southern part of the country. However, it has inevitably cut through rocky hills in the Centre Province to pass through dark tunnels before reaching Yaounde.

C. AIR TRANSPORT:

This means of transportation is the most recent of all the means to appear in Cameroon. It was in the early 1930s that small planes started touching down on Cameroonian soil. The Aero-club in Douala at the time received its first plane in 1934, and after the Second World War air transport gained more importance in the territory.

At independence the country was already linked to Europe by air through the French Air Transport Company, currently known as UTA. Later *Air-Afrique* was created. It is a multi-national air company made up essentially of the countries which had French influence. An agreement by the countries involved was concluded in Yaounde in 1961. Cameroon was a member of the new air company. Air-Afrique effected journeys to Europe and other African countries, and within Cameroon it served the towns of Douala and Yaounde

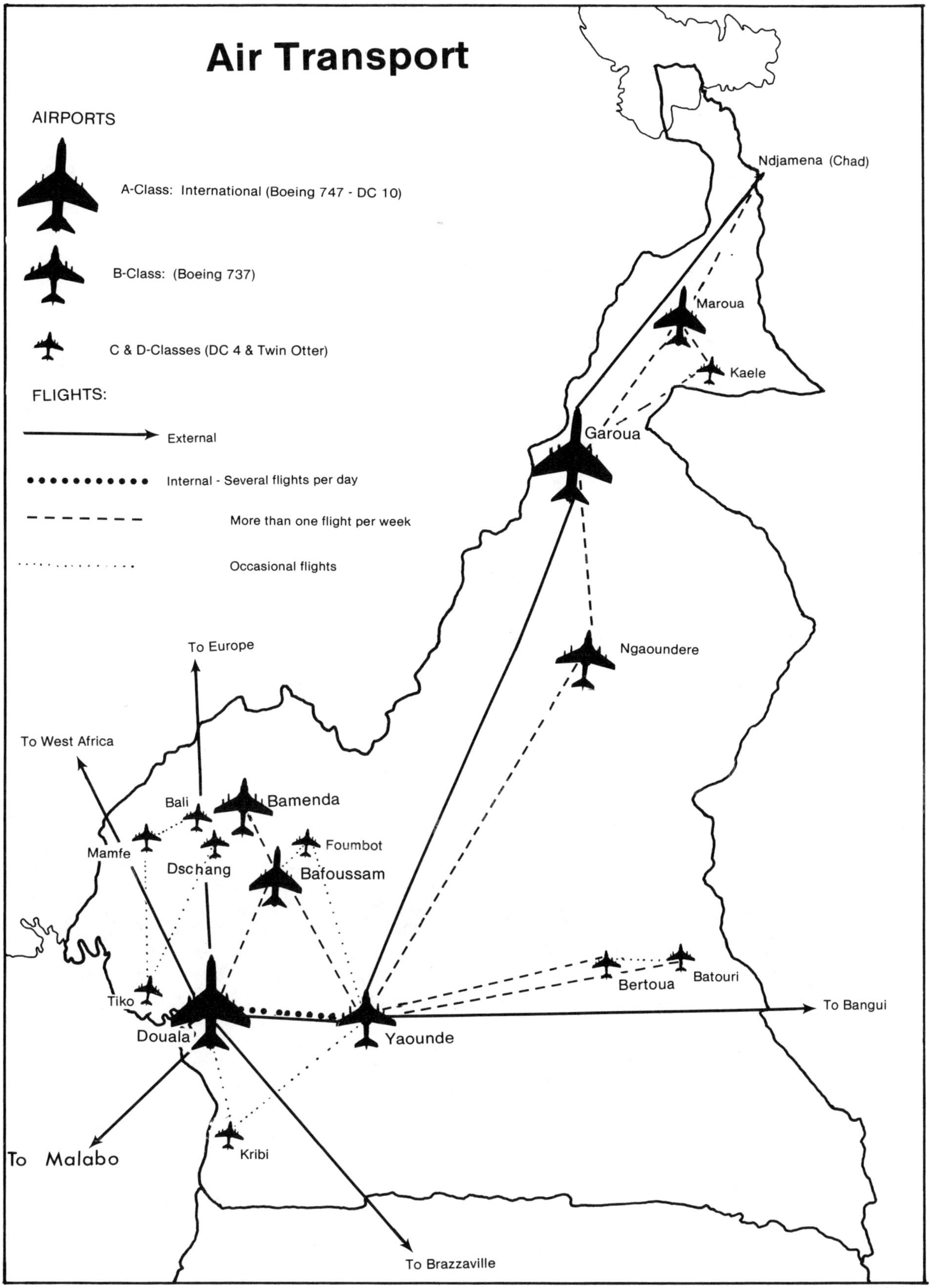
Air Transport
AIRPORTS
A-Class: International (Boeing 747 - DC 10)
B-Class: (Boeing 737)
C & D-Classes (DC 4 & Twin Otter)
FLIGHTS:
External
Internal - Several flights per day
More than one flight per week
Occasional flights
Ndjamena (Chad)
Maroua
Kaele
Garoua
Ngaoundere
To Europe
To West Africa
Bali
Bamenda
Mamfe
Foumbot
Dschang
Bafoussam
Bertoua
Batouri
Tiko
Douala
Yaounde
To Bangui
To Malabo
Kribi
To Brazzaville

and provided a link with the north. Journeys to other areas within the country were in the hands of a small company, Air Cameroun.

THE CAMEROON AIRLINES COMPANY:

Cameroon withdrew from the multi-national air company, Air-Afrique, on January 22, 1971. On June 4, 1971, the National Assembly sat and voted a law to create the Cameroon airlines Company. The company was thus constituted on July 26, 1971 and it started with a capital of 1,500 million francs CFA, 70% of this amount came from the Cameroon government and 30% came from the French Air Company, Air France.

The inaugural flight of this company, Cameroon Airlines, was effected to Paris on November 1, 1971 in a chartered Boeing 707. From that time on the story of air transport in Cameroon has always been linked to this national company, Cameroon Airlines (CAMAIR). This does not mean that foreign air companies are excluded. Other international flights to Cameroon also involve the following companies: Air Afrique, UTA (France), Aeroflot (USSR), Sabena (Belgium), Swissair (Switzerland), Iberia (Spain), Ethiopian Airlines, Nigerian Airlines, Air Zaire, Air Gabon, Air Centrafrique, British Caledonia, and many others.

Since its creation, the capital of Camair was raised to 2,900 million francs by February 1, 1973 and the company acquired its own planes: a Boeing 707 for long distances to Europe; three Boeings 737 for medium distances and a Twin-otter carrying 19 passengers serving the small aerodromes of the country. The company acquired a jumbo jet Boeing 747 in 1981 and since then it has continued to expand its services both within Cameroon and abroad.

AIR SERVICES BY CAMEROON AIRLINES:

Three main links operated by Cameroon Airlines are:

1. The line between Cameroon and Europe.
2. The lines to other African countries.
3. The internal lines of Cameroon.

1. The line between Cameroon and Europe:

This line leads to Paris then to Marseilles, Rome in Italy and Geneva in Switzerland. Another line also leads to London. With the expansion of the services new links are being sought in Europe and America.

2. The lines to other African States:

These lines link towns in different directions. Cameroon Airlines serves Bangui in the Central African Republic in the east, Libreville and Brazzaville in the south and Lagos in the Republic of Nigeria. Ndjamena lies at the terminus of the north-south internal line and does not need special planes to serve it. The journeys to the west link the capitals of countries. From Dakar Camair planes fly to Addis-Abeba passing progressively through Monrovia, Abidjan, Accra, Cotonou, Lagos and back to Douala, from where it goes to Kinshasa, Bujumbura, Nairobi, Addis-Abeba and then to Dakar.

3. The internal lines of Cameroon:

The Boeing 737 assures internal flights between Douala, Yaounde, Ngaoundere, Garoua, Maroua, Yagoua and Ndjamena in Chad. Towards the western and north-western parts of the country Camair flies to Tiko, Bafoussam, Dschang, Koutaba, Bali, Bamenda and Mamfe. To the east the airline serves essentially Batouri and Bertoua.

All the lines served by Camair do not bring the same amount of proceeds. Some of them are quite profitable, for example, those to Europe and a few internal lines such as the Douala-Yaounde, Douala-Bafoussam-Yaounde. The others are deficient in passengers and consequently bring losses to the company. Considering, however, that it is a para-public undertaking, the company's aim is not only to maximise profits, but the proceeds should at least be able to pay for the costs involved in rendering the services.

AIR TRANSPORT INFRASTRUCTURE:

Given the generally improving conditions in air transport around the world and in Cameroon, Cameroon Airlines has had to improve its infrastructure. The runways of most aerodromes have been lengthened, the buildings changed, and some airports have been completely overhauled to receive bigger planes. Consequently, there exist airports of different standards. In order of importance we have:

1) DOUALA AIRPORT:

The new airport of Douala is the most modern of all the airports in the country and was the first international airport of Cameroon. This airport was officially inaugurated on June 27, 1977 by the head of state. It has a runway of almost 3 km (2,850 m). The old airport, adjacent to the new one and still in use, has a runway of 2.3 km.

2) GAROUA AIRPORT:

This airport is the second international airport after Douala. Given the flat nature of the land around Garoua the runway of this airport is capable of being stretched for a much longer distance than that of Douala airport. However, Garoua receives fewer planes than even the Yaounde airport which is not an international airport.

3) YAOUNDE, BAMENDA, BAFOUSSAM, MAROUA AND NGAOUNDERE at present are the airports capable of receiving medium distance planes such as the Boeing 737. Yaounde airport has a runway of only 1,737 m and the concern now is to make it an international airport.

D. RIVER TRANSPORT:

Most of Cameroon's rivers are not navigable for the simple fact that they are interrupted by rapids and falls. One can mention a few of these falls appearing on some river courses. On the Wouri River are the Ekom Falls, on the Sanaga are the Nachtigal rapids and on the Menchum River are the Menchum Falls. Some of these rivers are very shallow because their beds get filled up with sand and silt brought down from the hills.

Some of Cameroon's rivers are navigable in the rainy season when their volume and depth are increased by floods. Even during this season only certain sections are navigable and hardly the whole course.

The principal navigable rivers in Cameroon include:

a) The Benue, which is navigable from mid-July to mid-September. The navigable section extends from Garoua down to Port-Harcourt in Nigeria, covering a distance of over 1,200 km. This makes Garoua a river port.

b) The Dibamba is navigable for 60 km and used for the floating down of timber logs. c) The Sanaga is navigable for 60 km up to Edea and for a further 85 km up to the Nachtigal Falls.

The Abi falls in Mbengwi.

d) The Wouri is navigable for 24 km upstream from Douala.

e) The Nyong is also navigable for 60 km up to the Dehane falls and after Mbalmayo it is again navigable up to Akonolinga for about 150 km.

f) The Ngoko is navigable from Mouloundou up to its confluence with the Ngoko upstream for many kilometres.

g) The Manyu and Ndian rivers are permanently navigable throughout the year. The Ndian is navigable from Rio-del-Rey by small vessels. The Manyu, on the other hand, is navigable from Mamfe down to Calabar during very high waters and the traffic is usually very high in the heart of the rainy season when the Mamfe-Ekok road becomes impassable.

h) The Chari is also navigable from Kousseri down to Lake Chad for the whole year.

On the termini of the navigable sections, river ports develop. Typical examples include Garoua on River Benue and Mamfe on River Manyu.

PORTS:

Cameroon counts four maritime ports, namely; Douala-Bonaberi, Kribi, Tiko and Bota (Limbe), and one important river port at Garoua.

Douala-Bonaberi Port:

Two ports, Douala and Bonaberi, handle most of the imports and exports of the Republic of Cameroon. Douala port is located on the left bank of the Wouri river. It has a quay of more than 300 metres long of deep water. On the right bank, facing the Douala port, is the Bonaberi port. Both of them are found at the limit of ocean navigation. This limit is set by the long bridge across the Wouri river linking the town of Douala with Bonaberi.

The port of Bonaberi has a quay of over 350 m. The Douala-Bonaberi port forms the main port of the country, handling over ninety percent of the country's foreign trade. It exports mainly timber and agricultural produce, such as, cocoa and coffee, and receives imports which include essentially vehicles and other manufactured goods. The Bonaberi port is noted for handling bananas destined for export. The main problem faced by the Douala-Bonaberi port is silting which calls for huge sums of money to be spent on dredging.

Traffic on the Douala-Bonaberi port has been increasing at a great pace, passing from 62,524 tons in 1911 to 3,533,499 tons in 1981. Very important is the fact that in recent years the port has increasingly handled the external trade of Chad and the Central African Republic.

Douala Port.

Tiko Port:

This port is accessible through the Bimbia creeks. Like Bonaberi, it served for a long period since colonial days as a banana port, but today it also handles exports of wood and rubber.

Bota (Limbe) Port:

The port is favourably situated but equipped only with rudimentary facilities. The two ports, Tiko and Bota, both handled the imports of the former West Cameroon State. With the reunification of the two federated states, the ports declined drastically almost to insignificance due to the shift of their economic activities to the Douala-Bonaberi port. Besides government interference in shifting some of the activities, the construction of the Reunification Road (Tiko-Douala), created a preference for road transport to water transport for many goods. Bota port in particular is a naturally deep sea port which, unlike Douala port, does not experience the phenomenon of mud, sand, and silt accumulations and will hardly need the expensive task of dredging. Presently it only handles the export of cocoa, coffee and kernels from the South-West Province. Studies are underway for the creation of a major port in the deep sea waters near Limbe at Cape Limbo.

Kribi Port:

The port of Kribi is located at the indentation zone of the coast at the mouth of the Kienke River. Accessibility up to the port is not very easy for large ships. The port of Kribi handles the export of timber logs from the southern part of the country and cocoa from the region around Ebolowa.

RIVER PORTS:

Garoua Port:

The port of Garoua is busy only for three months: July, August and September. Even within this time a sudden fall in the water level is capable of making it difficult for water vessels to move.

Until the opening of the railway line up to Ngaoundere, traffic on the Benue-Niger route was the easiest and cheapest means of transporting goods to the north. It is to be noted that since the 1960s the port of Garoua has been declining because of increased competition from road and rail transport. The port also saw its trade stagnate between 1967 and 1970 because of the Nigerian Civil War. In addition to these problems, trade using the port of Garoua requires payment in non-franc foreign currency for the fact that most of the shipping is in the hands of Nigerian-based companies.

It is important to note that navigable reaches of some rivers end in small ports, such as Yabassi on the Wouri River, Edea on the Sanaga, Mbalmayo and Abong-Mbang on the Nyong River, Kousseri on the Logone River and Mbonge on the Ndian River.

TELECOMMUNICATIONS:

INTELCAM, the International Telecommunications Corporation of Cameroon (a corporation jointly owned by Cameroon and some foreign bodies) has the sole power to exploit telecommunication systems in Cameroon.

Cameroon has 60% of the shares, La Compagnie Francaise des Cables Sous-Marins et Radio has 30%, and Cable and Wireless Ltd. of Britain has 10%.

INTELCAM aims at studying, installing and exploiting all submarine, radio-electric, terrestrial, spatial and other systems of telecommunication which may enhance the telephone, telex, international telegraph and also external radio transmissions and television of the country.

The country is served with both internal and

Aerial view of the Zamengoe Satellite ground station near Yaounde.

external telecommunication systems. A ground telecommunications station by satellite was set up at Zamengoe, a few kilometres away from the city of Yaounde since 1973. This station makes it possible for Cameroon to be linked to the outside world by phone.

There is a television system. Radio stations operate in all the provincial chief towns, including the national station in Yaounde. The provincial stations use English, French and local dialects to reach their audience. The country is well served with national and international newspapers. The *Cameroon Tribune* written in both English and French is particularly popular.

SUMMARY

Transport is particularly important in the field of development. It is generally recognized that the improvement of transport forms perhaps the most valuable single contribution towards economic, social and political development. Areas that are inaccessible and too distant from markets and ports have equally experienced a type of economic stagnation.

More importantly, improved and extended transport facilities are necessary to the widening and linking of the market in areas already settled, and in stimulating further production for internal and external trade in a country: cheap and easy transport links markets, brings additional buyers and sellers into contact with one another. Furthermore, easy transport allows exchange of goods between rural areas and urban centres — thus, the rural farmer rises above subsistence level, consuming locally grown food produced in his village and also making use of manufactured goods produced in towns. Town residents receive enough food supplies from the rural areas.

Improved means of transport reduces the expensive and tedious task of human porterage. It has been estimated that human porterage is some nine or ten times as expensive as transport by water, motor vehicle, or rail. This is so because frequent repetition of transport movements are demanded in a situation where only limited amounts can be carried by the individual in each operation; and such movements cannot be used for long distances.

Important developments have taken place in Cameroon following the general improvement in the various means of transportation. Export crop production has expanded while trade and cultural exchange between north and south have increased, thanks to improved means of transportation and free movement of the population. Northern peoples migrate in large numbers to work in the tree-crop farms and cities of the south, while many southern traders, technicians and civil servants settle to work in northern cities. The food demands of these migrants have contributed to the increasing flow of foodstuffs between north and south. The cattle trade between north and south has also increased.

CHAPTER 19: COMMERCE

INTERNAL TRADE

The level of economic progress of a nation can best be evaluated by considering the volume of exchange of goods and movement between the towns and the countryside, and between one region of the nation and the other. Added to this, is the rate of modernization of the means of transportation and distribution of the goods involved in the commerce of the country. A growing town with a developing countryside shows, for example, that the means of transporting building materials and other goods to the villages are improving.

The population of Cameroon, being essentially rural, demands that there be much exchange of products between towns and countryside, and between one region of the country and the others.

We can therefore distinguish many categories of traders within the country who are involved in the field of commerce at different levels.

a) The Peasant traders:

The sales of farm products by subsistence farmers to town residents occupy a major position in the internal trade of the country. Farmers supply the big towns with local foodstuffs, such as, beans, cocoyams, maize, cassava, yams, plantains, potatoes, groundnuts; fowls and animals, such as, goats, pigs and cattle.

After the sales, farmers buy salt, clothes, kerosene, utensils and a few luxuries like ornaments, beverages, etc., which they carry back to their villages. These are the peasant traders whose financial stand is usually very weak.

b) The Petit-traders or Market-boys:

These are those who own stores in built-up markets or in strategic corners of the towns and deal only in retailed manufactured goods. These petit-traders stock assorted goods, especially necessities and also clothes which they sell directly to consumers.

Besides this group, there is a more viable class that owns medium-size shops. This group includes mostly the Greek, Indian and Lebanese traders and even some rich Cameroonians.

c) The Big enterprises:

These are large undertakings owned mostly by foreigners. Some of them are old colonial companies such as "R and W King" and a number of them are owned by foreign capitalists.

They are engaged in the sales of expensive goods like vehicles, machines and building materials. Examples of such enterprises include SOCADA, SCOA, CAMI-TOYOTA.

Some of them who deal in less expensive goods usually sell them only in bulk. To these enterprises, can be added the super-markets such as MONO-PRIX, SCORE, VAPSAN, and many others which are found mostly in the big towns of the country.

INTER-REGIONAL TRADE

This concerns the exchange of products from where they are produced in abundance to where they are scarce.

Cattle is carried from the Adamawa to the south; groundnuts, cotton and leather articles from the north of Cameroon are carried to the south and in return the traders carry manufactured goods and other products of the south to sell in the northern part of the country.

As regards cash crops, three types of agents are responsible for their collection from the countryside to the town, and if necessary, to the ports or factories.

First, there are the individual traders who buy cocoa, coffee, etc., from the farmers directly. Second, there are the local cooperative societies which buy from the farmers and pass over to the National Produce Marketing Organization which then exports them. Thirdly, corporations, such as Bastos Company, buys products from local producers in order to process for direct consumption.

Internal trade is not without its problems. Built-up markets in towns have frequently been faced with fire accidents which have always amounted to serious losses. Within the last few decades municipal markets in Douala, Tiko, Kumba and Bamenda have been faced with serious fire accidents which have always been accompanied by heavy losses.

EXTERNAL TRADE:

Since independence there has been some marked progress in the external trade of Cameroon. Also important is the fact that Cameroon has, to some limited extent, been able to diversify its trading partners. With a good record of political stability, fiscal soundness and a liberal investment code, Cameroon has increasingly become an international market, especially for foreign traders eager to dump their country's mass-produced goods.

Exports

Cameroon exports mostly agricultural and forest products of which cocoa (the leading export crop by value) and coffee occupy a significant position. Other agricultural products exported by Cameroon include cotton, rubber, bananas, palm pro-

Yaounde Central Market.

ducts, tobacco and groundnuts. Non-agricultural exports include timber, crude oil, aluminium ingots and aluminium products, etc.

Foreign customers who receive Cameroon's exports include France, the Netherlands and the United States of America. These are the main buyers of Cameroon's exports, but minor ones include the Federal Republic of Germany, Italy, Japan, the United Kingdom, and many other African countries. Cameroon relies mainly on the export of primary commodities.

Imports

Cameroon imports mainly finished manufactured goods which are ready for immediate use. As Cameroon's desire to make progress increases, so has its purchases and level of indebtedness to foreign governments and agencies. Imports include heavy machinery, iron and steel goods, motor vehicles, electrical apparatus, medicine, pharmaceutical equipment, chemical products, books and writing materials, textile products, etc.

The main suppliers of these products are France (which supplies nearly one half of the country's imports), the United States of America, West Germany, Japan, Italy, the United Kingdom, the Netherlands, Belgium, Luxemburg, Nigeria, etc.

To understand the significance of the trade figures below, one will need to examine many factors which come into play in determining the balance of trade between Cameroon and each of its trading partners.

One notices, for instance, that balances with some more developed countries such as Britain, Japan and more especially France, are always on a deficit. This is partly so because Cameroon imports almost all her necessities (manufactured consumable goods and equipment) from these countries whereas they do not import as much from Cameroon.

The case of France is particularly unique. Through cooperation agreements signed between Cameroon and France since independence, the French have been able to secure a healthy portion of the Cameroonian market. Moreover, the terms of trade and the fact that Cameroon pays for French goods using the franc and not another currency always places France on the advantage.

PROBLEMS OF CAMEROON'S EXTERNAL TRADE:

1) The shortage of exports: There is an insufficient production of the various export commodities.
2) There is a predominance of primary products over industrial products. The instability of world prices of primary products creates uncertainty and subjects all export revenue estimates to the deterioration in world prices. The prices of raw materials are not as remunerative as those of industrial products, thereby always resulting in poor terms of trade for the country.
3) Cameroon's external trade is dependent on a few countries and on a few products.

EXTERNAL TRADE

PRINCIPAL COMMODITIES (million francs CFA)

Imports c.i.f.	1980	1981	1982
Cereals	5,779	4,411	5,683
Wheat Flour	1,462	1,725	933
Beverages	3,359	3,425	4,017
Cement and clinker	5,883	8,224	5,305
Alumina	4,179	10,021	5,414
Petroleum products	43,619	46,906	6,645
medicine	8,085	9,760	11,933
Plastics	7,898	9,117	9,543
Rubber, synthetic rubber and rubber goods	5,850	5,466	4,450
Paper and allied products	7,071	8,904	8,920
Synthetic textiles	2,345	2,935	7.983
footwear	2,225	2,2526	3,294
Iron and steel	32,439	32,405	34,411
Non-electric machinery	46,310	52,935	57,864
Electric Machinery	18,454	18,855	26,892
Road and transport equipment	24,192	33,380	36,062
Air transport equipment	554	824	1,089
Total (incl. others)	216,804	386,089	394,581

Exports f.o.b.	1980	1981	1982
Cocoa	40,333	42,388	36,926
Coffee (arabica)	21,310	16,956	17,996
Coffee (robusta)	46,100	42,896	27,283
Bananas	5,350	1,871	6,676
Rubber	1,621	1,178	964
Groundnuts	n.a.	n.a.	52
Tobacco	3,270	3,399	3,322
Cotton fibre	7,349	10,386	10,711
Cotton Fabrics	3,375	3,953	6,042
Palm nuts and kernels	812	188	120
Palm oil	949	1,217	1,722
Cocoa pulp	10,978	5,978	2,882
Cocoa butter	6,692	5,279	4,123
Logs	22,335	17,839	14,878
Sawnwood	4,661	3,797	3,699
Aluminium	4,527	3,722	4,750
Aluminium products	5,473	5,437	5,519
Crude Petroleum	83,583	91,985	127,937
Total incl. others)	296,981	290,851	306,314

PRINCIPAL TRADING PARTNERS (million francs CFA)

Imports c.i.f.	1980	1981	1982
Belgium/Luxembourg	8,074	7,974	10,583
France	145,264	159,330	174,873
Germany, Federal Republic	27,512	21,271	27,544
Italy	16,120	17,891	16,253
Japan	17,886	22,541	24,274
Netherlands	10,939	15,448	9,807
UDEAC*	13,159	12,561	3,388
UMOA†	7,637	3,597	3,116
United Kingdom	9,243	13,180	14,121
USA	17,785	24,804	29,890
Total (incl. others)	337,607	386,089	394,581

Exports f.o.b.	1980	1981	1982
Belgium/Luxembourg	4,725	2,666	2,407
France	60,619	58,595	71,040
Germany, Federal Repubic	16,671	17,502	16,522
Italy	22,695	15,114	14,022
japan	7,369	7,779	11,193
Netherlands	53,298	43,185	64,281
UDEAC*	13,159	12,413	15,624
UMOA†	3,204	2,490	1,734
United Kingdom	5,874	1,686	2,065
USA	83,675	114,366	135,166
Total (incl. others)	296,856	303,307	358,777

*** Union douaniere et economique de l'Afrique centrale (Customs and Economic Union of Central Africa), comprising Cameroon, the Central African Republic, the Congo, Gabon, and Equatorial Guinea.**

† Union monetaire ouest-africaine (West African Monetary Union), comprising Benin, Burkina Faso, the Ivory Coast, Niger, Senegal, and Togo.

Source - *Africa South of the Sahara 1986* - 15th edition - Europe Publications Limited.

a) Dependence on a few countries: Countries of the European Common Market constitute Cameroon's major trading partners but there are strong predominant links with France. It follows that in the external trade, the dependence of Cameroon on certain countries obliges the country to suffer the inconveniences of tied transactions, whatever the currency situation of the country. As a result, no internal measures can adequately check the widespread price increases as well as the constantly lowering purchasing power of Cameroonians whose meagre income cannot keep pace with the inflated prices of imported manufactured goods. Although Cameroon has managed to diversify the destinations of its exports, France continues to be the major source of imports.

b) Dependence on a few products: This concerns mainly exports. As a matter of fact, in spite of the diversity in the country's export products, Cameroon's earnings in this area are derived mainly from cocoa beans, coffee and timber. These three products accounted for 57.4% of Cameroon's export earnings in 1979/80 as against 57% in 1975/76, 53.7% in 1970/71 and 51.6% in 1965/66.

The above economic conditions are, however, changing. It is expected that with the increasing rate of industrialisation the amount of manufactured goods imported from the western world will reduce and Cameroon would be able to increase its industrial exports.

Moreover, the exploitation of Cameroon's petroleum has altered the previous situation whereby numerous tons of petrol were imported from other countries, and may lead to a more favourable balance of trade.

THE U.D.E.A.C. AS A TRADE BLOC.

ORIGIN:

Early before independence, the leaders of the Central African Republic, Chad, Congo (Brazzaville) and Gabon agreed to continue jointly many of the economic services they were about to inherit from the Federation of French Equatorial Africa.

They agreed to set up a customs union. Three years later, in 1961, a partial free trade area came into existence between the Equatorial Customs Union (UDE) and Cameroun. (All five countries were also members of the Franc zone).

In December 1964, the presidents of these countries signed a treaty, the Brazzaville Treaty, creating the Central African Economic and Customs Union (UDEAC), which replaced the Union Douarniere Equatoriale (UDE). Believing that they received insufficient benefits from the union, the inland countries, Chad and the Central African Republic in April 1968, withdrew and formed a new union with the Congo Leopoldville, now Zaire. The Central African Republic rejoined UDEAC as a full member after lengthy negotiations later in 1968, but Chad decided to maintain the status of an observer until December 1984 when it again joined the union as a full member.

In December 1983, the Republic of Equatorial Guinea (the only non-French speaking country) formally joined the Central African Economic and Customs Union, bringing to six the total number of member states.

OBJECTIVES:

The main objective of UDEAC is to encourage coordination and cooperation among member states as a source of economic strength. Economic integration among member states would enhance development. This integration is sought by,

1) Encouraging free trade and movement of people within the union,
2) Setting up of common external tariffs,
3) Planning for regional establishment and distribution of large-scale industries to avoid wasteful competition.

It was initially stated that both large-scale industries (e.g. oil refineries, steel mills, and chemical industries) were too expensive for each of these member states to establish alone, and joint effort was therefore required. It was also argued that even consumer goods, could not be profitably produced without assured outlets beyond the limited markets of individual states. Factory industries involve mass production of goods, and mass production of goods demands a large market. So far none of the UDEAC countries is thickly populated enough to provide an adequate market for factory goods. Consequently, a trade bloc of the central African countries would enhance industrial development and avoid wasteful competition.

Custom duties on goods imported from outside the UDEAC area are levied at the port of entry and redistributed among the member states. Whatever Cameroon loses, such redistribution is more than compensated for by the increased use of Cameroonian ports and transhipment facilities and by the growing market for Cameroonian industrial products in the other member states.

Solidarity Fund:

Because some member states (the coastal states) are favourably placed as far as the location of industries is concerned, and consequently derive greater benefits from the activities of the union, a solidarity fund was established to compensate the disadvantaged states.

PROBLEMS:

The smooth functioning of the union is hampered by a number of problems which include:

1) Poor and inadequate roads from one country to the other.
2) Some countries are ravaged by internal political upheavals, while others such as the Central African Republic (under Bokassa) have had their economies mismanaged.
3) There is a low level of trade among the member states. Individual states are deeply engaged in trade with France, their former coloniser.
4) Despite the number of states involved, the whole of UDEAC constitutes a relatively small economic area in terms of income and population, i.e. limited market of only about twenty million inhabitants.
5) Competition from Nigeria and Zaire:

A further problem arises from the fact that the UDEAC zone lies between two of Africa's economic giants, Nigeria and Zaire, each of which has a much larger internal market and a better industrial base. These countries are well placed to compete in the markets of UDEAC countries. Zaire's major industrial centre, Kinshasa,produces goods which compete with those produced in Brazzaville, Congo. On the other side Nigerian products in Cameroon (and even in Chad) are cheaper than those produced in Douala, thereby undermining the industrial base of Cameroon. The situation has only been partly helped by such trade inconveniences as differences in currency and custom tariffs.

The Brazzaville Treaty was revised in 1974 to increase the level of economic integration. The main amendments were connected with the new Development Bank of States of Central Africa, and the promotion of multinational companies within the union.

PART FIVE: REGIONAL STUDIES

CHAPTER 20: THE SOUTHERN PLATEAU REGION

This region is bordered to the north by the Adamawa plateau. From the latitude of Yoko the land starts rising into the Adamawa. To the west it is bordered by the coastal plain. The Nachtigal falls on the Sanaga mark the end of the plateau to the west.

The Southern Plateau is quite vast and extends to the southern and south-eastern boundaries of the country. It is an extensive crystalline plateau with hills and rocky massifs. The zone of the south-east forest is a vast peneplain with a monotonously undulating relief.

The Southern Plateau has a type of equatorial climate which is characterized by two rainy and two dry seasons (double rainfall maxima e.g. Yaounde).

It is a region entirely covered by thick and evergreen forest. North of Yaounde, in the localities of Saa, Obala and Batchenga this forest has been subjected to much cutting. Towards the south-east it is less attacked by human activities. Towards the north of this region the forest gives way to savanna vegetation.

The plateau is entirely occupied by Bantu and Pygmies. The western part of this plateau is more populated than the eastern part. The relative inaccessibility, and to a lesser extent, the dense nature of the forest, account for the sparse population in the eastern part. More to the north of this south-eastern forest, the population increases, especially as one comes closer to the Savanna zone where settlements such as Batouri, Bertoua, Abong-Mbang and a few minor ones are found.

The thick equatorial forest influences the way of life and the activities of the populations found in the region. Their settlements are small and form isolated groups of villages in the forests.

AGRICULTURE:

Here, intensive cultivation is very prominent. Smaller acres compared to those of the grassland areas are cultivated for food crops and more care is usually given to these farms because the soils easily suffer from leaching.

The main food crops here include cassava, cocoyams, yams and plantains.

Peasant cultivation in this region, especially of cash crops, is increasing. The main cash crops produced on this Southern Plateau include cocoa and coffee. Cocoa is the leading cash crop and it is cultivated in the localities of Abong-Mbang, Akonolinga, Bafia, Bengbis, Obala and many other smaller villages. Coffee and tobacco are also cultivated in the eastern part of the region. Tobacco is grown on large farms mainly by the Cameroon Tobacco Comapny while sugar cane is grown and processed at Mbandjock by SOSUCAM and CAMSUCO.

ANIMAL REARING:

The presence of the forest which excludes grassland vegetation for grazing, and the frequency of the trypanosomiasis disease (caused by the tse-tse fly), both make this region very unfavourable for cattle rearing. It is only in the northern part of this plateau, at the fringes of the Adamawa, that cattle can be reared. Only smaller animals, such as, goats and pigs, can be found here. Fowls are numerous and are reared mostly under modern techniques, especially in the urban centres in order to meet the towns' considerable needs of chicken and eggs.

As regards wild animals, the forests are a home for them, sheltering mostly buffaloes, chimpanzees, elephants, gorillas, monkeys, etc. Crocodiles and other water animals are numerous. Two important national forest reserves with many animals are found in this area — the Dja and Pangar reserves in the Upper Nyong and Lom-et-Djerem Divisions respectively.

Another item of paramount importance to the economy of this region is the timber wealth. It is enough to turn back to Chapter 17 to see the importance of forests in the country's economy. Only the Coastal and the Southern Plateau regions of Cameroon are endowed with timber wealth, but the heart of it is the Southern Plateau.

Forest exploitation had been going on for long in the localities of Yaounde, Mbalmayo, Sangmelima and Moloundou and sawmills are also multiplying in these towns. An immense portion of the forest is yet not being exploited due to lack of

THE SOUTHERN PLATEAU REGION

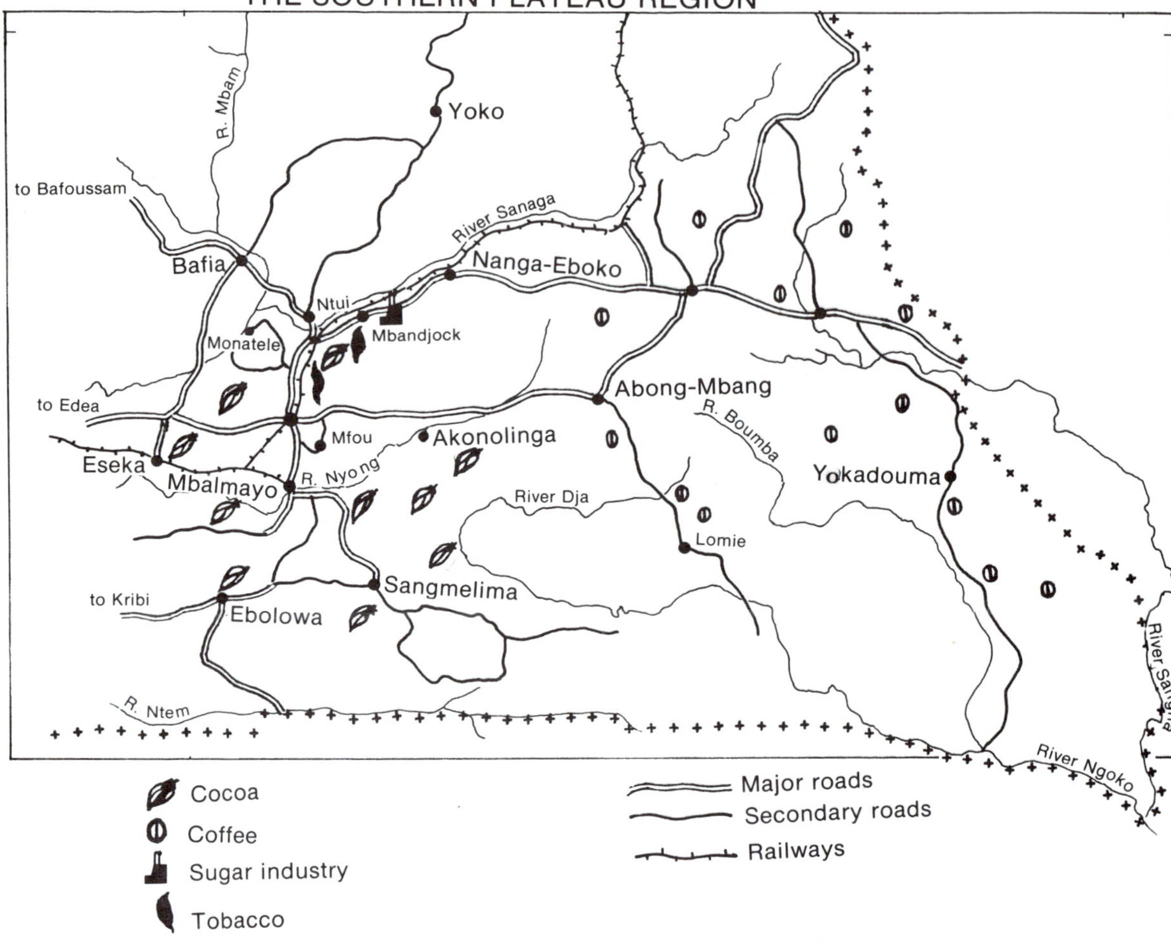

access roads.

Many corporations are springing up in this region to boost its economy. The major ones include CENADEC, CAMSUCO, SOSUCAM, SOFIBEL, MIDEBOM, STC, MIDEVIV, SODECAO, ZAPI of the East.

CAMSUCO: The Cameroon Sugar Company (Camsuco) was established in 1975 and its factory at Mbandjock was inaugurated by the Head of State on the 9th December 1977. It is the second agro-industrial sugar company of Cameroon and the first being SOSUCAM also located at Mbandjock.

Its main goal is the cultivation of sugar cane to process into granulated and cube sugar. This industry has helped enormously in the employment of many Cameroonians. By 1979 CAMSUCO had a total of 3,641 people on its list of employees.

MIDEVIV: The Food Development Authority

Poultry farm at Mvog-Betsi in Yaounde.

(MIDEVIV), is a government undertaking charged with setting up "green belts" around big towns; studying and encouraging the cultivation of food crops and fruit trees in these "green belts" with the main purpose of supplying the towns with the yields.

Among the many measures adopted to achieve its goals, it has created a centre for the distribution of banana and plantain suckers and cocoyam tubers at Mbankomo near Yaounde; at Ntui the MIDEVIV has a centre for the promotion of maize and groundnut cultivation.

The MIDEVIV in recent years has done much in the big towns (especially Yaounde) in the supply of local foodstuffs. It accomplishes this task by establishing a quick means of evacuating the products from their areas of cultivation to the towns for sale. The activities of MIDEVIV have also gained root in the other regions of Cameroon.

CTC: The Cameroon Tobacco Company (CTC) is responsible for the cultivation, fermentation and commercialization of tobacco in its areas of operation. Its activities are concentrated at Batchenga in the Centre Province, and more especially in the East Province where numerous exploitation structures are found. Over 9,000 tobacco farmers work under the framework of this company which supplies tobacco grains to them and also employs nearly 20,000 workers. The farmers receive technical advice from the CTC through its agency — the Tobacco Exploitation Service (TES) — a French undertaking which retains one third of the activities of Cameroon Tobacco Company.

The tobacco that is produced is sold to the world cigar producers, especially those based in France and America. Part of it is used for fabricating the SANAGA CIGARS.

ZAPI: The Integrated Priority Action Zones (ZAPI) was formed by decree No. 72/442 of September 2, 1972. It originated from the elaboration of the Second Five-year Economic, Social and Cultural development Plan of Cameroon. This corporation was based in the Centre and East Provinces, located on the Southern Plateau. Its goal is to provide technical advisory services to farmers for the production of cocoa and coffee. It also coordinates peasant activities by marketing their foodstuffs in urban centres.

Since the creation of the SODECAO in the Centre Province (which absorbed the work of ZAPI) the activities of ZAPI had since then been limited to the East Province and consequently called 'ZAPI of the East.'

Besides the economic activities of ZAPI, it also fulfills some social functions by collaborating with specialized services of health and hygiene. It brings women together for training in various aspects. It also provides financial credits to workers.

TOWNS

YAOUNDE: Situated on many small hills in the heart of the Southern Plateau, Yaounde is one of the earliest settlements of the Ewondo. This settlement gained its first importance in 1889 when the Germans arrived from Kribi to make it their military post. This was followed by the clustering of shops such as "John Holt" and "King." Later in 1921, after the expulsion of the Germans, the French made Yaounde the seat of their government. Its position was economically strengthened in 1927 when the railway line made its arrival in this town.

With a population hardly exceeding 7,000 inhabitants in 1933, this town grew rapidly immediately after independence to about 100,000 inhabitants in 1963; 313,706 inhabitants in 1976, and 656,700 persons in 1986. Yaounde is the second town of Cameroon after Douala.

It presently fulfills many functions. Paramount is the administrative role. It is the political capital of the country, the provincial capital of the Centre Province and the chief town of the Mfoundi Division and therefore houses mainly administrators, foreign diplomats and civil servants.

It is also the main intellectual centre of the country with many higher educational institutions, such as, the University of Yaounde and its affiliated schools. Besides these institutions there are numerous private and public colleges and schools.

Economically, it only comes after Douala, but its role as the major urban centre on the vast plateau is significant. It is a transit and redistribution centre for the many urban centres in this region.

Industrial activities here are still very low.

A partial view of Yaounde city.

However, there is a brewery, a cigarette factory and many sawmills.

The town is accessible by road from all directions; by rail and also by air. The beautiful town of Yaounde is constantly undergoing reconstruction work to bring it up to a modern city in the near future. Flanked by forest and savanna this town has considerable climatic advantages over Douala.

BERTOUA: The Baya are said to have migrated to their present site under their leader, Ndiba. After his death, he was succeeded by his son Mbartoua who reigned during the times of the Germans. Mbartoua resisted strongly against the Germans and in 1903, he was assassinated on the road to Betare-Oya. The name Bertoua comes from the distortion in pronunciation and transcription of Mbartoua by the Germans.

Bertoua is a small town compared to many other settlements in Cameroon. Situated some 343 km south of Ngaoundere this town had a population of 15,000 people in 1976. In the East Province, Bertoua has long been a smaller town than Batouri but after independence its role increased administratively faster than that of Batouri. It passed from the chief town of the Lom and Djerem Division to the Capital of the federal inspectorate of administration for the East, and finally to the capital of the East Province in 1972.

The administrative role is yet the principal function of Bertoua and consequently, many government facilities are being attracted to the town. Commerce and industry are still developing and it is hoped that, in future, Bertoua will count among the major towns of the country.

CHAPTER 21
THE COASTAL LOWLAND REGION:

PHYSICAL FEATURES:

As earlier mentioned in Chapter 2, this region is characterized by the flatness of the landscape, except where the Cameroon Mountain stretches into the sea.

Composed mainly of the sediments brought down from the highlands of the interior, and also of weathered volcanic soil, this region is agriculturally very useful. However, the zones which are covered by swamps and sandbars are not very fertile, for example, the area of Rio-del-Rey drained by the Ndian river and the Douala basin drained by the Wouri and Mungo rivers.

CLIMATE:

Climatically, this region presents distinct characteristics. It is the region of the Equatorial Maritime Climate. The temperatures are constantly high, usually above 25° C. The rainfall is generally very heavy and atmospheric humidity throughout the year is high. Despite the existence of the two seasons (dry and wet), all the months of the year receive at least some amount of rainfall. It is the only region of the country which receives maritime influence. Coastal areas have temperatures that are moderated by sea breeze whereas the areas situated far away from the coast are comparatively very hot.

VEGETATION:

The Coastal Lowland region, like the Southern Plateau region is the heart of the forest vegetation but unfortunately for the Coastal Lowland region, it is losing much of its forests due to human exploitation for the cultivation of crops and for settlements. Only the swampy and less accessible forest areas are not being exploited.

POPULATION:

Again, like the Southern Plateau region, it is the region of the Bantu and Pygmies. Due to the economic viability and job opportunities offered by this region in its numerous plantations, industries and commerce, it serves as an attraction region, drawing people from the hinterland. Almost every human settlement here is characterized by a predominance of in-migrants. Douala, for example, is one of the settlements where in-migrants out-number the native population. The population concentration is high in the towns but the forests are almost void of human habitation.

THE ECONOMY:

This region could be termed the *nerve centre* of Cameroon's economy where important agricultural, commercial and industrial enterprises are concentrated.

AGRICULTURE:

This is the region where plantation farming of the large capitalistic scale is found. The fertile volcanic soils on the lower slopes of Mt. Cameroon and on the entire Mungo region, coupled with the favourable climatic conditions, make this region good for the cultivation of many cash and subsistence crops. There are banana, rubber, oil palm and tea plantations established by big enterprises. Apart from these large capitalist plantations there also exist smaller peasant farms which are mainly for cash crops.

The major agro-industrial undertakings here include the C.D.C. (Cameroon Development Corporation), PAMOL, SOCAPLAM, HEVECAM, O.C.B. (Organisation Camerounaise de la Banane).

THE C.D.C.

During the thirty years rule of the Germans in Cameroon (1884-1914) they imported and planted cocoa, coffee and oil palms on the rich soils at the foot of the Cameroon mountain. These agricultural undertakings were interrupted during the First World War and by 1918 the Germans were defeated and their property passed over to the British who held the then Southern Cameroons.

By 1924 the plantations created by the Germans were put up for sale, and many of them were purchased by their former German owners. At the outbreak of the Second World War the plantations were again confiscated from the German entrepreneurs and placed under the control of the *Custodian of Enemy Property* whose seat was at Lagos in Nigeria. This lasted from 1939 to 1946 and on the 1st of January 1947 the plantations were declared national property whose products were for the welfare of the citizens, and they were placed under a new body, the C.D.C. — Cameroon Development Corporation.

The C.D.C. was later reorganized on the 28th of

THE COASTAL LOWLAND REGION

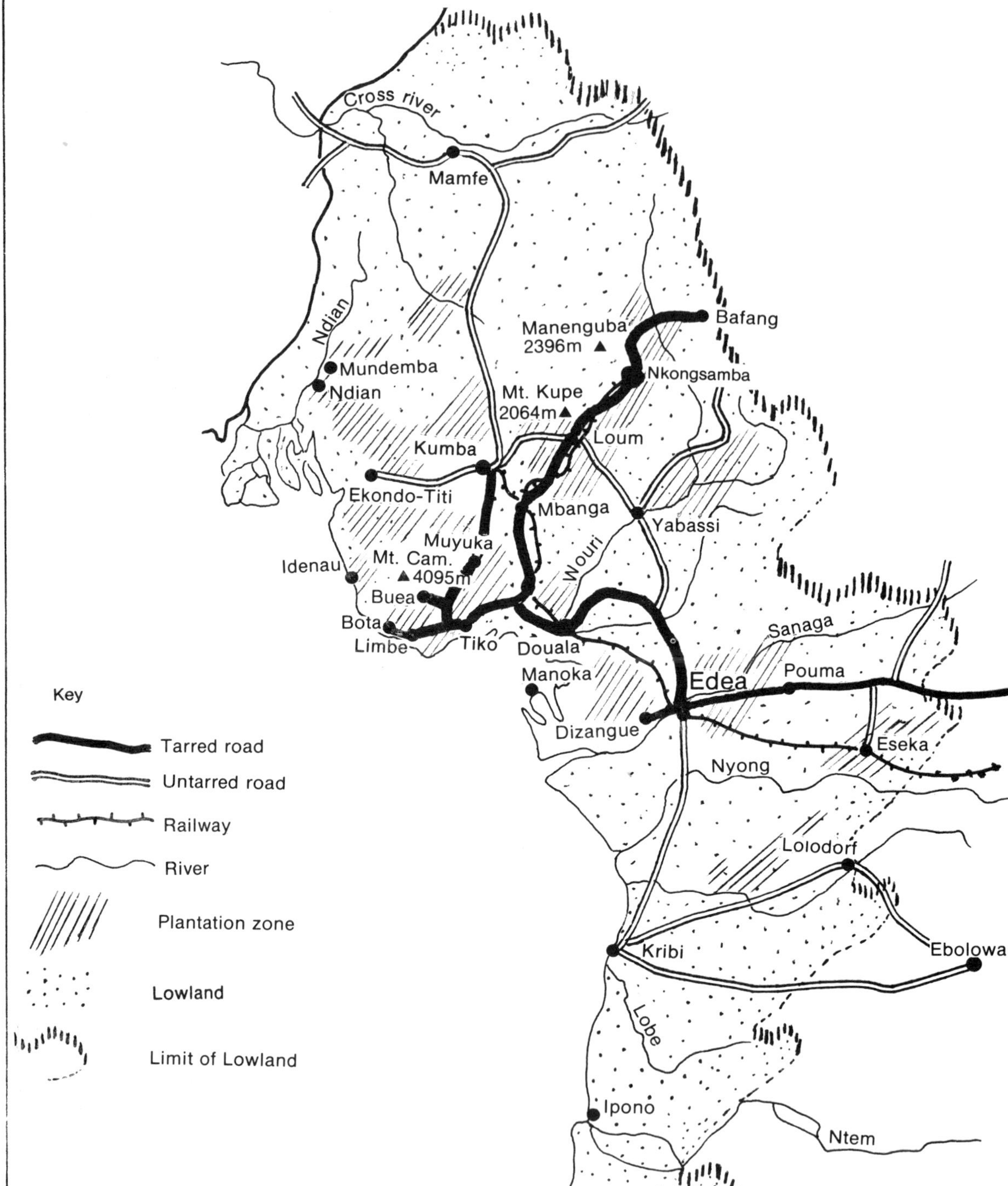

September 1973 by Decree No. 73/597. The main purpose of the corporation is to acquire, develop and operate plantations of tropical crops.

By 1975 the corporation was controlling extensive lands of an area of 80,000 hectares, by 1979 the total area covered mainly by cultivated crops was 35,578 hectares.

Today the CDC is extending operations to Ndu in the N.W. Province, Djuttisa near Dschang in the West Province, Kompina and Penda Mboko in the Littoral Province.

The C.D.C. operates extensive rubber plantations at Missellelle. Tiko, Mabeta, Mukonje, Meanja, Tombel, Mbonge, Malende and Kompina. Banana plantations are found mainly at Tiko and Molyko, and tea at Tole and Ndu. Oil palms are the most extensively cultivated at Bota, Meanja, Tiko, West Coast, Limbe and Mondoni. Pepper is cultivated at Meanja and Bimbia.

Most of these plantations are equipped with factories to process the products.

The C.D.C. also renders social services to its workers in the form of medical care, games, education, and housing. The corporation also opens and maintains roads to the various residential camps.

PAMOL (Plantations Pamol du Cameroun) is engaged mainly in the cultivation and processing of oil palms. The company has extensive plantations at Ndian, Mundemba, Lobe, Ekondo-Titi and Tiko.

Their two mills located at Lobe and Ndian are involved in processing the harvested palm nuts. This company also offers housing, educational and medical facilities to its workers. The population of Ndian is served with an up-to-date hospital owned by Pamol.

SOCAPALM (Societe Camerounaise de Palmeraies) was created in 1968 by a Presidential Decree and was charged with planting and exploiting oil palms at an agro-industrial scale.

With a capital of 6,120 million francs C.F.A. provided jointly by government and other international financial institutions, it had to realise mighty projects at Mbongo and Eseka which comprise large farms and two oil mills. Other farms operated by SOCAPALM are located at Edea, Kienke, Song-Ndong and Dibombari. The Dibombari plantations are also equipped with a mill.

HEVECAM and SAFACAM: These two corporations are engaged in the cultivation and exploitation of rubber. HEVECAM operates in the Ocean Division around Kribi while SAFACAM operates in Dizangue — just a few kilometres from the town of Edea.

The list of agricultural corporations dealt with above by no means exhausts the number in this region.

LIVESTOCK AND FISHING:

The role of this region in the promotion of livestock rearing is minimal because of the general lack of grazing lands, and the presence of mosquitos which attack cattle and spread the trypanosomiasis disease. However, goats and pigs are scattered all over the area. Poultry are more common and kept mostly around the towns.

Fishing, on the contrary, is well developed in this region. It is here that both subsistence and industrial fishing are practised in the sea and rivers. Industrial fishing by large companies is done in the sea while subsistence fishing by individuals is done using traditional methods, both in the sea and in the rivers.

FOREST EXPLOITATION:

Unlike in the Southern Plateau region where most of the forest is still unhampered, the forests of the Coastal Lowland region for long have been subjected to much exploitation for export as well as for local use of its wood.

The forests of the South-West Province, especially in the Fako, Meme and Ndian Divisions, have been greatly reduced for the establishment of plantations and for settlements. Here, a network of forest roads exists. These roads were created by lumbering companies.

Due to the intensity of forest exploitation in this region, most of its towns are characterised by the presence of sawmills.

INDUSTRIES:

It is no exaggeration to say that this region controls the industrial installations of the country. Almost all the heavy industries of Cameroon are located here. The industries of this region range from light to heavy. It is enough to turn to the map on industries to see the types of industries located in this region. Of these numerous industries we will discuss in detail the oil refinery located at Cape Limbo and the paper pulp factory, which are among the most significant.

THE OIL REFINERY (SONARA)

The National Refinery set up at Cape Limbo near the town of Limbe is yet the only crude oil refinery of the country. The total land surface covered by the refinery, the maritime installations and the social infrastructure is over 54 hectares. The site at Cape Limbo was chosen because of the natural deep sea port which allows heavy tankers to draw near to the shore.

Production: The refinery refines crude oil to produce the following commercial products: Butane, Gasoline (ordinary and high octane), Kerosene, Jet fuel, Gasoil and Fuel oil.

Gas oil and Fuel oil are a mixture of these different products extracted from underground. The first step is to separate the mixture into these various products by a type of distillation called *atmospheric distillation* which resembles an alcohol distillery but a little more complex. Some products coming out of this process are ready for use, others such as gasolines require further treatment in the second unit called *catalytic reformer* which causes a chemical change to give high octane gasoline. Some products still require further treatment to remove impurities, especially sulphur, before commercialisation is done. All these are treated in a third process called *hydrodesulphurisation* of the distillates.

Capacity: The refinery has a capacity of 2,000,000 tonnes of crude oil per year. This capacity is sufficient to satisfy the needs of the Cameroonian market for several years and to further strengthen the country's economy through sales of the excesses.

Petroleum Refinery at Cape Limbo.

Reception and stockage of oil: To stock crude oil and finished products before expedition, several tanks are built, which cover a large surface area. All the tanks have a capacity of more than 500,000 m^3 of crude oil.

Apart from the area for loading road tankers, the reception of crude oil, and the expedition of finished products both require special installations which constitute the second phase of the project called *maritime installations*. Crude oil is brought by tankers loaded at sea and having capacities reaching 100,000 tons. The tankers cannot reach the shore and thus unload at a buoy at a distance of 2,500 m from the shore. The buoy is linked to the refinery by two sealines composed of pipes placed on the sea floor which allow tankers to pump in the crude oil or receive the fuel oil which it exports.

Finished products are sent to Douala on a coaster which will moor at a wharf along a 400 m jetty made on 100 metres of stone bed and 300 metres of concrete sleeve, prefabricated and placed on the sea bed.

All the exploitation and managerial activities of the refinery are controlled by the National Refining Corporation, SONARA.

Reception and storage tanks at the sea-shore.

THE PAPER PULP INDUSTRY (CELLUCAM)

The Paper Pulp industry (CELLUCAM) based at Edea constitutes a giant step towards the industrialization of Cameroon. It is the single largest wood factory in black Africa and goes a long way to project the image of Cameroon in the Paper Pulp industry, both in Africa and the world. CELLUCAM used a sum of 75 billion francs CFA to equip this industry with high technological installations, housing and training facilities.

Quite often paper and pulp mills are integrated units though paper making and pulping require in effect two separate mills with different requirements for their location.

The factors that influenced the location of the paper pulp industry at Edea are numerous.

1) Many pulp industries tend to be located near the source of their raw materials — wood, which is bulky and of low value. Modern pulp factories are large and usually equipped with elaborate, costly and durable plants so that they necessarily have to be set up in places with extensive forests that can supply enough suitable wood for many years. For this reason the government in 1968 commissioned the French Centre for Tropical Forest to undertake an inventory of 100,000 hectares of forest in the area where CELLUCAM now operates. The

results proved that the vast equatorial forest surrounding the town of Edea can assure the industry of reliable sources of wood for many years.

2) Pulp mills are commonly found on navigable or "floating" rivers because rivers can provide cheap water transport by floating to lumbering areas. The Sanaga at Edea could be used for transporting some small logs but the problem with most tropical woods is that they are very heavy and can sink.

3) The pulping industry is a heavy consumer of electricity and it is often attracted to power sites on rivers. The Cellucam industry no doubt had to be set up just at the mouth of the Edea H.E.P. station.

4) Large quantities of water are consumed in the pulping process. Some chemical processes require several thousand litres for each tonne of pulp produced. The water has to be fresh and clean. Cellucam makes use of 600 litres of water per second and this water is later purified before being emptied into the Sanaga river to avoid killing water creatures and plants.

5) Moreover, the site of Edea, some 80 km from the Port of Douala, coupled with a good road and railway line, make the link from factory to port very easy for the transportation of equipment.

Production:

Production started in March 1980 and on the 16th of April 1980 it was inaugurated by President Ahidjo. The challenge that remains to be met now by Cellucam is to make the investment economically profitable.

Presently, the industry produces annually about 122,000 tonnes of pulp which is being exported in its entirety but it is hoped that in future it will be transformed into paper and related products.

Future Projects:

The intended paper mill will produce about 40,000 tonnes of paper annually which include both printing and light wrapping papers. This phase will meet the demand for paper and paper products in Cameroon and the UDEAC countries. Besides the installation in the near future of a paper manufacturing mill Cellucam plans to develop into an integrated wood manufacturing

The paper Pulp factory (CELLUCAM) at Edea.

enterprise. This will also include the building of a saw mill which will process between 25,000 to 30,000 tonnes of squared lumber from logs.

COMMUNICATIONS:

This region is the most privileged of all the regions in the country, because it is adequately served with road, rail, air and ocean transport facilities. No other region has all these means.

The major towns here are linked with well surfaced roads. These roads originate from Douala and radiate to Limbe, Kumba, Nkongsamba and Edea. Communication with the remote areas is not easy and the problem is aggravated in the rainy season by the sticky nature of the soil which leads to very slippery roads.

By rail, this region has a relatively dense network. By air, it has the first international airport of the country and internal airports at Bessongabang, Tiko, Kribi and Nkongsamba. Communication by water is significant. Maybe, one would have stressed that it is the maritime location of this region that is at the background of its prosperity. The large estuary created by the Wouri, Dibamba and Mungo rivers allowed for enough penetration of Europeans far into the land in the last century. That alone brought the people of this region in contact with outside civilisations. Then, in recent times, the maritime position opens this region to foreign trade. No doubt, the main economic centre of the country — Douala, is found here.

TOWNS:

DOUALA, the biggest town of Cameroon appears to be situated along the sea-side but in reality it is spread out on the left bank of river Wouri, some 24 km from the sea.

Its population as recorded during the 1976 population census was 458,426 inhabitants. Estimates for 1986 place the population figure at over one million. The population of Douala town is quite diverse, composed mainly of in-migrants from the western part of the country who come in to seek jobs in the industries. Cameroonians from the other regions of the country and non-Cameroonians also move to this town to work, not only in the industries but also in government and private offices, or to sell their goods. Douala is the major doorway into the country, both by air and sea. It is one of the most important air and seaports located along the coast of West Africa.

Economically, this town is termed the "economic capital" of Cameroon. Because of its site as a seaport, it handles over 90% of Cameroon's foreign trade. The main products exported through Douala port are timber, cotton, bananas, cocoa, coffee, sheet and crepe rubber, bauxite and petrol. Imports through this port include textiles, clinkers, aluminium from Guinea, flour, iron, vehicles, machines and also building materials.

Industrially, this town hosts more than half of the country's factories. These numerous factories produce fertilizers, cement, salt, soap, metals, chemicals, shoes, glass, beer and many other products. The fishing industry is centred at Douala because fishing at industrial level can only be done in the neighbouring ocean waters.

Douala also has administrative and educational functions. It is the chief town of the Littoral Province and as a result many government offices are found in the town. Many educational institutions are found in this city. Besides the numerous primary and secondary schools, the town also hosts a branch of the Yaounde University, the Advanced School of Commerce.

NKONGSAMBA, situated some 145 km north of Douala, Nkongsamba is the third town of Cameroon after Douala and Yaounde with a population of 71,298 inhabitants in 1976. It is also the biggest town of the Mungo Division. Found at the foot of the Nlonako Mt., this town extends along the valley of a small and winding stream. It saw its growth after 1914 when the First World War interrupted the construction work on the railway. The accidental terminus later gave birth to this big town. Nkongsamba is important for coffee trading because it is located in the centre of a coffee growing region.

LIMBE: Limbe was founded in 1858 by the English Missionary, Alfred Saker. From Fernando Poo, Alfred Saker, the English Baptist Missionary, arrived and was struck by the beauty of Ambas Bay, with its islands and the towering mass of the great Cameroon Mountain.

He landed with his crew on Mondoleh Island, but immediately chose the rocky lands on the left bank and at the mouth of the Limbe River — a mountain stream of clear water flowing into the sea close to the town.

King William of Bimbia, then a well known friend of Saker and the Baptists, sold outright to them the whole of Ambas Bay with its islands and a strip of coastline 16 km (10 miles) long to a depth of 8 km (5 miles) from the seashore.

Saker asked Queen Victoria to allow the settlement be named after her. The Queen accorded her consent and thereafter the settlement was called, Victoria. This name lasted nearly a hundred and twenty-four years, and was later changed in 1982 to Limbe.

Limbe served as the most important sea port during a good portion of the colonial era. It was

A partial view of Douala City Centre.

the main port of the West Cameroon state.

The role of Limbe as a sea port dwindled almost to insignificance due to the extension of the Douala port and the construction of the Tiko-Douala road. In recent years, however, Limbe is being revived, mainly because of the large oil refinery at Cape Limboh.

Limbe is the chief town of the Fako Division and as a result it fulfills administrative functions. Its site at the seashore makes it a good seaside resort area with beautiful sand beaches at Miles 6, 8, and 11. It had a population of 27,016 inhabitants in 1976. The Botanical Garden with well maintained trees and shrubs makes this town a beautiful place to visit.

KUMBA: Kumba is the most populous town in the South-West Province and the chief town of Meme Division. It is a busy town, functioning mainly as a commercial centre. Its relatively high population of 59,000 inhabitants (in 1976) is engaged mainly in trading and a large proportion of these traders are foreigners. The bulk of these foreigners come from Nigeria and consequently repatriate a good fraction of the proceeds from their businesses. This fact partly explains why, despite the high level of economic activities going on in this town, it continues to maintain an old look with many semi-permanent buildings built of *caraboard.* Another possible factor may be the general lack of commitment and little dynamism on the part of the local population. Other non-indigenous inhabitants come from the western highland region of the country.

As the most centrally located town of the province, Kumba stands at the main cross-roads of the province and serves as a redistribution centre for the following administrative Divisions: Manyu, Meme, Ndian and Fako. It is also situated at a railway terminus.

Kumba has many saw-mills. Like Nkongsamba, it is a coffee and cocoa trading centre. It is also an educational centre with many secondary schools.

BUEA: The town of Buea is situated at the foot of Mount Cameroon at an altitude of 1,000 m. Endowed with a cool and fresh climate, this town attracted the German colonizers to make it the capital of the country. This lasted from 1901 to 1909. Unfortunately, the town saw its administrative role degenerate from a country's capital to a state capital of the West Cameroon State (1961 -1972) and in 1972 it still reduced further to the capital (or chief town) of one of the administrative provinces of Cameroon.

Buea is intermittently covered by mist for a greater part of the year. Despite the special attractions — cool temperatures, a wide range of fresh vegetables, sometimes fresh milk and cream, and a grand view of the coastal lowlands — Buea is often regarded as a depressing station because of its foggy and overcast weather. Having also been associated with formal bureaucratic life, Buea is considered a rather 'dormant' town, whereas Limbe, the closest significant town, has for several decades maintained a lively and buzzing vitality, which attracts many people. However, Buea still attracts a good number of tourists to the old German relics and also the Cameroon Mountain.

Buea has a modest population (24,580 inhabitants in 1976). With the exception of the Tole tea factory situated at the outskirts of the town, Buea is void of any major industry.

EDEA: The town of Edea is built on the banks of the Sanaga River and on the Douala-Yaounde line where the railway crosses the road. It is one of the medium-size towns of Cameroon (25,493 inhabitants) whose population is engaged mainly in industrial work.

This town has a gigantic dam on the Sanaga River from where hydro-electric power is generated. Closely attached to this power station is the heavy aluminium smelting industry which consumes more than half of the power from the plant. In 1980 the town again received another big industry — the Paper Pulp industry, CELLUCAM, which uses timber to produce paper pulp.

Other significant settlements in the Coastal Lowland Region include Mamfe, Mundemba, Loum, Muyuka, Tiko, Mbanga and Tombel.

CHAPTER 22: THE WESTERN HIGHLAND REGION

PHYSICAL FEATURES:

Although we had earlier seen the relief and structure of this region it is worthwhile mentioning again here that this region is comparatively more elevated than the Adamawa region.

There are numerous plains inserted between very high peaks. Some of the major plains in this region include the Mbaw and Ndop plains which receive sediments from the hills and become very fertile. They are of great agricultural value and therefore are usually centres of important agricultural undertakings.

In certain areas, especailly in the North-West Province, the elevated areas end in abrupt escarpments so that many waterfalls are found in this region. The Bamenda escarpment which overlooks the town, and the Sabga slope are clear examples of such areas where streams are seen tumbling down from the hills.

Since the western region is that of ancient vulcanicity, there exist interesting features of crater lakes, such as the Baleng, Awing, Nyos, Oku, Wum and other smaller ones; basaltic plains which are very fertile, for example, the Foumbot area; volcanic cones and plugs which are also numerous in Foumbot. In recent years, the volcanoes in this region have shown a lot of activity. In the nights of August 15 and 16, 1984 explosions of highly toxic gases in the Monoun Lake near Foumbot killed 37 people. Two years later, in the night of August 21, 1986, another similar explosion occurred in Lake Nyos, near Wum, killing over 1,534 people and a large amount of cattle.

Some of the isolated hills or inselbergs in this region are of hard crystalline rocks. These crystalline rocks are generally responsible for the escarpments and inserbergs which exist in this area.

CLIMATE:

This hilly region of the country has a fresh climate with low temperatures and abundant rains. The area enjoys a climate of two distinct seasons: the dry and rainy seasons. The rainy season which is longer than the dry season starts from mid-March to mid-November. The dry season, on the other hand, starts from mid November to mid-March. The average annual rainfall in this region is 2,000 mm.

Another characteristic of this climate is that the mornings, especially during the dry season, are cold and frosty while the afternoons are hot and dry. During the dry season average day temperatures go up as far as 28° C while in the rainy season the average temperatures go as low as 15° C. Some of the depressions inserted between hills where movement of air is not very free, have high temperatures and even a lot of mosquitoes (mosquitoes like hot areas). The Ndop plain, for example, is hotter than the surrounding highlands. The atmosphere in the dry season is very dusty while in the rainy season the skies are clear.

Given the altitude, rainfall and temperatures, the vegetation that is bound to thrive is savanna. The vegetation on the hills is savanna while the valleys have gallery forests. Towards the southern part of the region the savanna gives way to secondary forests which result from the destruction of the primary forest.

POPULATION:

This is the region of highest population concentration in the country because the natural environment offers very fertile soils for agricultural activities and very conducive climate for human habitation.

The western highland region is the home of the semi-Bantu people of Cameroon. The Bamileke who form the largest tribal grouping here are actually centred in the West Province, and closely followed are the Bamum who are concentrated in Foumban.

Towards the North-West province, which falls in this region, one comes to an area which is inhabited by three main groups. The Tikars who form the majority here include Nsaw, Kom, Bafut and all the inhabitants of Ndop, and the majority of the inhabitants of Bui, Menchum and Donga and Mantung Divisions.

Another group, the Widikum, has its members centred mainly in Mezam and Momo Divisions. The Bali are in Mezam Division.

The third group, the Fulani, mainly cattle rearers (Bororos, Hausas) are scattered all over in this region.

THE ECONOMY

AGRICULTURE:

This is the region where the natural landscape has been seriously transfomred by the hard-working semi-Bantu people who inhabit the area. The fertile volcanic soils in most of the areas especially in the region of Foumbot, have greatly attracted many people for farm work.

The main cash crops produced in this part of the country include Arabica coffee which is found almost everywhere; cocoa is produced only in limited areas such as in the Upper Nkam and in the Menoua Divisions. Tea is cultivated at Djutissa in Dschang and at Ndu in the Donga and Mantung Division, quinquina at Dschang and tobacco in the Mbouda and Foumbot areas.

On the western highlands the inhabitants generally have more than enough to consume from their subsistence crops so that the excesses are carried to the neighbouring towns. Intensive farming is practised mostly on the Bamileke highlands to produce subsistence crops which are composed mainly of cereals. Maize constitutes the major cereal in this region followed by beans, groundnuts, guinea corn and rice.

Tubers here include cocoyams, cassava, yams and potatoes. Plantains do very well in this region. Vegetables, especially tomatoes and cabbages also grow very well here.

In the North-West Province the vast flat lands of Mfonta in Bafut are occupied by modern farms on which both subsistence and cash crops are grown. The Mfonta program is a settlement program for young farmers.

The Rural Training Centre (RTC) Mfonta which is run by the Presbyterian mission trains young farmers and animal breeders who in the course of their training, which lasts some five months, acquire tuition on general agriculture and animal breeding.

The centre, at the end of the course, gives out loans in aid to the farmers to settle.

Noteworthy is the role played by the National Centre for Studies and Experimentation in Agricultural Mechanisation (CENEEMA) based at Bambui. This body aims at mechanising agriculture, and in accomplishing this task, hires out tractors and other agricultural machines to farmers located at a radius of 50 km from the centre.

Besides hiring out machines, CENEEMA also carries out research to produce machines aimed at satisfying local needs, such as cassava graters, coffee pulpers, maize shellers, oil presses, groundnut crackers, etc.

The place of rice is becoming more and more significant in this region due to the coming of many rice corporations, and many projects undertaken by the government to increase rice production. Vast rice farms are usually found on the plains, such as, the Mbo plain between the Littoral and West provinces; Ndop and Mbaw plains in the North-West province. The valleys of some rivers have for long been used for rice farming as is the case of the Menchum river valley, and the Noun valley near Foumban. Elsewhere upland rice is also found. The major corporations in this region involved in the cultivation of rice include the SODERIM and the UNVDA.

THE SODERIM:

The Mbo plain Rice Development Corporation: This corporation was created in 1977 to continue the work already started by the Mbo Plain Rice Development Authority (MIDERIM). The SODERIM was created with the primary aim of increasing and improving the production of rice, and secondly, to check rural exodus into the already very crowded neighbouring towns. SODERIM operates mainly in the Mbo plain on an area of over 30,000 hectares.

The operation zone of SODERIM is at the border of the West and Littoral provinces. In the West province it operates in the Mungo Division. The Mbo plain is drained by the Nkam River which forms the boundary between the west and littoral provinces. The broad valley of this river is always flooded in the rainy season. These floods are then used in the whole plain for rice cultivation.

Rice production here is also favoured by the presence of a ready market which is made up of the neighbouring big towns of Nkongsamba, Douala, Dschang, Bafang, Bafoussam, and an adequate road infrastructure to evacuate the harvested rice.

THE U.N.V.D.A.

The Upper Noun Valley Development Authority formally created in 1970 is charged with the developmment of rice cultivation in the headwaters of the Noun Valley above the Bamendjim dam, in order to supplement the incomes of the local populations of the region. This operation covers the whole of the Ndop plain extending down to Bangolan. By 1978 an area of about 800 hectares had already been put under rice cultivation with an average output of 2 - 2.3 tons of rice per hectare.

Presently, the zone under cultivation is being increased drastically to improve the living conditions of over 6,000 rice farmers in this area.

The U.N.V.D.A. also buys *paddy* (unhusked rice) from the local farmers for transformation and commercialisation. The measure to hull and commercialise rice had been undertaken by the U.N.V.D.A. in order to reduce the costs of storage, transportation and also the problem of sales usually faced by individual farmers.

Other activities of the U.N.V.D.A. include the installation of rice hulling mills, the creation of vast rice nurseries to provide farmers with high quality rice, and the spreading of new rice varieties, especially the "Tainan V" and the "Mangakely."

A vast and modern rice farm at Ndop run by the U.N.V.D.A.

The U.N.V.D.A. project is financed jointly by the Cameroon government with an initial capital of 595 million francs, F.A.C. with 250 million francs, and a loan from the CCE (Caisse Centrale) — a French financing organisation, of 900 million francs CFA.

The project is doing much to improve the area of its operation by constructing small irrigation dams, maintaining roads, opening up of poor drainage areas, foot paths and irrigation paths.

On the western highlands the cash crops include mainly Arabica coffee cultivated on individual family holdings; vast coffee plantations exist at Foumbot and Santa; tea is at Ndu and presently being extended to the hills of Santa and Dschang; tobacco mainly at Mbouda and quinquina in Dschang.

Besides the rice production corporations we also find the W.A.D.A., U.C.C.A.O., N.W.C.A. Ltd. and CENADEC which are engaged in agricultural activities.

W.A.D.A.:

The Wum Area Development Authority has as its objectives, the improvement of the living standards of the rural populations within the environs of Wum. It accomplishes this task by providing advisory services to the farmers, providing financial agricultural credits to farmers, giving assistance to cooperative farmers. Assistance to farmers is done by providing them with trained oxen that can be used for ploughing, and agricultural inputs, such as, tractor services, fertilisers and chemicals, seeds and seedlings.

W.A.D.A. helps in ameliorating handicraft and technical services by the maintenance of simple agricultural tools and furniture.

U.C.C.A.O.:

The Union of Arabica Coffee Cooperatives of the West (U.C.C.A.O.) organises the collection, processing, packaging and sales of Arabica coffee. The Union has over 100,000 farmers as members who are grouped into six cooperatives. The centres of the various cooperatives are at Dschang, Mbouda, Foumban, Bafoussam, Bafang and Bangante. The various cooperative centres are coordinated by the central head-office at Bafoussam.

CENADEC:

The National Centre for the Development of Cooperative Enterprises (CENADEC) is concerned

THE WESTERN HIGHLANDS

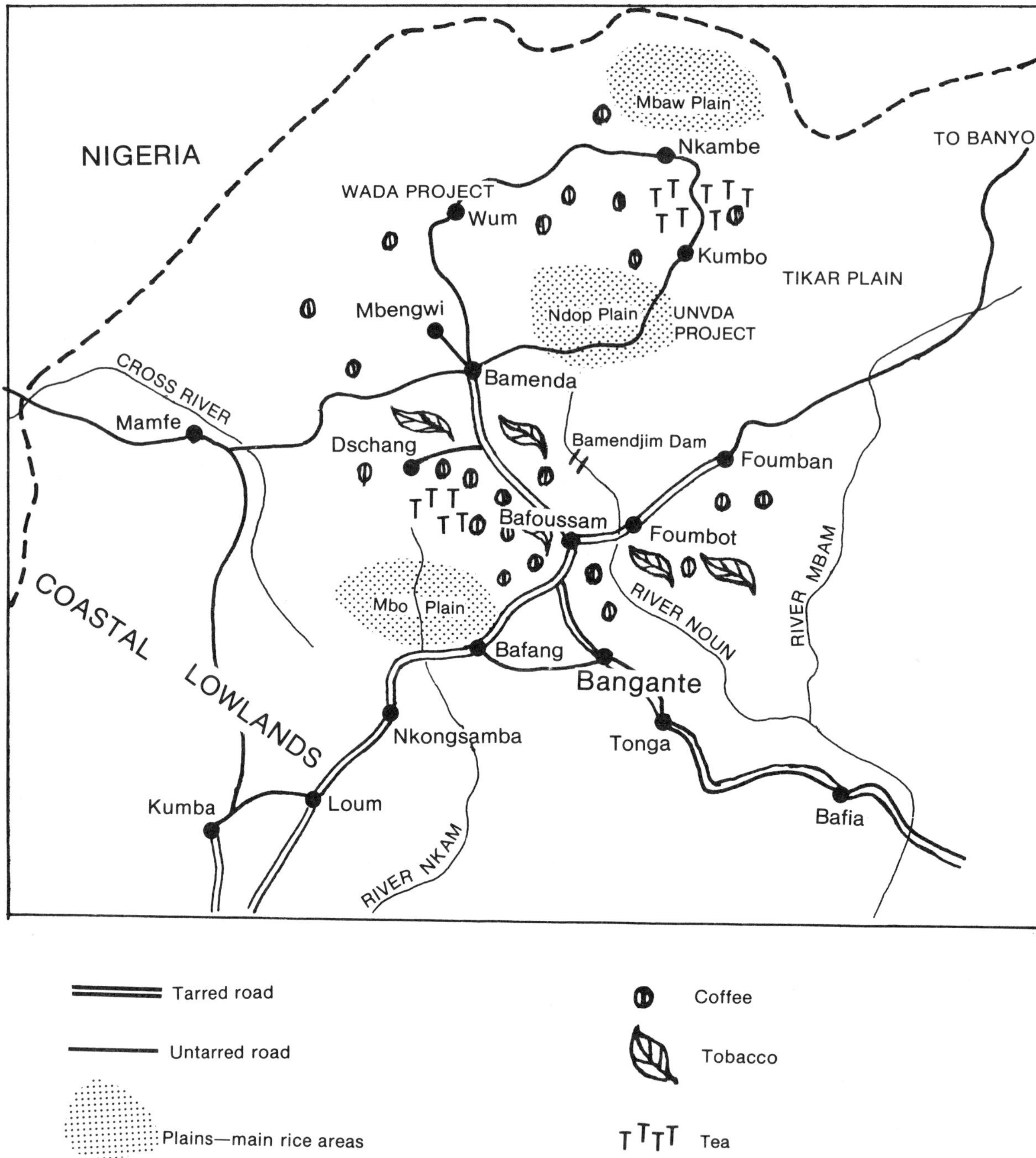

with the growth of cooperatives, since they constitute a means not only of economic and social progress but also an instrument for the development of the populations, especially those in the rural areas.

The CENADEC in pursuing its goals intervenes in the promotion and development of cooperatives; it controls and assists in the management of the cooperatives; it promotes and planifies the work of the cooperatives. It also provides administrative and financial help to the cooperative societies.

The CENADEC is presently active in the Centre, South, South-West and in the North-West provinces. In the North-West province it is concerned with consolidating the supervision and management of the existing cooperative societies. In addition, it runs a cooperative college in Bamenda.

COOPERATIVES AND MUTUALITY:

One realizes that the North-West Province is the leading centre of the cooperative movement in Cameroon. This is evidenced by the presence of many cooperative societies and the only cooperative college in the country — the National Cooperative College Bamenda. Produce and Marketing cooperatives have for long dominated in the coordination of farmers' activities in this province. In 1978/79 for example, 116 primary cooperative societies and three consumer cooperatives, with a total of 30,232 members, were regrouped into 11 cooperative unions and became affiliated with the North-West Cooperative Association (NWCA) ltd.

The N.W.C.A. ltd. is the only licensed produce buying agent in the province. It deals directly with the National Produce Marketing Organisation.

The cooperatives control the harvesting, processing, transportation, storage and commercialisation of the produce in order to reduce exploitation of the farmers by middlemen.

ANIMAL REARING:

The western highlands are not only famous for the cultivation of subsistence and cash crops but also for the rearing of animals. The SODEPA operates a cattle ranch with about 7,000 animals at Dumbo in the North-West province on an area of 30,000 hectares. The place of this region as far as livestock is concerned can best be understood if one looks back at the table on livestock in Cameroon in 1978.

The Bamileke lands which are densely peopled and where each family fences its plot, do not have cattle except small livestock such as goats and sheep. Towards the Bambutus and Bamenda highlands which are almost bare of people there is a lot of cattle rearing.

The Kounden zootechnical station near Fomban runs piggeries and poultry farms.

INDUSTRIES:

This region is generally poor in industries. The rice producing areas are equipped with rice-hulling mills. Only local crafts industries exist in most of the towns; however, it is an important sector of the country for the tourist industry.

INFRASTRUCTURE:

The western highland region has only two means of transport: land and air transport.

Land transport exists everywhere but the roads usually deteriorate and become impassable when the rains arrive. Most of them are untarred. Only one line from the littoral province (that divides at Bafoussam to Bamenda and to Foumban) is tarred. The Bamenda-Bafoussam-Foumban tarred road is only part of the transafrican road, which has to be continued from Bamenda through Mamfe to Nigeria and from Fomban to the Central African Republic. Apart from this single line, the poor conditions of the untarred roads are usually aggravated by the hilly nature of the landscape. The examples of the Kom hills and those hills surrounding the Mbo plain present landscapes that naturally hinder road construction. The hills hamper railway construction. Thanks to the existence of laterite in some zones. It is often scooped and spread on the surfaces of some of the roads to reduce the slipperiness occurring in the rainy season.

There are presently five major airports in this part of the country at Bafoussam, Koutaba, Dschang, Bali,and Bamenda.

TOWNS:

BAFOUSSAM: Located at an altitude of 1,450 metres some 275 km from Douala and 363 km from Yaounde, the town of Bafoussam is the principal town of the West Province.

Half a century ago Bafoussam contained just a handful of people but today it has grown into the seventh town of Cameroon with a population of 62,239 inhabitants. This rapid growth can be attributed, primarily to the dynamic nature of the indigenous people and secondly to the functions attached to it by the government.

A review of the oral history about the creation of this town would be of some interest in understanding the growth of the town.

After the expulsion of the Germans, the French in their desire to enforce the colonial control over her zone, decided to establish an administrative post as early as 1926 on the western plateau. Choosing the site was difficult because none of the traditional chiefs was willing to give up his land for the purpose of an administrative post. The then chief of Banjoun, Fotso II chose the present site of Bafoussam — a site that did not bind him since the said land did not belong to him. Chief Mambou, then of Bafoussam was opposed to the colonial decision to set up a post on his land. He organised his people and they burnt the equipment of the eight officials who arrived. The colonial authorities in turn captured Mambou and detained him for the action. Chief Fotso II

Bafoussam Town.

who later saw the progress of Bafoussam regretted his decision to allow the post to be set up on a different person's land.

From headquarters of a subdivision in 1925 Bafoussam became a district in 1959; seat of the Administrative Inspectorate of the west in 1961 and finally in 1972 it was made headquarters of the West Province.

It is not only the administrative role that has evolved, but the population too has expanded. From the oldest estimates of 1938, the population of Bafoussam was set at 4,800 inhabitants. In 1958 it had risen to 11,000 inhabitants. From then on the growth has been so rapid that by the 1976 census it had come up to 62,239 people. This drastic increase was due to troubles which drove the populations of the most afflicted rural zones to the towns.

During those sad years people escaped from Bamougoum, Bamenka, Batie, Baham, Bandekop, Bahonan, Bandjoun and Bangou and packed themselves in the town of Bafoussam, fleeing the insecurity of the countryside. With the return of calm many people who had been deprived of their territories, houses and wealth decided to settle permanently in the town.

Economic development, of course, also kept pace with the administrative and demographic growth. This town serves as a market centre for the entire population of the environment and also as the headquarters of U.C.C.A.O. The industries that are presently found in this town include a brewery, a tobacco processing factory and many local crafts centres.

Situated at the converging point of roads coming in from Douala, Bamenda and Foumban, this town is chiefly accessible by road. It is also accessible by air but the traffic here is still very scanty. It is hard to imagine the railway arriving in this town because of its highlands, but if it does, the impact will be very positive.

At the present pace of development this town is expected to be one of the leading metropolises of the country in future.

BAMENDA: The town of Bamenda was founded in 1901 when the first German station, built in Bali in 1890, was transferred to Mendankwe village.

Situated in the centre of a mountainous region,

the settlement of Bamenda is placed on a depression below an escarpment from where a complete view of the town can be obtained.

The town which had a population of 67,184 (in 1976) is the sixth town of Cameroon after Douala, Yaounde, Nkongsamba, Maroua and Garoua. It has three main sections, namely: Mankon, Bafreng, and "Station". Mankon is the centre of activities and contains the largest portion of the town's population. It is the centre of commercial activities. Bamenda Old Town (or *Abakpa)* situated at the foot of the escarpment, used to be the main market centre of the whole region and was a mixed community of Hausa and members of the surrounding villages. Today, it is almost the slum portion of Bamenda. Bafreng, the second portion, is separated from the main part of the town by a small stream that finds its way down from the escarpment through a very narrow valley. It is a fast growing area whose lands are being taken up for commercial and residential houses. Overlooking these two sections is the "Station", which is the administrative part of Bamenda with many government offices. It is situated at the edge of an escarpment. Residing in this area are top-ranking civil servants and administrators.

The town of Bamenda is the centre of attraction for the neighbouring villages. It is a major crossroads for the entire North-West province. The roads come in from all directions. An airport in Bafut is only a few kilometres from Bamenda town. Located in the region are other fast growing urban centres which reinforce the growth of Bamenda. They include, Kumbo, Fundong, Mbengwi, Ndop, Wum and Bali. The heaviest traffic leading into the town is through the Bafoussam — Bamenda road.

Bamenda has, in the last few years, experienced some rapid growth. As a provincial capital of the North-West Province, and a divisional chief town of Mezam Division, this town houses many government offices and consequently many civil servants.

Bamenda is one of the major educational centres of the country with many private and public secondary schools. Its role as an industrial centre is negligible. It has only local crafts workshops with no major industrial installations.

Bamenda Town.

CHAPTER 23: THE ADAMAWA REGION

PHYSICAL FEATURES:

Placed in the centre of the country, the Adamawa region starts from latitude 6° N to 8° N and is separated from the Western highlands by a small *col* around longitude 11° 4″ E, just west of the Mambila Mountains.

The Adamawa is a raised crystalline block with higher altitudes in its northern part than in the south. The average height of this plateau is 1100 metres. The ancient crystalline massif is covered in certain areas by recent volcanic outpours of basaltic rocks. This plateau is also incised by many rivers which flow down from it. The Adamawa ends in the north with an abrupt cliff below which starts the vast Benue plain. The region receives enough rainfall of about 1,200 mm per year; the temperatures are fresh and the sky is usually clear.

POPULATION:

This region is insufficiently inhabited by human beings because of several reasons. The Adamawa plateau constitutes part of the ancient lands of Africa which have suffered several phases of erosion throughout many centuries. These erosional processes have rendered the soil infertile. Secondly, the soils of this area experience the phenomenon of *laterisation* to produce *hardpans* which are not good for agriculture. Last and most important was the Fulani invasion which drove the Bantu from this region further south. These Fulani did not, after their triumph, establish any dense settlements but used the vast areas for animal grazing. The absence of conducive natural conditions for agriculture, the relative isolation of the region, especially the south-eastern slopes, have excluded permanent settlement of any ethnic group so that most of those who know this area are only herdsmen who never like sedentary life —nomadism is a major characteristic of their life-style. Though sparsely populated, the Adamawa is still capable of having more people. More than 60 percent of the population here are moslems and stock-rearers who keep moving from hill to hill. More than 80% of the people here are illiterate.

THE ECONOMY:

LIVE-STOCK RAISING:

Livestock raising is obviously the leading activity in this highland region, which is endowed with good pastureland. Besides the available pastureland, the presence of the animal-loving Fulani (Bororos) makes the area an important animal rearing zone.

Many animal rearing corporations are based here and most of them practise modern techniques of ranching and cross-breeding of imported and well selected local breed. Here, many veterinary centres are found. The Wakwa animal rearing station near Ngaoundere, selects and demonstrates modern methods of cattle ranching and also tries to associate animal rearing with crop cultivation. It also cross-breeds local and imported cattle for the purpose of developing a species that is better adapted to the region and that can give a greater yield of both meat and milk. This is the major operation zone of SODEPA — a government corporation for the development and exploitation of animal products. This corporation was created in 1974, and has as its major objectives, the acquisition, creation, administration, exploitation and development of all animal rearing enterprises. It also commercialises animals, and transformed animal products and by-products. In the south-east of the plateau at Ndokayo, SODEPA runs a cattle ranch with about 6,650 cattle on an area of over 40,000 hectares. Towards the northern part at Faro there is another ranch which is smaller than the Ndokayo ranch.

By 1978 the amount of cattle in the Adamawa region alone was estimated at 1.4 million out of the 3.4 million cattle for the whole country for that year. The Fulani sell these cows to the southern part of Cameroon, especially in the towns of Yaounde and Douala and even out of the territory to Gabon and Congo.

Transportation of the cattle is done mostly on foot and by train. In some few cases the already slaughtered cows are transported by air to the towns of Douala and Yaounde.

AGRICULTURE:

By importance, this activity comes only after animal rearing in this region. A comparative study of the surface areas cultivated per inhabitant in the various regions of Cameroon shows that the farmer in the Adamawa has the least surface area, despite the vast lands that exist here. Only the Baya, Mboum and the Dourou tribes of this area do farming on small plots cultivated by individual families for yams, cocoyams, cassava, maize, millet and groundnuts. Besides the individual

THE ADAMAWA REGION

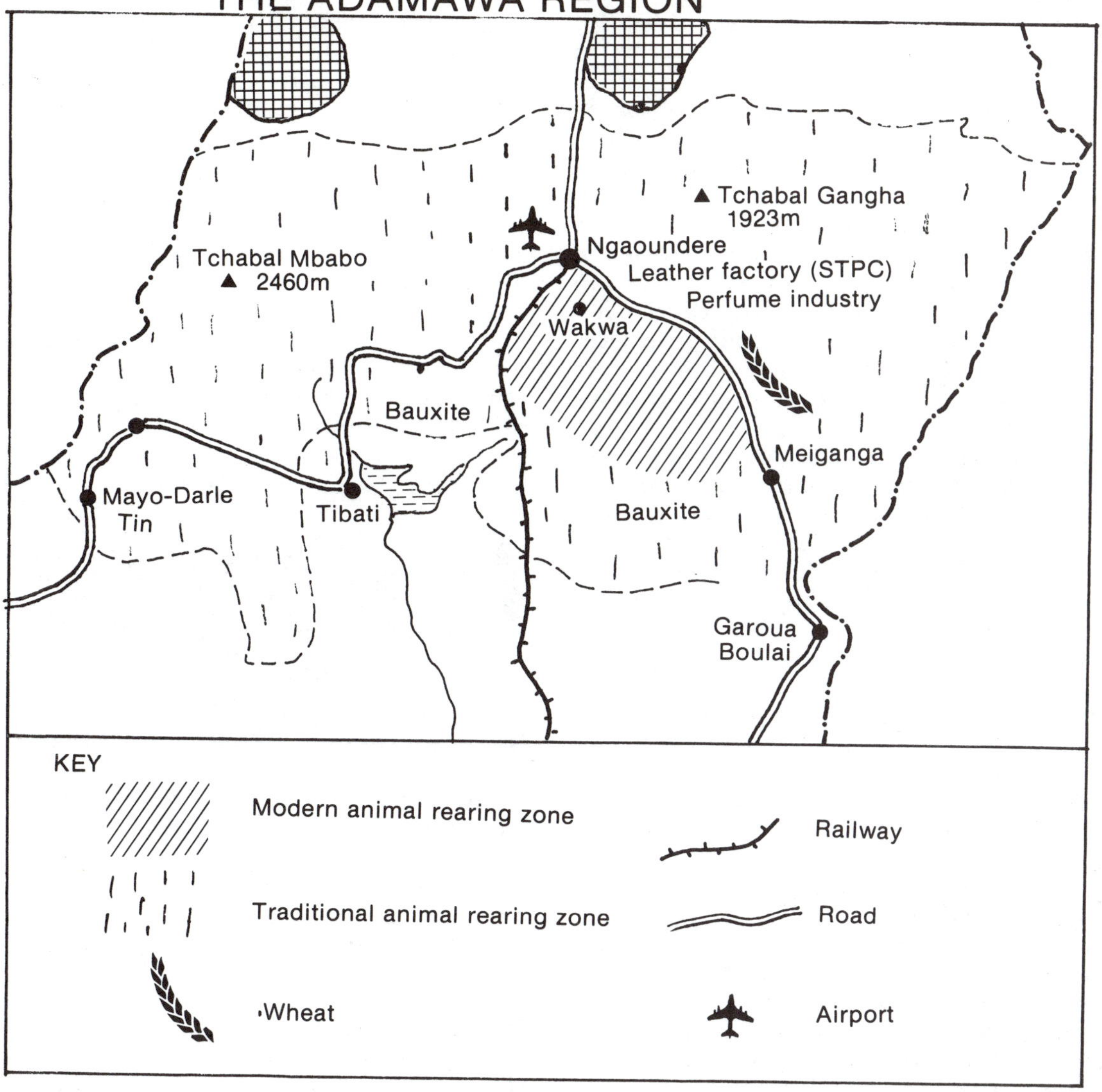

family farms, few large plantation farms are springing up for the cultivation of groundnuts, millet and wheat.

Trials for wheat on the Adamawa plateau at Wassande led to the establishment of the Wheat Development Corporation — SODEBLE —which undertakes the farming of wheat on an area of over 10,000 hectares. It intends to set up a flour mill in Ngaoundere, which will have a daily capacity of 800,900 quintals. Presently, the only flour mill in the country is at Douala and uses imported wheat.

TRANSPORT NETWORK:

Presently, the Adamawa is linked to the other regions by road, rail and air. The region shows a general insufficiency in roads and the few existing roads are of very poor standards.

The region receives three major roads from the south which converge at Ngaoundere before continuing to the north. Up to Ngaoundere the roads are of poor state probably because this section is also served with a modern railway line, which many passengers usually prefer to the roads.

From Ngaoundere, where the railway has its terminus, the only means of transportation to the north on land is by car and it is from here that the road is well tarred. By air, this region is poorly equipped. Only the Ngaoundere airport is of reasonable size.

INDUSTRIES:

The Adamawa region is equally poor in industries. Among the industries of significance the perfume industry can be mentioned. The S.T.P.C. has also set up a big factory at Ngaoundere for treating the skins of cattle, sheep and goats for the manufacture of leather.

The prosperity of the Adamawa region depends primarily on the inhabitants of the area who necessarily have to change their nomadic way of life to settle and be engaged in more farm work because this will lead to the growth of more urban centres. Secondly, the sooner the exploitation of the bauxite deposits at Minim-Martap and Ngaoundal begins, the better for the region, but it should be noted that extractive industries never attract much labour force and consequently, we do not expect any big agglomeration to grow because of the mines.

The Mbakaou Dam: Another thing of economic importance may be the Mbakaou dam constructed on the Djerem River to augment the volume of the Sanaga river for the H.E.P. station at Edea. This dam was completed in 1970 and it holds about 2 million cubic metres of water upstream. A dam without any factory is of little economic value. To say that the waters will be used for irrigation is probable but very doubtful with the highly nomadic animal-rearers of the region.

TOWNS:

The Adamawa is characterised by the general absence of important attraction centres. The only significant town is Ngaoundere.

NGAOUNDERE:

This is the principal town of the Adamawa plateau and it is situated at an altitude of 1,100 m. This town got its name from a small neighbouring granitic peak — the Ngaoundere hill. It is essentially a moslem town which originally belonged to the Mboum people. The Foulbe (Fulani) arrived and settled very peacefully, but after some time they overthrew the Mboum. Since then, the town has grown under the auspices of the moslem Foulbe.

Today, Ngaoundere, with a population of 39,000 inhabitants is the administrative chief town of the Adamawa Division and the capital of the Adamawa Province. The population may not increase very fast because its hinterlands are sparsely peopled.

Ngaoundere is at an important route junction. The roads from the south converge at Ngaoundere before continuing to north Cameroon. It is at the terminus of the trans-Cameroon railway line which comes from the south. It is also a stop-over centre for planes bound for the north and from the north to the south.

Collection of hides and skins in the Adamawa for the processing factory at Ngaoundere.

The combined effects of these means of transport have given Ngaoundere a new stimulus for growth. Ngaoundere is also growing fast into a commercial centre. The decentralisation program for the Yaounde Universtiy has brought part of the faculty of Science to Ngaoundere to give first-stage students in chemistry and biology professional training in the fields of stock raising and animal, food and chemical industries.

CHAPTER 24: THE NORTHERN LOWLAND REGION:

The region of North Cameroon starts after the Adamawa from around latitude 8° N and extends right to lake Chad, covering an area of about 90,000 km^2.

The physical, social and economic aspects of North Cameroon make the region very peculiar and distinct from the other regions.

RELIEF:

From the physical landscape we recall that the relief is not homogenous but shows an alternation of very low plains and highlands: the Benue depression in the south, the Mandara Massifs in the centre and the vast sedimentary basin in the north.

The Benue Depression in the South: The Benue depression is situated at the northern piedmont zone of the Adamawa highlands and it is limited in the north by the Mandara highlands. The average altitude of this depression is 250 m and the lowest portion is occupied by the valley of the Benue River. Here, we find thick layers of alluvium which were laid down in this region during the quaternary era.

These alluvial deposits would have been greatly used for agriculture if this zone were densely inhabited by people. Unfortunately, the low population density here does not make use of this advantage.

Towards the Adamawa the depression has high isolated peaks such as Hossere Vokre (2049m), and in the south-west of Garoua are the Alantika Mts, (1885 m).

The Mandara Massifs in the Centre: Some few kilometres north of Garoua the land rises into a number of massifs which group together to form the Mandara Mts. The major portion of this elevation is situated in the west at the frontier of Cameroon and Nigeria. The Mandara Mts. separate the Benue depression from the northern plain, otherwise called the Chad plain. The isolated inselbergs to the east of the main massif are essentially extensions of the Mandara structure. This zone slopes progressively eastwards into what we call the Diamare plain.

The elevated portions of north Cameroon are also more populated than the plains.

The Northern plain (Chad Plain): In the north-east of the Mandara Mts. the land slopes gently in the direction of Lake Chad to form the Chad plain. This plain actually extends over a larger area in the Republic of Chad than in Cameroon. The Cameroonian portion extends eastward only as far as the Logone River. A large part of this plain is flooded in the rainy season.

The northern plain, like the Benue depression, is sparsely populated. This northern plain is sometimes referred to as the *Yaeres.*

CLIMATE

The northern lowland region has a tropical type of climate which is characterised by two seasons, wet and dry. The length of the dry season increases, whereas rainfall reduces as one moves northwards. For a good part of the year, November to June, this region is swept by the north-east harmattan, which is hot and dry. At this time little rain falls; extremes of temperatures result from the dry air, which allows both intense insolation during the day and pronounced radiation during the night.

HUMAN AND SOCIAL ASPECTS:

From the human aspect, the peoples of the north are very different from those in the south and have quite different ways of social life.

The ethnic composition of the northern population is diverse (see chapter 8) but greatly mixed up and spread out very unevenly over the region.

This population is concentrated on the Mandara Mts. The sparsely populated areas are usually characterised by the presence of national wildlife reserves.

THE ECONOMIC ASPECTS

The northern region is a continental region which for long has escaped European influence but today the development of this region has been embarked upon fully.

The north was neglected by colonialists until after the end of the colonial period because of:

a) Its continentality and inaccessibility:

The distance from Douala port was too far and the only natural route to the north from the coast was by River Benue but one had to pass through Nigeria.

b) Low prices of North Cameroon Goods:

Cotton, groundnuts and millet, which were produced in the north were cheaper and of less economic importance than the crops cultivated in the south such as cocoa and coffee.

Besides, transportation costs also discouraged the colonialists and as a consequence of these factors, it was not very profitable to trade with the

THE NORTHERN LOWLAND REGION

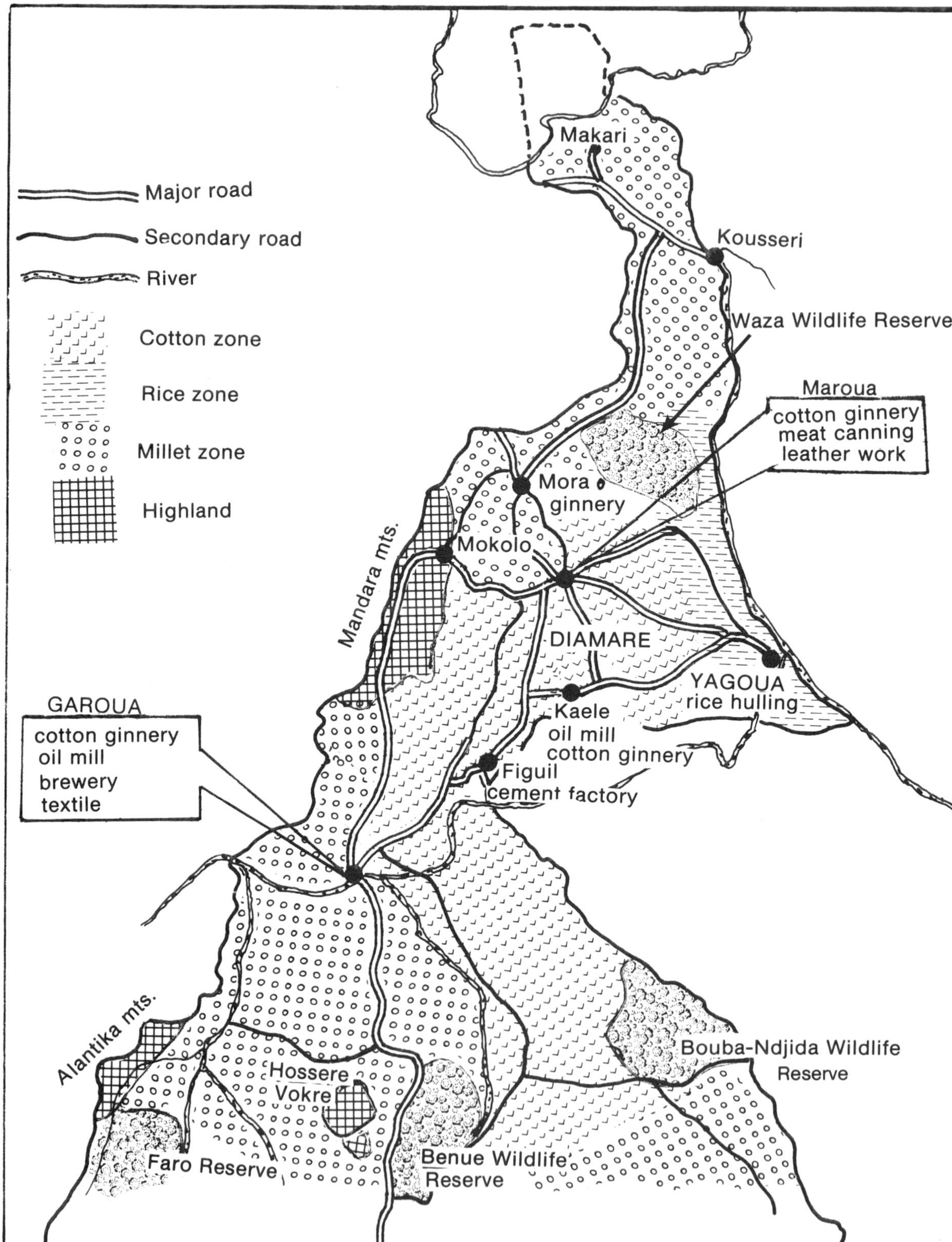

north.

c) Human problems:

The peoples of north Cameroon were not easily aroused by aspects of western civilisations. Some groups had their own civilisation and traditions. Some societies were materially well advanced and civilised and therefore detested any other form of civilisation.

LATE DEVELOPMENT:

Even though most of the economic realisations were after independence, we have to note that as from 1947 a series of agricultural, pastoral, infrastural, educational and many other social projects were already planned and started in this region.

Cameroon, being a member of the French Union at that time had to benefit from the development programs which France had for its overseas territories.

I) AGRICULTURE AND PASTORAL FARMING:

Three main projects were introduced —

a) The cultivation of commercial cotton by the CFDT in 1950 (CFDT — Compagnie Francaise pour le Developpement de Fibres Textiles)

b) It was also at this time that the Logone —Chari rice project was started by S.E.M.R.Y.

c) There was the organisation of veterinary centres for cattle with the establishment of a veterinary school at Maroua.

II) INDUSTRIES:

With high cotton production came cotton ginneries at Kaele, Maroua, Garoua, Mora and Touboro. The CICAM (Cameroon Cotton Industry) spins and weaves cotton in a textile complex set up at Garoua. The CICAM industry in Garoua is presently being extended to produce bedsheets and synthetic cloth in future. There is a cotton oil mill at Maroua and another one is being set up at Garoua. There is a groundnut oil mill at Pitoa, rice hulling factories at Yagoua and a meat canning industry in Maroua-Salack. The Brasseries du Cameroun brewery in Garoua presently produces both beer and soft drinks and many tons of ice blocks. CIMENCAM (Cameroon Cement Industry) inaugurated its factory at Figuil, 93 km from Garoua on the 23rd November, 1971. This factory has a furnace with a yearly production capacity of 46,000 tons of Portland Cement. Limestone and clay which constitute the essential raw materials are got from Figuil and Bidzar. Only gypsum is imported. There are numerous handicraft centres in the small towns and more particularly in Maroua.

Women walking to a community well to bring drinking water near Garva, north Cameroon.

The Lagdo Dam:

The construction of the Lagdo dam on the Benue river helps to strengthen the economy of north Cameroon. Besides providing hydro-electric power to north Cameroon towns, the dam is used for irrigation in the rice farms, and for fishing.

III) INFRASTRUCTURE:

A more convenient connection with the south was started in 1953-54 with the construction of the western road through Foumban-Banyo to Ngaoundere; and later by the eastern road through Bertoua, Meiganga, Ngaoundere and then to the north. Lately, the Ngaoundere-Kousseri road has been tarred up to Maltam. Branches of this road linking important towns and economic zones have also been improved.

As regards air transport, aerodomes for DC3 and DC4 planes were constructed at Garoua and Maroua before independence but at independence that of Garoua was already receiving DC6 planes. Presently, Garoua receives larger planes such as the Boeing 737 and 747.

The Garoua airport has been transformed into the second international airport of the country after Douala.

IV) SOCIAL INSTALLATIONS:

Great strides are being made to change and improve the living and social standards of these rather conservative peoples of this part of the country.

The first moves started with the establishment of the Government secondary schools at Garoua, Bongor, etc. Presently there are many colleges and high schools.

Dispensaries and hospitals are also being built in most of the big towns of the north.

V) TOURISM:

The place of tourism in north Cameroon has already been discussed elaborately under the tourist industry. Suffice it to say that the tourist industry has a bright future in north Cameroon. Only transport and lodging facilities still need to be increased in this domain.

Cotton spinning in Garoua.

SOME IMPORTANT AGRO-INDUSTRIAL PROJECTS:

S.E.M.R.Y.: The S.E.M.R.Y. (The Corporation for the Expansion and Modernisation of Rice Cultivation in Yagoua) is a corporation which is concerned with rice production in the regions of the Logone River where favourable conditions exist.

The alluvial plain here is usually flooded but these floods need to be controlled through modern techniques by digging canals and dykes to be used on the "Hardes" — area considered unsuitable for cultivation and abandoned.

The local farmers too need to be trained to practise modern farming methods on this vast plain which extends from Yagoua towards Pouss in the north and towards Moulvoudaye in the west.

S.E.M.R.Y. also aims at improving the local cultivatiom methods of rice in order to increase the yields and quantity available for consumption, and also to ease the transportation of the product. It also organises the sales of unhulled rice (Paddy).

The initial target of S.E.M.R.Y. at its inception was to transform 4,300 hectares of land through full irrigation control to produce 17,000 tons of paddy by the end of the start-off period.

The goal was attained and further programs were embarked upon which today give very high yields.

Presently S.E.M.R.Y. operates three main units variously termed S.E.M.R.Y. I, II and III.

S.E.M.R.Y. I, has as its centre Yagoua and its area of operation covers 5,300 hectares.

S.E.M.R.Y. II, is centred at Maga and has an area of 7,000 hectares to cover. The zones to be covered are in the localities of Guirvidig, Maga and Pouss.

S.E.M.R.Y. III is centred at Kousseri and has an area of 600 hectares to cover in the localities of Logone Birni, Kousseri, Gulfei and Gulfei Gana.

The activities of these three units are controlled and directed by the central head office at Yagoua.

SODECOTON:

The Cotton Development Corporation is designed to promote cotton production in north Cameroon. All activities related to the production, commercialisation and consumption are controlled by SODECOTON which replaced C.F.D.T. in 1974.

It also trains and gives technical assistance to local farmers, transports cotton, carries out ginning processes and takes care of the sales of fibre to the local industry such as CICAM. It labels cotton seed oil as, Huile de Diamaor, and grundnut oil as, Huile d'Abbia.

The first areas for cotton were the surroundings of Kaele, Maroua, and Guetale but in the early fifties the crop reached the regions of Logone and Chari. It has also spread to the Benue depression.

The SODECOTON trains its senior personnel at Guetale and its junior staff at Makebi, Pitoa, Guider and Bere.

CEREAL BOARD:

It was founded in 1974 and was aimed at taking charge of the cereal market of the northern province in order to control the stock with the view to avoiding price fluctuations.

TOWNS:

GAROUA: This beautiful inland town is placed on the right bank of the Benue river and situated at an altitude of 235 m. It has experienced a rapid growth in recent years. With a population of 69,285 inhabitants it is the fifth town of Cameroon. Serving as the administrative headquarters of the vast North province, this town assembles people from the immediate surroundings and the other parts of the country. It is also the administrative chief town of the Benue Division. The town also has a good number of industries which include a brewery, textile and cotton ginning and spinning industries.

The town is accessible in the rainy season through the Benue river which makes it the major river port of the country. It is also accessible by road and air.

MAROUA: This is the most populous town of the northern region with a population of about 70,000 inhabitants. It is the fourth town of the country. Its population is mainly Foulbe who are engaged in trade and in the making of local crafts out of leather.

Besides serving as a market-town it is also the administrative centre of the Diamare Division and the provincial capital of the Far North Province. Its road and air links are the two ways of easily reaching the town. Its few industries are meat canning and local crafts.

PART SIX: POLITICS AND DEVELOPMENT.

CHAPTER 25: PRINCIPLES FOR GOVERNMENT POLICIES IN CAMEROON.

In almost every country in the world today governments have an unquestioned responsibility for stimulating progress in most spheres of economic, social and political development. In economic development, government has a number of functions which include: building public services, influencing attitudes, shaping the economic institutions which influence resource utilization, influencing income distribution, controlling the quantity of money, controlling fluctuations, ensuring full employment, and influencing the level and areas of investment.

Very important in Cameroon, in particular, and Africa as a whole, the way in which political changes take place has profound effect upon the direction and scope of economic development, which in turn, affect future prosperity. For whatever the amount of natural resources, historical or spatial basis of development, and whatever the theories developed by economists, sociologists or geographers for a particular place at a particular time, in the final analysis, it is governmental action (or inaction) that determines the direction and scope of economic policy. The building of roads, railways, dams, new settlement areas; the motivation for regional economic cooperation between states and the maintenance of a calm political atmosphere depend essentially on governmental ability (or inability) to act justly and wisely.

Any government which squanders public wealth under heavy concealment, allows social services to decay, or favours one ethnic group against the others, certainly does not, by such behaviour encourage development. In fact, such governments only succeed in creating a blinded and confused citizenry unable to understand why their country gets so impoverished. The final results of such bad governments are usually political chaos, coups, and civil wars, which in no way, promote development.

It will not be possible to do full justice to this vast and complex issue within an introductory book of this nature. However, let us explore, in outline, some of the frequently mentioned principles followed by the economic planners of Cameroon.

The curious student who asks whether Cameroon is a capitalist, or socialist (or communist) nation receives an answer which lies somewhat in the middle. The government of Cameroon applies a lot of socialist principles to be called a socialist state and a lot of capitalist principles to be called a capitalist state but the use of either of these terms to describe the whole economic system will be grossly misleading. Cameroon does not adhere to dogmatic or ideological principles in a foolish way. It seeks all the policy alternatives available and adopts those that are appropriate to the problem depending on the situation, no matter whether such policies turn out to be capitalistic or socialistic. Thus, a *mixed economy* is the economic system to be found in Cameroon.

Three basic economic principles have often been mentioned in relation to Cameroon's economic system, namely: Planned liberalism, Balanced Development and Self-Reliant Development.

1) PLANNED LIBERALISM:

Planned liberalism brings the advantages to be anticipated from the spirit of free enterprise, the sense of efficiency, and responsibility, in private initiative. At the same time, the economy is consciously planned, directed, stimulated and controlled by the state, the quarantor of general interest.

According to the policy of planned liberalism, the government identifies development options and establishes its priorities, given its limited financial resources. The priority projects are contained in National Development Plans, usually established for a period of five years. Those projects that will help to consolidate the economy, create employment for Cameroonians, and ensure some level of balanced development of all the regions and all the peoples of Cameroon are given preferential treatment. Large-scale industry, the mining of minerals, the marketing abroad of the principal products, education, energy, transport infrastructure and forestry are handled by the state because these activities concern the essential resources of the whole nation and should be handled wisely. The state then becomes a major investor. Both Cameroonian and foreign entrepreneurs then operate within the guidelines and regulations set up by the government.

On the other hand, small-scale industry, trade, restaurants, farming and many other forms of minor economic activities are in the hands of private individuals. Cameroon therefore, recognizes the importance and permits the existence of the private sector with some adequate level of planification. Cameroon's planification still guarantees private ownership of property as well as the security of individuals and their wealth.

2) BALANCED DEVELOPMENT

A complementary concept to planned liberalism is what the public authorities have always termed *Balanced Development*. The concept of balanced development does not, in any sense, imply a situation of perfect balance in levels of development between the regions or peoples of Cameroon. It is a concept that seeks to promote social justice given the consideration that Cameroonians are a united people of equal status irrespective of ethnic and geographical background. It is a concept that seeks to redress inequalities that resulted from selected attention given to different regions and peoples by missionaries and colonizers. In short, the concept stems from the fact that it is "unfair to concentrate development efforts in only one part or area of the country." Thus, we note the following key points:

- **Balanced regional development:** The different regions of the country are important in one way or another. None of the ten administrative provinces has to be neglected by the leaders of the country, and each of these regions has an equal obligation to contribute to national development.
- **Balanced social Development:** Each Cameroonian citizen, irrespective of age, sex and ethnic background, receives a remuneration or share of the national cake corresponding to his or her effort in national construction. To achieve this goal the government seeks to equip every administrative division with, at least, enough secondary schools, transport infrastructure and medical centres to raise the social standing of the citizens. Thus, the disciplines and faculties of the University of Yaounde are being decentralized and relocated in the various provinces to develop into new universities.
- **Balanced economic Development:** Between towns and villages, between farmers, industrial and service workers, between civil servants and workers in the private sector, differences in levels of income and standards of living are reduced to raise the mass of the people and not a privileged few. The drive to supply electricity to small towns and even down to the villages seeks to encourage economic development of all the people, both rural as well as urban residents. Balanced economic development also ensures that none of the sectors of the economy is neglected.
- **Balanced Cultural Development:** Cameroon has a diversity of cultures. Positive aspects of these numerous local cultures are encouraged as well as modern techniques from abroad.

3) SELF-RELIANT DEVELOPMENT:

Much has been said and written about self-reliant development. Unfortunately, a lot of the literature fails to bring out its real meaning. It is important, first of all, to remember that self-reliance contains more than just the idea of self-suffieiency in food production.

Self-sufficiency and self-reliance are policies usually devised to safeguard a nation from the dangers of too much dependence on the outside world for the supply of essential commodities. In reality, no country in the world is self-sufficient but there are many that are self-reliant. Even the so called developed nations depend heavily on the less developed countries for the supply of raw materials and energy resources, but they are self-reliant because of their technology and the fact that they have control of their economies.

Trade with the outside world, especially with the rich capitalist nations, could adversely affect the internal economic situation in the following ways:

a) Cheap imports from abroad could disrupt domestic industries.

b) Tariffs in foreign lands could disrupt industries devoted to export.

c) Embargoes or monopolies abroad could cut off essential supplies or result in high prices.

To stem off such dangers many leaders of less developed nations have talked about self-reliant development. But what do they mean by *Self-Reliant Development?*

As the term implies, self-reliant development simply refers to a situation where the citizens of a nation can count on their resources and more particularly on their capabilities to achieve development. The development of a nation is brought about by the people themselves, working hard and intelligently. Self-reliant development, in other words, *is the capacity to generate one's own development.*

The introduction of technology or new methods of production can bring development. Technology could be imported and adapted to local conditions. Agriculture provides the surest means to finance the new technology from abroad. Foreign aid, loans, imported technicians and large-scale industries set-up in less developed countries do not bring development if the citizens only stand by as spectators waiting to buy whatever comes out of the industries. By participating in hard and intelligent work, by devising appropriate policies and by having good leadership a nation can become developed and such a nation can be said to be self-reliant. There is self-reliant development when the citizens have the capacity to introduce new technology into the production of subsistence goods. It is not enough to be self-sufficient in the supply of natural resources or to have foreign technicians working to produce goods in a nation.

SELF-RELIANCE AND INTERNATIONAL TRADE:

It is important to note that trade in itself is not bad. The way it is conducted between poor and rich nations is what matters. The extent of participation in international trade is not a measure of self-reliance. Many people refer to a situation of self-sufficiency with little participation in international trade and not to technology. Thus, they conclude that a self-reliant country does not need to participate in international trade. The crucial question is whether a nation can participate confidently in international trade without allowing its economy to depend solely on that trade.

Some countries are able to participate in international trade without allowing themselves to be "shocked" by changes that occur in the world market while others cannot withstand minor changes in the world market. Nations which depend solely on the export of some primary products are particularly vulnerable, no matter whether they participate more or less in international trade.

Central to the notion of self-reliance is the ability to meet domestic demands from domestic sources while still participating in international trade without relying on it. Self-reliant development is complementary to political independence. Self-reliant development calls for the creation of an independent national economy:

- Production essentially for national needs.
- Control of the economy in national hands.
- Increasing diversification of foreign ties.
- Rural development.
- Improving income distribution to better the living conditions of the masses.
- Government or national control of a large part of the economy.

Thus, self-reliance presupposes development generated by internal forces and internal needs. Cameroonians therefore would be expected to be masters of their economic destiny, providing the bulk of the labour force — managers, technicians and unskilled labour. This is to be contrasted with a situation where foreign capitalists are invited to set up large-scale manufacturing industries, using unskilled cheap labour and producing essentially for export.

For self-reliant development to succeed, agriculture and technology must play a strategic role. If dependence on foreign capital is reduced, only the wealth produced in the countryside can be counted on to finance national projects of investment. Similarly, if external markets are to be reduced in importance, incomes of rural residents must be increased in order to provide local markets for new products. Self-reliance therefore means less dependence on foreign aid and loans, which have been seen to be disruptive of many developing economies.

Not only agriculture alone, but the citizens should be able to move beyond traditional modes of production. They should develop the capacity to introduce some technology in production and be able to manage their economy efficiently.

CHAPTER 26. GOVERNMENT AND ADMINISTRATIVE UNITS OF CAMEROON.

Cameroon has a strong and centralized form of government. The main source of important policies and direction for all government activity is Yaounde.

The Republic is divided into ten administrative *Provinces,* each placed under the jurisdiction of a governor appointed by the head of state.

The governors are responsible to the head of State. They are responsible for the maintenance of law and order, for coordination of the activities of the *Divisions,* and for exercising control over the civil service in the Province.

Each Province is subdivided into Divisions *(departements),* each named after a particular geographical feature. The Division is headed by a Divisional Officer *(prefet).* The Divisional Officer is the executive agent of central authority responsible for seeing that laws and by-laws on all subjects are observed in his administrative unit. He may call on the security officers to preserve law and order. He reports on economic and administrative matters to the provincial governor.

The Divisions are again divided into Sub-Divisions *(arrondissements)* headed by assistant Divisional Officers *(sous-prefets).* Districts are created within Divisions and Sub-Divisions . They are administered by District heads.

Administrative Divisions of Cameroon

I. FAR NORTH: Maroua
- 1. Diamare: Maroua
- 2. Kaele: Kaele
- 3. Logone-Shari: Kousseri
- 4. Mayo Danay: Yagoua
- 5. Mayo Sava: Mora
- 6. Mayo Tsanga: Mokolo

II. ADAMAWA: Ngaoundere
- 7. Djerem: Tibati
- 8. Faro-Deo: Tignere
- 9. Mbere: Meiganga
- 10. Mayo Banyo: Banyo
- 11. Vina: Ngaoundere

III. NORTH: Garoua
- 12. Faro: Poli
- 13. Benoue: Garoua
- 14. Mayo Rey: Tchollire
- 15. May Louti: Guider

IV. EAST: Bertoua
- 16. Upper Nyong: Abong-Mbang
- 17. Boumba-Ngoko: Yokadouma
- 18. Kadey: Batouri
- 19. Lom and Djerem: Bertoua

V. WEST: Bafoussam
- 20. Noun: Foumban
- 21. Bamboutus: Mbouda
- 22. Menoua: Dschang
- 23. Mifi: Bafoussam
- 24. Nde: Bangangte
- 25. Upper Nkam: Bafang

VI. LITTORAL: Douala
- 26. Mungo: Nkongsamba
- 27. Nkam: Yabassi
- 28. Sanaga Maritime: Edea
- 29. Wouri: Douala

VII. NORTH-WEST: Bamenda
- 30. Donga-Mantung: Nkambe
- 31. Menchum: Wum
- 32. Bui: Kumbo
- 33. Mezam: Bamenda
- 34. Momo: Mbengwi

VII. SOUTH—WEST: Buea
- 35. Manyu: Mamfe
- 36. Meme: Kumba
- 37. Ndian: Mundemba
- 38. Fako: Limbe

IX. CENTRE: Yaounde
- 39. Mbam: Bafia
- 40. Mefou: Mfou
- 41. Mfoundi: Yaounde
- 42. Lekie: Monatele
- 43. Nyong and Kelle: Eseka
- 44. Nyong and So: Mbalmayo
- 45. Nyong and Mfoumou: Akonolinga
- 46. Upper Sanaga: Nanga Eboko

X. SOUTH: Ebolowa
- 47. Dja and Lobo: Sangmelima
- 48. Ntem: Ebolowa
- 49. Ocean: Kribi

LOCAL GOVERNMENT

Local governments in Cameroon, for the most part, have limited independence from the central government. The provision of education, health facilities and many other important services at local level is determined by the ministries of the central government.

Local self-government, though less important, exists in urban as well as in rural areas. They have

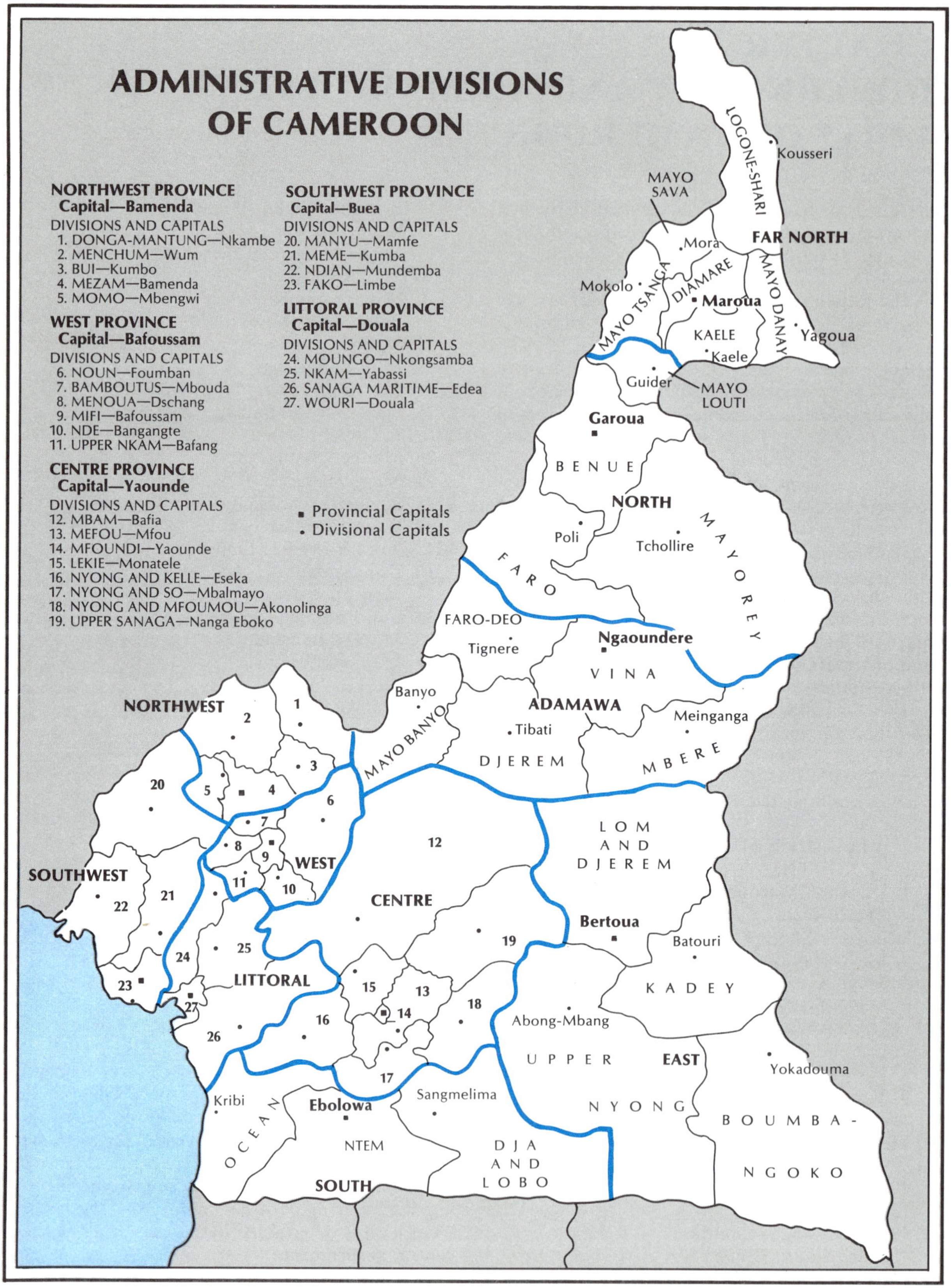
ADMINISTRATIVE DIVISIONS OF CAMEROON
NORTHWEST PROVINCE
Capital—Bamenda
DIVISIONS AND CAPITALS
1. DONGA-MANTUNG—Nkambe
2. MENCHUM—Wum
3. BUI—Kumbo
4. MEZAM—Bamenda
5. MOMO—Mbengwi
WEST PROVINCE
Capital—Bafoussam
DIVISIONS AND CAPITALS
6. NOUN—Foumban
7. BAMBOUTUS—Mbouda
8. MENOUA—Dschang
9. MIFI—Bafoussam
10. NDE—Bangangte
11. UPPER NKAM—Bafang
CENTRE PROVINCE
Capital—Yaounde
DIVISIONS AND CAPITALS
12. MBAM—Bafia
13. MEFOU—Mfou
14. MFOUNDI—Yaounde
15. LEKIE—Monatele
16. NYONG AND KELLE—Eseka
17. NYONG AND SO—Mbalmayo
18. NYONG AND MFOUMOU—Akonolinga
19. UPPER SANAGA—Nanga Eboko
SOUTHWEST PROVINCE
Capital—Buea
DIVISIONS AND CAPITALS
20. MANYU—Mamfe
21. MEME—Kumba
22. NDIAN—Mundemba
23. FAKO—Limbe
LITTORAL PROVINCE
Capital—Douala
DIVISIONS AND CAPITALS
24. MOUNGO—Nkongsamba
25. NKAM—Yabassi
26. SANAGA MARITIME—Edea
27. WOURI—Douala
▪ Provincial Capitals
• Divisional Capitals
Kousseri
LOGONE-SHARI
MAYO SAVA
FAR NORTH
Mora
Mokolo
MAYO TSANGA
DIAMARE
Maroua
MAYO DANAY
Yagoua
KAELE
Kaele
Guider
MAYO LOUTI
Garoua
BENUE
NORTH
Poli
Tchollire
MAYO REY
FARO
FARO-DEO
Tignere
Ngaoundere
VINA
ADAMAWA
Banyo
MAYO BANYO
Tibati
DJEREM
Meinganga
MBERE
NORTHWEST
WEST
SOUTHWEST
CENTRE
LITTORAL
LOM AND DJEREM
Bertoua
Batouri
KADEY
Abong-Mbang
UPPER NYONG
EAST
Yokadouma
BOUMBA-NGOKO
Kribi
OCEAN
Ebolowa
Sangmelima
NTEM
DJA AND LOBO
SOUTH

councils whose members are elected to run local affairs.

Urban councils are found in towns and are headed by mayors. In major towns, for example, Douala, Yaounde, Nkongsamba, Bafoussam and Bamenda, the head of the council is a special Government Delegate appointed by the president of the Republic. In the rural areas, the councils are headed by Municipal Administrators.

Apart from the units and administrators mentioned above, traditional government is in the hands of chiefs (or *fons)* and Lamibe. Villages had traditionally been ruled by powerful leaders known to be able to maintain law and order in a very effective way. They also performed diverse governmental roles before the imposition of western forms of government. Today, the Ministry of Territorial Administration recognizes the importance and the legitimacy of these rulers.

CHAPTER 27.
CAMEROON'S INTERNATIONAL RELATIONS.

Cameroon's relations with the outside world are broad and characterized by a sense of anti-ideology "because true vision must make room for adjustments to the hard realities of world politics." Cameroon cooperates with most nations of the world irrespective of their political creed.

Mark Delancey (1986: 195) summarizes the principles governing Cameroon's foreign relations: "nonalignment; anti-imperialism and especially opposition to the racist, white-dominated regime of South Africa; African solidarity and cooperation; domestic economic growth and development through collaboration with foreign assistance and investment; and the maintenance of national sovereignty ..."[1]

The nation's leaders have always stated that Cameroon does not interfere in the internal affairs of other nations and does not expect other nations to interfere in Cameroon's internal affairs. Even though Cameroon seeks cooperation with countries irrespective of their ideological leanings, the case of the Republic of South Africa with its discriminatory apartheid system against blacks presents the only exception.

1. Relations with Other African States

Cameroon enjoys good relations with almost all African nations, especially the six neighbouring states. As a bilingual country, Cameroon acts as a link between English and French-speaking African countries. Cameroon is also a member of the following regional groupings:

a) UDEAC: The Central African Economic and Customs Union to which Cameroon belongs has already been discussed at length in Chapter 19.

b) The TRANSAFRICAN HIGHWAY COORDINATING COMMITTEE to which Cameroon belongs, was formed in 1971. Other member states include the Central African Republic, Kenya, Nigeria, Uganda and Zaire. Its chief function is the coordination of planning and construction of a trans-African highway.

c) LAKE CHAD BASIN COMMISSION: Founded in 1964, its members include Cameroon, Chad, Niger and Nigeria. These are countries which share the lake. The commission coordinates the development of the Lake Chad basin waters for agriculture, animal husbandary and fisheries.

d) ORGANIZATON OF AFRICAN UNITY: Cameroon is also a member of this intergovernmental organization which groups sovereign African states.

2. Relations With Other World Regional Organizations

a) UNITED NATIONS ORGANIZATION (UNO). Cameroon is a member of the UNO including some of its specialized institutions such as UNESCO, UNICEF, FAO, WHO, IMF, IBRD, etc.

b) AFRICAN, CARIBBEAN AND PACIFIC STATES. Following the Lome Convention signed by 57 African, Caribbean and Pacific states, Cameroon belongs to that group of ACP states having preferential trade links with the European Economic Community (EEC).

3. Relations with France

Cameroon's relations with Europe have always tilted in favour of France. As Ndiva Kofele-Kale, the Cameroonian scholar, points out, France's privileged position in Cameroon is derived from a historical bond forged over a long time and sustained by a complex web of emotional, psychological and legal ties.[2] To add to this, as a European power, France was present at the coast of west Africa and engaged in the African slave trade which seriously reduced the population of Cameroon; as a colonising power struggling for territories in Africa, she received a large part of Cameroon as a League of Nations' mandate territory. Cameroon was an *associate territory* in the French Union. On the eve of independence, the two countries signed a series of cooperation agreements and since then, France has remained the predominant European trading and investment partner of Cameroon. However, in recent years, Cameroon has developed many economic ties with trading partners other than France.

4) Relations with the United States of America

The United States of America has always wanted Africans to believe that she was never interested in acquiring portions of land (or colonies) in Africa but the activities of the American Presbyterian Missionaries on the coast of West Africa testify to the fact that she was also engaged in the 'scramble.' Her efforts were only thwarted

---

1. Mark Delancey "Cameroon's Foreign Relations." In *The Political Economy of Cameroon,* by Schatzberg and Zartman (eds.) P. 195.
2. Ndiva Kofele-Kale. "Cameroon and Its Foreign Relations." In *African Affairs* vol. 80 no. 319 April 1981. P. 200.

by the strong foothold already gained by the Germans and other European nations in this part of the world. Consequently, American involvement in the continent, until recently, has simply been negligible (except in Liberia and through her corporations). One then wonders what has caused the sudden upsurge of American interest in Africa, and more especially in Cameroon, during the last few decades. First, the identification of millions of black Americans with Africa as their "mother land" and their concern for racist and imperialist activities on the continent have led many of them to embark on studies focused on Africa.

Second, the American public perception of the Soviet Union as a threat to "western interests" on the African continent gradually led the U.S. government to turn its attention to Africa, especially to countries that are strategically located and those that are rich in natural resources.

In the early 1960s, the U.S. began the American Cooperation program in Cameroon which did not really take shape until 1975. In the 1980s, the United States, under Ronald Reagan's administration, adopted a more aggressive approach in the 'search for a place' in Cameroon and this was done through a number of ways, notably through increased aid and cultural ties. In late 1984, the American warship, USS FAIRFAX COUNTY, paid a three-day visit to the port of Douala and its captain maintained that the visit "was in line with the cooperation links that exist between Cameroon and the United States of America."

American officials have always referred to the political stability, economic development and more particularly, Cameroon's emphasis on agricultural development to become self-sufficient in foodstuffs, as the reasons for developing close ties with Cameroon. Sceptism about such explanations is based on the fact that at the time that Cameroon was a lot worse off there was no desire to help and moreover, the help is not directed to those African nations who need it most. The fact of the matter, however, is that where there is crude oil Americans are to be found: their dollars, banks and other articles of trade.

Where American interests and investments are deeply involved there is a strong desire to create ties and also spread anti-Soviet (or anti-communist) sentiments which often brings political instability. This increased American involvement, therefore, is greeted with cautious optimism by Cameroon's elites committed to the policy of non-alignment.

5. Relations with the Eastern Bloc Nations.

The USSR has no deep historical ties with Cameroon especially as far as slave trade, colonialism and imperialism are concerned. Diplomatic relations opened with the Soviet Union shortly after 1972. The Soviet Union has since then provided scholarships to Cameroonian students who have pursued courses in different fields in that nation. Trade links, though limited, are increasing.

Links with the Peoples Republic of China have even been more forthcoming, especially after China stopped supporting the UPC party. The PRC asisted Cameroon in operating rice farms in different parts of the country in the early 1970s, and opened a hospital at Mbalmayo. A few Cameroonian students have also won scholarships to study in Chinese institutions. In the capital city, Yaounde, the Chinese have constructed a large magnificent cultural centre which also serves as the headquarters of Cameroon's political party.

Cameroon also has diplomatic ties with a number of eastern bloc countries, which include among others, Vietnam, Poland, North Korea and Cambodia.

CHAPTER 28
THE NOTIONS OF "THIRD WORLD" AND "DEVELOPMENT"

Many scholars have written about the problems of the third world. Those who seek to make a living on third world studies have put out voluminous literature outlining third world characteristics as if it were one homogenous mass, despite the differences in income, values and cultures from one nation to the other. As a common characteristic of such literature, life in the third world is described in the most demeaning terms, portraying living conditions as unbearable. The third world countries have variously been called 'backward', 'underdeveloped', 'undeveloped', 'less developed', or 'developing' countries. Sometimes the people of such countries are said to be unhappy, poor, and primitive. To support their claims, statistics of income per capita and other material goods are manipulated, and estimates of the level of happiness among inhabitants of such regions are made, even though such statistics say nothing of the distribution of income within society nor of the level of general well-being. Through European and American movies and pictorial magazines widely distributed throughout the world, reports of starvation and poor living conditions are carefully selected from the less developed world and presented to the world. This has been a narrow, biased and highly ethnocentric Euro-American tradition to evaluate nations solely on physical quantification of man-made goods, the result of which has been to attribute superiority to the 'First and Second Worlds'.

With an abundance of such literature from the 'first world' people have come to make very general and sometimes very vague, wrong associations and false assumptions of living standards of either the so called First World or Third World countries. For example, most inhabitants of Africa, Asia and Latin America believe that there are no miserable living conditions in the 'developed countries' while many inhabitants of America and European countries believe that most of Africa is jungle and starving people.

WHAT ESSENTIALLY IS THE THIRD WORLD?

The term "third world" is a remnant of Cold War thinking, which divided the world into two rigid blocs, the Capitalist Western World and the Communist Eastern World, with all countries not allied to the United States of America or the U.S.S.R. lumped in the remaining third category. This political use of the term was very imprecise because some countries, such as South Africa, could not fit well in any of the categories. Other nations were important political allies of each of the two main powers but did not even share the same political ideologies. With such imprecision and ambiguities the use of the term gradually lost meaning, then shifted from political to economic, and today it is generally used to refer to states in Africa, Asia and Latin America that are considered not highly industrialized.

It is important to recall that most of the countries in Africa, Asia and Latin America, achieved independence after the Second World War and to differentiate them from the older nations of Europe, they were collectively referred to as, *Emergent Nations,* expressing a "western" ethnocentric outlook as if the peoples of these nations had no history and no past before achieving independence. The term soon gave way to the expression "underdeveloped societies" to distinguish them from the "developed societies." Such categories did not show that progress, if any, was taking place at all in the "underdeveloped" countries. The problem the terms "developed" and "underdeveloped" faced was that many of those so called "underdeveloped societies" were already making remarkable progress at the same time that some of the "developed nations" were undertaking development projects to become more developed. Consequently, there was a need to look for another term, "developing societies". This search for new terms has not yet ended. More recently, the United Nations (UN) has used the expressions "Less Developed Countries" (LDCs) and "More Developed Countries (MDCs) to differentiate between those countries that have not made much economic or industrial development from those that have made some significant achievements. These UN expressions contain the essential fact of development as a continuing process in both societies.

Whatever term western writers use, it is essential to note that there is a persistent effort to contrast living standards of western nations with those of other nations to be found in Africa, Asia and Latin America even though there may be as many similarities as there are contrasts. Even when these contrasts in living standards are mentioned, another common charcteristic of the literature is to be noticed. The literature often contains wrong comparisons. Western writers concentrate on the poor people and overlook the upper strata in the "LDCs", and for the western nations they dwell on the living standards of the rich people and almost completely ignore the poor people. This is exemplified in the writings of Denis Austin who looks at the LDCs and MDCs. He writes that, "There are not three worlds but two, a developed world and an underdeveloped world —a world of those who have and those who have not. The haves not only possess more goods; they have a richer life, enjoying political stability, old age, full stomachs, freedom from anarchy, warm houses in winter, cool houses in summer, the advantage of travel, and the ravels of technology."*

Such an assessment is grossly misleading because it compares the poor people in LDCs to the rich portion of the population in the technologically advanced countries. It is not a parallel comparison. The millions of people languishing in misery in the crowded *ghettoes* of American and European cities are never discussed in such writings. No account is made of the thousands of homeless people lying in the subways and on the streets of American cities. They are also overlooked. Yet, these are shocking scenes to any newly arrived African or Asian to the United States of America.

In most Euro-American writings such countries as Cuba with little or no real poverty (and without millionaires) are classified among "third world" countries while many western nations with millions of homeless, hunger-stricken, drug-addicted and really poor people are classified among the "First World" nations.

The third world is not one mass. It would be highly misleading to portray the "third world" as an unchanging world of widespread poverty. Most third world countries have made, and are still making significant progress, yet there has hardly been a change in the number of world economic categories. Cameroon is not anything near what it was a decade ago; even between one year and the other there are great changes, and so are many other non-western countries.

WHAT REALLY IS DEVELOPMENT?

It must be realised that physical or technological development is but one aspect of national development. A balanced view of development should comprise the cultural, personal and physical aspects of that society. Thus, we may define development as, "the sustained improvement of an entire society and social system towards a better and more just life." The distinction between this broad concept of development should always be made from the now limited concept of economic development which considers only material progress. Walter Rodney (1982:3) notes that "development in human society is a many-sided process." We should try to identify the different aspects of development.

— Physical development includes, man-made goods produced by the use of technology.

— Cultural development includes, the values, norms and traditions of a society.

— Personal development comprises the psychological directions of individuals.

Development should be of continuing utility to mankind and the process of developing should be thought of as a continuing process and not as one that is static. This explains why such countries as Japan and the United States of America continue to engage in research to become more developed. In other words, they are still developing.

THE LEVEL OF DEVELOPMENT.

a) Physical Development: In order to judge the level of physical development of any nation, one needs to assess the positive function and utility of natural and man-made resources in relation to the citizens of the nation. Natural resources of mineral, water, oil, soil, air, plant and animals are provided by nature. The mere presence of natural resources within a nation is not development.

Man-made resources result from the application of technology. If natural resources are harnessed to provide food, shelter, health and protection for the citizens; that is, if resources are employed to produce positive consequences for citizens and for other peoples with whom citizens interact, there is physical development. This is what economists term economic growth. Technology, the utilisation of knowledge for useful ends, is an important requirement for man-made resources. Technology is required for the production of (1) communication equipment (e.g. television, telephone, radio), (2) transportation equipment (e.g. cars, planes, trains), (3) military hardware (e.g. bombs, guns, missiles), and all other goods and services for public consumption. Let us examine carefully how the use of technology may be directed against the process of development.

---

*Denis Austin, "Prospero's Island" in Scott W. Thompson (ed.) *The Third World: Premises of U.S. Policy. p. 55.*

Communication Equipment: If the modes of communication are controlled by high-level national or private authorities so that the ordinary person in the street cannot express his or her views, this has a negative effect on the intellectual development of the nation. For example, the typical images of Africa on American television reveal only the exotic and negative aspects: starving people, huts and naked people. Their usual show, *Tarzan,* creates lasting images in the minds of Americans of how Africans in the jungles interact with wild animals. Thus, one easily comes across seemingly well-educated Americans asking very bizarre questions, such as, "Are there tarred roads in Africa? Is there electricity in Africa?" On the other hand, images of America held by Africans are essentially of skyscrapers, cities with well-paved streets. Many photographs by westerners carefully avoid the homeless people in the streets, the numerous dilapidated buildings of *the ghettoes* and the violent criminal activities on human life with the use of guns. Moreover, the printing presses have turned out textbooks adopted in educational systems, which perpetuate neocolonialism, imperialism and racism. Because truth and reality are distorted we cannot claim that this is development.

Military Equipment: Technological progress has produced weapons used for both international aggression and internal subjugation of some portion of the national population. With technological progress, blacks were uprooted from Africa and ensalved in the western world for more than four hundrd years; thousands of people were decimated by bombs in Japan at Hiroshima and Nagasaki. Some individuals and nations ("the merchants of the devil", as they are sometimes called) increase their income by specializing in the production of weapons and inciting conflicts within weaker nations in order to sell weapons. Today, the world is under the scare of nuclear extermination. These examples only show how military technology has not served a useful purpose to mankind, yet many would say, this is development.

Many countries — Korea, Brazil, Mexico, Kuwait and Indonesia, are actually more advanced in many aspects of economic and social life than some European countries, yet they are classified as "third world" nations because they do not form part of the western world. At best, western writers would make subtle subgroups within the third world for these countries that have made significant progress.

b) Cultural Development: By culture we mean the body of all non-genetic information passed from generation to generation. Cultural development is another important component of development. The culture of a society is usually reflected in its values, norms and traditions. The economic and political institutions of any human society are structured to enhance the way of life of that society. Thus, two institutions are important here: the economic institutions and the political institutions.

Economic Institutions: They embody those cultural values placed on the creation, governing, and distribution of goods, whether in a traditional (subsistence), market or planned economy. All these three types of economy involve participation and decision-making in the production of goods and services. In a traditional economy everyone is involved in decision-making; in the market economy both consumers and producers, and in the planned economy a few people manage the economic institutions in the interest of the whole society. Two types of goods are produced in any society, "fun items" necessary for pleasure — cars, television sets, etc.; and "life-sustaining items" necessary for basic human survival — housing, food, water supply, electricity, etc. In many highly industrialised nations production is geared essentially towards "fun items" while "life-sustaining items" are controlled by a few corporations and supplied at exhorbitant prices.

In some instances, where "life-sustaining items" are even in large supply, their distribution is grossly unequal and marked by discrimination. These large inequalities become even more acute, for example, in South Africa and in the United States of America, where racial background are taken into account. Equally repugnant in the United States is the fact that females with similar academic qualifications as males are paid differently, usually lower than men in the same jobs and occupations. In Africa, the Republic of South Africa, with its apartheid system, has always been classified by Americans and Europeans as one of the most developed African countries even though it is the most underdeveloped in human terms. If any nation discriminates against a portion of its population or does not create conditions that improve the well-being of all or most of its people, it is uncivilized and underdeveloped.

The political institutions: This is another cultural component in any appraisal of national development. In most Euro-American discourses on political institutions, nations that are led by those not elected by the people and more specifically through a multi-party system, are characterised as authoritarian and such leaders are variously termed absolute monarchs, dictators or tyrants.

the way in which these types of governments conduct their decisions is traditionally termed illegitimate. Marx Weber, the German Sociologist, pointed out that there are more than one way by which legitimate leaders emerge. As a matter of fact, many societies in the world have, over the centuries, selected and continue to select their legitimate leaders not necessarily through mass multiparty elections. Churches, organizations, villages and many nations have always selected their leaders in a number of ways and their authority has always been accepted by the subjects.

The quality of any government or any leader should not be solely judged in terms of election criteria but in terms of the usefulness of that government and the dedication of the leader to his people. Cameroon, and many other newly independent nations, have had very popular leaders who have proved dedicated and responsive to the people's problems. Many industrialized nations have governments that are mainly concerned with rewarding only that portion of the population that brought them to power while neglecting those who did not elect them. Such governments that are only concerned with portions of the population cannot be termed "developed". Governments of all systems have responded differently to their people. The quality of a government depends essentially on how it responds to its people's needs. Any government that comes to power through an approved procedure by the citizens and is of good service to its people is "developed." On the other hand, any government that seeks to be of service to only a portion of the population is "underdeveloped."

c) Personal Development: Personal development refers to the inner feeling — increased skill and capacity, greater freedom, self-discipline, self-esteem and a sense of responsibility. As mentioned above, this comprises the psychological direction of individuals. It is really doubtful if a rise in income alone or material wealth equally increases human happiness and one's self-esteem. Many would agree, however, that a person's contentment with his life depends essentially on his sure sense that he is a valuable being, that his life has some worth and that he belongs to a community of human beings who care about him. If he knows that there is no community on which to depend for recognition and respect his life would appear to be useless. Self-esteem is an important component of good life. Self-esteem — a sense of worth and self respect, of not being used as a tool by others for their own ends — is necessary for good life and people of all societies need it. They may term it dignity, honour, recognition, but it refers to the same thing.

In many homes in the western world (or the first world) where television has replaced human companions, where the telephone enables people to transact business at a distance with little or no human contact, and where hand guns are frequently used to take away human life at random, people have tended to be individualistic and confined to themselves. The American economist, Everett Hagen, has observed his country and asked the critical question, "Is economic development worthwhile?" After examining the unhappy state of many Americans and their dependence on sleeping pills, pills to relieve headaches, potions to relieve stomach aches, tranquilizers and anti-depressant medications, he concludes that material development does not necessarily imply happiness. He adds that "almost one half of American adults must relieve the anxiety of their lives or mitigate their state of depression by taking tranquilizers, sporifics or anti-depressant medications..." Any society where individuals have very little self-esteem, where killing is so rampant and the rate of suicide so high, is not civilized and cannot, therefore, be termed "developed".

CONCLUSION

This chapter has been included to re-evaluate the narrow traditional view of development which is often based solely on technology and an increase in man-made materials without relating such development to the service of citizens. Increased man-made materials are not development if they are not beneficial to the individuals who make up that society. Economic development or economic growth is only one aspect of the broad concept of development. Development, in its full sense, considers the cultural, economic (or physical), and personal aspects. Unfortunately, because of the significance western nations attach to material values, worthiness or the term "developed" has been conferred only on those countries which possess economic and technological power. Other concepts associated with development but not always included are: the level of industrialization, the state of technology, the abundance and utilization of natural resources, the character of the people, and political tradition.

It should also be noted that economic well-being and happiness appear to be largely a relative matter. There is little doubt that the rich in any given society are generally well off and more respected than the poor, but there seems to be no clear connections between happiness and material well being when one makes cross-national comparisons. For example, the Republic of South Africa is one of the richest countries in Africa but the level of happiness there is certainly lower than in most African countries. In his article, *The Reality of Underdevelopment,* Denis Goulet, the American Philosopher/ Political scientist, talks of how the rich nations discover that they too are underdeveloped. He states:

> Large numbers of Peace Corps volunteers, upon returning to the United States, have examined their own society with critical new eyes. After their experience in Africa, Asia or Latin America, they can compare the treatment accorded old people in the United States with that dispensed to the aged in societies where the extended family provides satisfying roles for grandparents. Not surprising, many judge treatment of the aged in the US to be cruel: "like throwing obsolete people on the junk heap".*

---

* Denis Goulet, "The reality of underdevelopment" in Michael P. Todaro, *The Struggle for economic Development,* p. 8.

REVISION QUESTIONS

1. Write briefly on the following:
   a) Early Christian Missionaries in Cameroon.
   b) The partition of German Kamerun.
   c) Ruben Um Nyobe, the nationalist leader of Cameroon.
   d) Ahmadou Ahidjo — his achievements and failures.
   e) The Plebiscites of February 11, 1961 and reunification.

2. Cameroon has been one of the most stable and most prosperous countries of Africa.
   a) Explain how this came about.
   b) In your opinion, do you expect this to continue for the next decade? Give reasons.

3. a) With the aid of a sketch map of Cameroon, show the position and extent of the following:
      i) The coastal lowlands
      ii) The Southern Plateau
      iii) The Adamawa Plateau
      iv) The western highlands
      v) The Northern lowlands
   b) Describe briefly the characteristic features of the surface area of any two of these regions.

4. Write an account of volcanic activity in Cameroon.

5. Distinguish the main rock types found in Cameroon.

6. Write an account of the climate of Cameroon showing clearly the influence of latitudinal extent, relief, distance from the sea (continentality), and the wind system.

7. Describe and account for the differences in the total rainfall and seasonal distribution on the coastal lowlands and the northern lowlands of Cameroon.

8. Compare and contrast the climate of Yaounde with that of Douala.

9. With the aid of a sketch map, describe and account for the distribution of vegetation types in Cameroon.

10. a) What is a Soil Profile?
    b) Explain briefly how the following factors influence soil formation:
       i) Parent material
       ii) Climate
       iii) Plants and animals
       iv) Topography (slope)
       v) Time

11. a) With reference to specific examples, discuss the uses that are made of the principal rivers of Cameroon.
    b) What factors limit the extent to which they can be used.

12. Write a short account of the Sanaga River, and explain its importance to Cameroon.

13. Name the drainage basins of Cameroon and briefly describe one of them.

14. With reference to specific examples, classify the lakes of Cameroon according to how they were created.

15. a) On a sketch map of Cameroon, mark the approximate areas occupied by the following peoples:
       i) the Douala
       ii) the Pygmies
       iii) the Bamum
       iv) the Tupuri
       v) the Bamileke
       vi) the Shuwa Arabs
    b) Select two of these peoples and show how their lives are adapted to the geographical conditions in which they live.
    c) Write a short history of either the Bamum or the Fulbe.

16. With reference to Cameroon, write briefly on any three of the following:
    a) Birth rate
    b) Fertility rate
    c) Death rate
    d) Population pyramid
    e) Sex Ratio
    f) Infant mortality
    g) Life expectancy

17. Describe and account for the distribution of population in Cameroon.

18. Choose one area of high population concentration and one other of sparse population and account for the population density.

19. a) Show on a sketch map the main sources and destinations of migrants in Cameroon.
b) Write a brief explanation of the direction of the movements of these migrants.

20. Population explosion is a major handicap to economic growth in Cameroon. Explain why you agree or disagree with this statement.

21. Discuss the Theory of Demographic Transition.

22. a) Account for the growth of urban centres in Cameroon.
b) What do you understand by the terms:
i) Urban Primacy
ii) Urban Hierarchy
c) Discuss five urban problems in Cameroon.

23. Choose two cities in Cameroon which you have studied: State clearly how similar and different they are in their site, situation and functions.

24. Discuss the following:
a) i) The concentric-zone model
ii) The Sector model
iii) The multiple-Nuclei Model
b) To what extent are these models applicable to cities (or towns) in Cameroon?

25. Discuss briefly, Walter Christaller's Central Place Theory.

26. a) What are the main features of traditional agriculture in Cameroon?
b) Distinguish between "subsistence crop farming" and "cash crop farming".
c) What are the main problems of agriculture in Cameroon?

27. Agriculture, and not oil, constitutes the backbone of the nation's economy. Discuss.

28. a) Name and locate the area of production of three export crops in Cameroon.
b) For any one of these crops, describe the physical conditions under which it is grown and the method of cultivation.

29. Explain where in Cameroon, and under what conditions, cocoa is cultivated.

30. Natural and human factors influence animal raising in Cameroon. Explain.

31. Outline the major problems of industrialization in Cameroon.

32. With reference to Cameron, discuss two of the following:
a) The growth of the tourist industry
b) Problems of road and rail transport
c) UDEAC as a trade bloc

33. The tourist industry can make significant contributions to the economic, social and cultural development of the peoples of Cameroon. Critically examine the validity of this statement.

34. Timber exploitation is a major economic activity in Cameroon.
a) On a map of Cameroon indicate the major areas of timber extraction and one port of timber export.
b) Discuss the development of wood-based industries in any one part of Cameroon.
c) Describe the methods adopted to ensure the continued supply of timber.

35. With reference to Cameroon, write on one of the following:
a) subsistence fishing.
b) industrial fishing.

36. Under the headings relief, climate, soils, drainage and economic activities, write a geographical description of either the Western Highland or the Northern Lowland regions of Cameroon.

37. Write briefly on the following:
a) Planned Liberalism.
b) Balanced Development.
c) Self-Reliant Development.

38. a) What is "development?"
b) Discuss the different aspects of "development".
c) The United States of America, Cameroon, France, The Republic of South Africa and Chad can all be termed "developing countries" or "underdeveloped countries". Explain.

GLOSSARY

aa: A type of lava-flow which solidifies forming a rough surface — the roughness resulting from the violent escape of gases from within the lava; e.g. the aa of Ekona.
Active Population: Those who work and receive salaries or wages for the work they do.
Age-Structure: The distribution of different age categories in a population.
Air mass: A large body of air that is characterized by a sameness of temperature and humidity.
Alluvium: Loose sediment deposited by a stream.
Anticyclone: An area of high pressure from which air spirals in all directions. Anticyclones circulate in a clockwise direction in the northern hemisphere and in a counterclockwise direction in the southern hemisphere.
Balance of Payment: The difference between a nation's imports and exports over a given period. Consists of goods, services, gold, capital.
Basalt: A dark-coloured, fine-grained igneous rock, formed by the solidification of lava.
Basement Complex: A mass of very ancient igneous and metamorphic rocks, usually (but not necessarily) pre-cambrian in age and very complex in structure, which underlies the oldest identifiable rocks in any region.
Bauxite: End product of the process of laterization; residue of aluminium oxides and hydroxides; principal ore of aluminium.
Beach: accumulation of loose sediment maintained by depositional processes; zone of transition between land and sea.
Bedrock: in-place rock underlying a region, as opposed to transported material.
Birth Rate: The number of births per 1,000 members of a given population in a given year.
Bush Fallowing/Shifting-field Cultivation: Bush fallowing is a farming system where the farmer abandons a piece of farmland to recover some of its fertility for a few years so that it can be used after some period for more crops. It is sometimes called *shifting-field cultivation.* In the strictest sense, in shifting-field cultivation the farmer at the moment of abandonment does not consider any possible return to the piece of land.
Caldera: An extra large volcanic crater, or a large depression which remains after the top of a volcanic cone has either erupted and blown away or been enlarged by subsidence.
Capitalism: An economic system in which the means and resources of production are privately owned and a major incentive for economic operations is profit.
Col: The low passage zone on a range of mountains.
Composite cone: A volcano composed of both lava flows and pyroclastic material.
Concentric-zone Model: The model developed by Ernest Burgess, who envisaged urban growth as a series of circles radiating from the central business district.
Confluence: The meeting point of two or more rivers.
Continental climate: Climate that is associated with the interior of a continent or vast landmass; i.e. far away from the moderating influence of the sea. *Continentality* therefore, refers to farness from the sea. Continental climate is characterized by very large temperature ranges.
Coriolis Force: This is the deflecting force of the earth's rotation on all free-moving objects. Except at the Equator, winds and all other moving objects, no matter what their direction may be, are deflected to their own right in the northern hemisphere and to their left in the southern hemisphere.
Crater: The rounded funnel-shaped hollow at the summit of a volcano.
Evapotranspiration: Transfer of water vapour into atmosphere by the combined action of evaporation from soil and transpiration from vegetation.
Ethnocentrism: The belief that one's culture represents the only true and good way, and the subsequent tendency to judge other cultures by these standards. (This is to be contrasted with Cultural Relativism —which is an attitude of objectivity toward other cultures, involving the attempt to understand each culture in terms of its own values, norms and standards).
Deficit Balance: A shortage of money, when expenditures are greater than income, or when liabilities are greater than assets.
Distributary: One of the numerous branches into which a river divides on its delta or on an alluvial fan.
Drainage Basin: The total surface area drained by a network of streams — the streams usually converging at one outlet.
Ferrallitc Soils: Soils that are rich in iron and silica but relatively poor in aluminium oxides.
Flood plain (zone): The flat, low-lying portion of a stream valley subject to periodic inundation.

Fumeroles: A vent in the ground in a volcanic region, emitting steam and other gases.
Geysers: A spring that throws up hot water with explosive force from time to time. Geysers are commonly found in volcanic areas. Most often, the water shoots up from the geysers in great columns, cloudy with steam.
Harmattan: A dry N.E. wind blowing from the Sahara towards the West African coast.
Headward Erosion: The cutting back upstream of a valley above its original source by rain wash, gullying and spring-sapping.
Hydro-electricity: (water power). Electricity produced by water power generation — usually from water stored in dams or natural fall of water.
Hydrography: The science of the water bodies.
Imperialism: The political and economic control of one society by another.
Inflation: A steady upward movement in the level of prices.
Inselberg: An island mountain; an isolated hill on a plain or plateau region usually caused by weathering.
Interpluvial Period: The periods between the pluvial periods characterized by little or no precipitation. (See Pluvial Period).
Kaolin: A whitish type of clay produced by a complex chemical breakdown of aluminium silicates.
Lagoon: A shallow body of water, particularly one separated from the open sea by islands or sand bars.
Land Breeze: A local wind blowing from land toward a body of water during the night in coastal areas.
Laterite: A brown, highly leached soil type found in the tropics that is rich in oxides of iron and aluminium.
Lapilli: Small fragmentary materials ejected from a volcanic mountain into the air during an eruption.
Latex: The milky fluid which comes out from rubber trees after tapping.
Lava: Molten or fluid rock that issues from a volcano onto the surface of the earth.
Leaching: The removal of materials in solution by the passing of water through the soil. Soluble material is removed from the upper layers of the soil and carried downward to the lower layers by percolating water.
Magma: Molten or fluid rock material beneath the surface of the earth.
Mandate Territory: An area established under Article 22 of the League of Nations Convenant, whereby the German Colonies were placed under the guardianship and administration of various Allied powers, appointed by and responsible to the League.
Meridian: A great circle drawn from any point on the earth that passes through the north and south poles.
Metropolis: A very large city having many smaller cities and settlements that are tied to it.
Monoculture: Farming in which the cultivation of a single crop predominates.
Multiple-nuclei Theory (model): A theory that modern cities are typically organized around a number of specialized areas, each of which influences the character and development of the surrounding neighbourhood.
Nationalism: A consciousness of shared identity among membes of a politically distinct territory.
Natural Increase (of population): The surplus of births over births in a population in a given time period.
Pedologist: A specialist in the study of soils.
Peneplain: An almost level plain which results from a long period of denudation.
Piedmont: Lying or formed at the base of mountains.
pH Scale: A common measure of the degree of acidity or alkalinity of a solution; it is a logarithmic scale ranging from 0 to 14. A value of 7 denotes a neutral solution, values below 7 indicate greater acidity and numbers above 7 indicate greater alkalinity.
Plateau: An elevated flat area, sometimes called a "table land".
Plebiscite: A vote by which the people of an entire country or district express an opinion for or against a proposal especially on a choice of government or ruler.
Pluvial Period: Periods of heavy precipitation in geologic time.
Radial Drainage: A system of streams flowing to all directions away from a central elevated area, usually a plateau or mountain.
Rate of Natural Increase: The rate at which a population is increasing in a given year due to surplus of births over deaths, expressed as a percentage of the base population. The rate does not include the effects of immigration and emigration.
Reafforestation, Reforestation: The planting of trees over an area previously forested but which has been destroyed by bush fires or cutting down.
Regolith: The layer of rock and mineral fragments that nearly everywhere covers the earth's land surface.
Relative humidity: The ratio of the amount of water vapour present in a quantity of air compared to the amount that could be held by the same air if it were saturated, usually expressed as a percentage.

Right Bank (of a stream): That bank of a stream which is to one's right as one faces downstream. The left bank is then to the left as one faces downstream.
Rock: A consolidated mixture of minerals.
Scoria: A mass of volcanic rock full of holes due to the expansion of imprisoned gases.
Sea Breeze: A local wind blowing from the sea during the day in coastal areas.
Smog: Originally used to describe a combination of smoke and fog; today, the term is used as a synonym for general air pollution (especially in urban centres).
Soil: A combination of mineral and organic matter, water and air: that portion of the regolith that supports plant growth.
Soil horizon: A layer of soil that has identifiable characteristics produced by chemical weathering and other soil-forming processes.
Soil Profile: A vertical section through a soil showing its succession of horizons and the underlying parent material. Mature soils are characterized by a series of horizontal layers called horizons, which comprise the soil profile.
Spit: A long narrow accumulation of sand and shingle stretching from the land into the sea.
Sugar Loaf: A rounded conical hill.
Transhumance: The seasonal movement of men and animals to fresh pastures.
Trust Territory (Trusteeship): An area administered by another country, appointed by and responsible to the United Nations Organization, with the intention of enabling that area to finally attain political independence.
Volcano: A hill or mountain built-up around the vent by the accumulation of the materials which come out of the vent.
Vulcanism or Vulcanicity: These terms are used to express all the phenomena associated with the movement of molten material, rock fragments, and gases from the interior to the surface of the earth; volcanic action.
Watershed (or Divide): The highland area from which rivers take their rise and flow to different drainage basins.
Weather: The state of the atmosphere at any given time and place. The conditions of temperature, pressure, wind, humidity, cloudiness and precipitation over a short period of time — usually several hours or days.
Weathering: The process by which rocks are decomposed or disintegrated as a result of exposure at or near the earth's surface.

BIBLIOGRAPHY

Alexander, Boyd, "Lieut. Boyd Alexander's Expedition in West Africa" *Geographical Journal,* 34, 1, July 1909, PP. 51-55.

Alexandre, Pierre. *Langues et Langue en Afrique Noire.* Paris: Payot, 1967.

______, "Proto-histoire du groupe beti-bulu-fang; essai de synthese provisoire," *Cahiers d'Etudes Africaines* (Paris), V. No. 20, 1965, 503-560.

______, "Rapport de mission ethnolinguistique au Sud Cameroun," *Cahiers d'Etudes Africaines* (Paris), II, No. 8, 1962, 630-634.

Alexandre, Pierre, and Binet, Jacques, *Le Groupe Dit Pahouin (Fang-Boulou-Beti).* (Monographies Ethnologiques Africaines Series). Paris: Presses Universitaires de France, 1958.

Alima, Jos-Blaise. "Cent peuples, une nation," *Jeune Afrique* (Paris), No. 623, December 16, 1972, 56-61.

______, *Les Chemins de l'unite. ABC Paris. 1977.*

Anene, J.C., *The International Boundaries of Nigeria 1885-1960: The Framework of an Emergent African Nation.* New York: Humanities Press, 1970.

______, "The Nigerian-Southern Cameroon Boundary: an Ethno-Political Analysis". *Journal of the Historical Society of Nigeria,* 2, 2 Dec. 1961, PP. 186-195.

Ardener, E., and Macrow, D.W. "Cameroon Mountain", *Nigeria,* 62, 1959. PP. 230-245.

Ardener, Edwin. *Coastal Bantu of the Cameroons.* London: International African Institute, 1956.

______, "Kamerun Idea," (Pt..I), *West Africa,* No. 2,147, June 7, 1958, 533.

______, "Kamerun Idea," (Pt. II), *West Africa,* No. 2,148, June 14, 1958, 559.

______, "The Nature of the Reunification of Cameroon." Pages 285-337 in A. Hazlewood (ed.), *African Integration and Disintegration.* London: Oxford University Press, 1967.

______, "The Political History of Cameroon," *World Today* (London), XVIII, No. 8, August 1962, 341-350.

Ardener, Shirley, and Warmington, Alan. *Plantation and Village in the Cameroons.* London: Oxford University Press, 1960.

Atlas de la Republique Unie du Cameroun, ed. J.A. Paris, 1979.

Azevedo, Mario (ed.), *Cameroon and Its National Character.* EUGA. Jackson, Mississippi 1984.

Bafoussam: A Mushroom Town: Imprimerie Nationale, Yaounde.

Bertoua: A Crossroad Town. SOPECAM, Yaounde.

Bederman, Sanford H., "Mount Cameroon, West Africa's Active Volcano." *Nigerian Geographical Journal,* 9, 2 Dec. 1966. PP. 115-128.

______, "Plantation Agriculture in Victoria Division, West Cameroon: An Historical Introduction," *Geography* (London), LI, No. 4, 1966, 349-360.

Billard, Pierre. *Le Cameroun Federal.* 2 Vols. Lyons: Imprimerie de Beaux Arts, 1968.

Carter, Jean. "The Fulani in Bamenda." *Journal of Tropical Geography* (Singapore), December 1967, PP. 1-7.

Chilver, E.M. "Native Administration in the West Central Cameroons, 1902-1954," Pages 113-118 in K. Robinson and F. Madden, *Essays in Imperial Government.* London: Oxford University Press, 1963.

Clarke, John I., "The Trans-Cameroon Railway". *Geography,* 51, 1, 230. Jan. 1966. PP. 55-58.

______, "The Transcameroonian." *Geographical Magazine,* 40, 15, July 1968. PP. 1268-1277.

Clarke, J.I. "Population distribution and dynamics in Cameroon" in Zelinsky, W., Kosinski, L.A. and Prothero, R.M. (eds.), *Geography in a crowding world.,* London, OUP. 1970. PP. 348-62.

Clignet, Remi and Jordan, Frank. "Urbanization and Social Differentiation in Africa: A Comparative Analysis of the Ecological Structures of Douala and Yaounde, in *Cahiers d'Etudes Africaines,* 11, 1971. PP. 261-297.

Colin, Elico, "La Capture Imminente du Logone et du Chari par la Benoue." *Atlas de la Geographie,* 35, 196. July 1926. PP. 383-384.

Comice Agro-Pastoral de Bertoua, du 5 au 8 Fevrier 1981. Yaounde, 1981.

Criaud, J. *Geographie du Cameroun.* Edition Saint Paul Moulineaux, 1976.

Dark, Philip. "Notes on the Eton of the Southern French Cameroons," *Man* (London), LVI, No. 132, September 1956, 124-125.

Debel, Anne. *Cameroon Today,* editions, J.A. Tours; Imprimerie-reliure Maison Mame, 1977.

Delancey, Mark W., "Cameroon's Foreign Relations," in Schatzberg, M.G. and Zartman, I.W. (eds.) *The Political Economy of Cameroon.* New York: Praeger Publishers, 1986. PP. 189-217.

Deloaroziere, R. "Cameroun, inventaire ethnique et linguistique du Cameroun sous mandat francais," *Le Journal de la Societe des Africanistes* (Paris), IV, No. 2, 1954, 203-208.

Douala: An Octopus-like Town. SOPECAM, Yaounde, 1981.

Duckworth, E.H., "Crater lakes in Bamenda Province" *Nigeria,* 37, 1951. PP. 65-79.

Dugast, Idelette. Inventaire ethnique du Sud-Cameroun. (Memoires de l'IFAN, Populations, No. 1). Douala: IFAN Centre du Cameroun, 1949.

Dugast, I., *Inventaire Ethnique du Sud-Cameroun,* IFAN, Yaounde, 1949.

Evans, F., *et al.* "Volcanic Eruptions on the Cameroons Mountain." *Geographical Journal,* 60, 2, August 1922. PP. 135-141.

Food and Agriculture Organization. *Soils and Ecology of West Cameroon.* (ETAP Report No. 52083-65-CMR.) Rome: FAO, 1965.

Forestry Ordinance No. 73/18 of 22nd May 1973 and its Decree of Implementation No. 74/357 of 17th April 1974.

Gardinier, David E. *Cameroon: United Nations Challenge to French Policy.* London: Oxford University Press, 1963.

Gifford, Prosser, and Louis, William Roger (eds.). *Britain and Germany in Africa.* New Haven: Yale University Press, 1967.

Gleave, B. Michael, "The West Cameroon Meat Scheme" in *Geography,* vol. 50, 1965. PP. 166-168.

Gouellain, Rene, "Douala: Formation et Developpement de la ville pendant la Colonisation," in *Cahiers d'Etudes Africaines,* 13, 1973. PP. 442-468.

Gwan, Emmanuel, "Migrational changes in West Cameroon", in Clarke, J.I., and Kosinski, L.A. (eds.), *Redistribution of Population in Africa.* Heinemann, London. 1982. PP. 124-132.

Guillaume, G.M.D. *Notes on the Cameroon Mountain.* Buea, Cameroon: Government Press 1966.

Haig. E.F.G., "The Cameroons Mountain: A General Conspectus". *The Nigerian Field,* 6, 3, July 1937, PP. 118-120; Oct. 1937, PP. 172-182.

Information Bulletin: A Publication of the Provincial Press and Information Centre, Buea. No. 6/81. 1981.

I.P.A.R., Geographie du Cameroun. CEPMAE, Yaounde. 1976.

Jennings, J.H., "The Eruption of Mount Cameroon. February-March 1959." *Geography* 44, 3, 1959. PP. 209-208.

Johnson, Willard R. *The Cameroon Federation: Political Integration in a Fragmentary Society.* Princeton: Princeton University Press, 1970.

Joseph, A. Richard, "Ruben um Nyobe and the 'Kamerun' Rebellion" in *African Affairs,* vol. 73, No. 293, Oct. 1974. PP. 428-448.

______, *Radical Nationalism in Cameroun: Social origins of the U.P.C. Rebellion,* Oxford: Clarendon Press. 1977.

Journaux, Bala Mbanga, Parisse. *Geographie du Cameroun.* Collection notre milieu, Caen, 1968.

Journaux, Bala Mbarga, Parisse, *Le Cameroun, L'Afrique, Le Monde.* Collection notre milieu, Caen, 1968.

Kaberry, Phyllis M. "Retainers and Royal Households in the Cameroons Grassfields." *Cahiers d'Etudes Africaines* (Paris), II, No. 10, October 1962, 282-298.

______, *Women of the Grassfields: A Study of the Economic Position of Women in Bamenda, British Cameroons.* (No. 14) London: H.M.S.O. Colonial Research Publications, 1952.

Kale, P.M. *Political Evolution in the Cameroons.* Buea, Cameroon: Government Printer, 1967.

Leeming, A.J., "A Historical Sketch of Victoria, British Cameroons," in *The Nigerian Field, Vol. XV, No. 4,* October 1950. PP. 184-189; *Vol. XVI, No. 1,* January 1951. PP. 37-45.

Lembezat, Bertrand. *Les Populations Paiennes du Nord-Cameroun.* Paris: Presses Universitaires de France, 1961.

Le Vine, Victor T., *The Cameroon Federal Republic.* (2d ed.) Ithaca: Cornell University Press, 1971.

______, *The Cameroons from Mandate to Independence.* Berkeley: University of California Press, 1964.

______, "Introduction to the Study of Cameroon: A Bibliographic Essay, *"Journal of Cameroon Affairs,* I, No. 2, March 1972, 10.

______, and Heri, M'Ballah. "Federal Republic of Cameroon," Pages 519-532 in Helen Kitchen (ed.), *The Educated African.* New York: Praeger, 1962.

Le Vine, Victor T. *The Cameroon Federal Republic* (2nd ed.) Ithaca: Cornell University Press, 1971.

______, *The Cameroons from Mandate to Independence.* Berkeley: University of California Press, 1964.

Loung, J.F. *Le Cameroun.* Hatier, Paris, 1973.

Morel, Y., *Tableaux Economiques du Cameroun.* Douala, 1978.

Mveng, Engelbert. Histoire du Cameroun. Paris: Presence Africaine, 1963.
Mytelka, L. Krieger, "Common Market with some Uncommon Problems", In *Africa Report.* October 1970. PP. 14-17.
National Year-book 1980. SOPECAM Yaounde, 1980.
Ndiva Kofele-Kale. *An African Experiment in Nation Building: The Bilingual Cameroon Republic Since Reunification.* Boulder: Westview Press, Inc., 1980.
Ndongmo, Jean-Louis. "The competitive migration fields of Douala and Yaounde, Cameroon", in Clarke, J.I. and Kosinski, L.A. (eds.) *Redistribution of population in Africa,* Heinemann, London. 1982. PP. 133-137.
Ngwa J.A. *A New Geography of Cameroon.* Longman Group Ltd. London: 1978.
North Province. SOPECAM Yaounde. 1980.
O'Grady, Ron, *Tourism in the Third World.* Orbis Books, MaryKnoll, N.Y. 1982.
Okediji, Peter A., "Humanistic Understanding of Development" in *African Urban Studies,* No. 4, Spring 1979. Michigan. PP. 7-13.
Podlewski, Andre, "Cameroon," in Caldwell, John C. (ed.) *Population Growth and Socioeconomic Change in West Africa.,* New York, Columbia University Press. 1975. PP. 543-564.
Podlewski, Andre Michel. *La dynamique des principales populations du Nord Cameroun entre Benoue et Lac Tschad* (Office de la Recherche Scientifique et Technique Outre-Mer, Cahiers: Sciences Humaines, III, 4). Paris: ORSTOM, 1967.
Ritzenthaler, Robert E., and Rîtzenthaler, Pat. *Cameroons Village: An Ethnography of the Bafut.* Milwaukee: Milwaukee Public Museum, 1962.
Rodney, Walter, *How Europe Underdeveloped Africa.* Howard University Press, Washington, D.C. 1982.
Safari in Cameroon. Imprimerie Chirat. Paris, 1979.
Science and Technology Review: Social Sciences Series; No. 1-2, Yaounde, D.G.R.S.T. January-April 1983.
Seino, M., *The History of Cameroon.* Bamenda, 1981.
South-West Province. Minfoc, Yaounde
Tandap. Lucas. "Urbanization and Population Redistribution Trends in Cameroon," in Clarke, J.I. and Kosinski, L.A., (eds.) *Redistribution of Population in Africa.* Heinemann, London. 1982. PP. 138-145.
"The Far-Flung Fulani," *Nigeria Magazine* (Lagos), LXXV, No. 75, December 1962.
Thomas, Michael F. "The Bamenda Highlands," *Geographical Magazine* (London), XXXIV, No. 4 August 1966, 282-295.
Tourism in Cameroon. Paris, 1977.
Welch, C.E., *Dream of Unity: Pan-Africanism and Political Unification in West Africa,* Cornell University Press, Ithaca 1960.

INDEX

A
aa, Lava 17, 193
Abapka 171
Abasakom 54
Abega, Martin 6
Achaf 54
Achain 54
Acquired Immune Deficiency Syndrome (AIDS) 127
Action Nationale 7
Adamawa Plateau 14, 16
Adventitive Eruption 20
Aeroflot (USSR) 144
Afo-A-Kom 127
Africa in Miniature 119
Afri-Stevedoring 136
Age Distribution 62
Ahidjo, Ahmadou 5, 6, 7, 8, 9, 11, 12, 13, 108, 161
Air Afrique 142, 144
Air Centrafrique 144
Air Gabon 144
Air Zaire 144
Ajung 54
Ake 54
Akosombo Dam 108
Akou 101
Akum 55
Akwa, King 3, 47
Akwa Yafe River 14
Alantika Mt. 16, 175
ALCAM 7
ALUBASSA 113, 114
ALUCAM 108, 113, 114, 115
Ambas Bay 162
American Cooperation Program 187
American Presbyterian Missionaries 186
Andegechung 55
Andokat 14
Andonalu 54
Annobon Island 16, 22
Anticyclone 24, 193
Apartheid 186
Arabica Coffee 92, 167
ARCAM 6
Ardener 47
Ardos 58
Armnesty International 9
Arrondissements 183
Asbestos 106
Asia Minor 58
Assale, Charles 7
Assimilation 5
Associated Territories 5
ATCAM 6
Atlantic Basin 38
Atmospheric Distillation 160
Avicenia 30
Awing 121, 165
Azonal Soils 36
Azores 24, 26

B
Babanki 53
Babinga 45, 50
Baboute 50
Badji 55
Bafang 51
Bafoussam, town 169
Bafut, Tribe 53, 55
Bagante 51
Bahr-el-Ghazal 43
Bajong 55
Bakilingi 18
Bakoko 47
Bakota 47
Bakundu 45
Bakweri Land Committee9
Balanced Development 117, 180, 181
Baleng Lake 51, 121, 165
Bali-Chamba 55
Bali-Gangsin 54
Bali-Gasho 54
Bali-Gham 54
Bali-Kontan 54
Bali-Kumbat 54
Bali-Muti 54
Bali-Nyonga 54, 55
Balong 45
Bambili 53
Bamboko 45
Bamboo 51
Bambutus 16, 38
Bamenda Town 170
Bamendjim Dam 42, 108, 131
Bamessi 53
Bamileke, Tribe, 51
Bamum Tribe 51, 53, 54
Bamvele 60
Bandjoun 51
Bane 50
Banglang Mt. 16
Banso 54
Bantu 45, 47, 55
Bassel Mission 55
Bassa 45, 47
Bassin Oku 54
Bastos Company 114, 149
BATA 113
Bautchi Plateau 43
Bawum 53
Baya 50
Beba-Befang 55
Bell, King Douala Manga 3, 5, 47
Benue Depression 16
Benue River 40, 41
Benue Wildlife Reserve 122, 123
Bertoua, town 156
Besonaku 54
Beti 50, 68
Beye'ele, 50
Birth Rate 60, 193
Bisiga 56
Bismark, Otto Von 3
Biya, Paul 13, 63
Blackpod 89
Bokassa 152
Bombs 18
Bonaberi 47
Bonaku 47
Bonanjo 47
Bonapriso 47
Bonebela 47
Bonga 129
Bongola River 40
Bororo (Mbororo) 57, 66, 101, 165, 172
Botanical Garden 163
Bouba-Ndjida Wildlife Reserve 121, 123
Boumba River 40
Boumnyebel 8
Brasseries du Cameroun 177
Brazzaville Treaty 152
British Caledonia 144
Bucharest 69
Buea, Town 163
Buffle Noir 123
Bugiri 55
Bulu 50, 68
Burgess, Ernest 79, 81, 82
Bush Fallowing 84
Buwe Bukari 53

C
Cadbury and Fry 89
CAFISTRO 131
Calabar, State 38
Cameroon Airlines 64, 144
Cameroon Democrats 12
Cameroon Development Corporation 9, 64, 68, 92, 95, 97, 99, 111, 116, 157, 159
Cameroon Electricity 108
Cameroons Indigenes' Party 12
Cameroon Labour Party 12
Cameroon Line 22
Cameroons Muslim Congress 12
Cameroon National Federation (CNF) 9
Cameroon National Union 11, 12, 13
Cameroon People's Democratic Movement 13
Cameroon Socialist Party 12
Cameroon Tobacco Company 95, 153, 155
Cameroon Tribune 147
Cameroon United Congress 11, 12
Cameroon Youth League 9
CAMI-TOYOTA 149
CAMOA 113
Campo Reserve 124
Campo River 40
CAMSUCO 87, 89, 96, 111, 116, 153, 154
Cape Limbo 106, 109, 115, 159, 160, 163
Capsid Pests 89
Caraboard 47, 163
Carp 132
Cassiterite 106
Catalytic Reformer 160
Catfish 132
CELLUCAM 114, 115, 120, 133, 136, 160, 161, 164
CENADEC 154, 167, 168
Central African Economic and Customs Union 152
Central Business District 81
Central Functions 82
Central Place Theory 77, 82
Centre Climatique 121
Centre for Studies and Experimentation in Agricultural Mechanisation 166
Centre Touristique 121
CERICAM 114
Cereal Board 179
Chad Basin 38, 40
Chad Lake 41, 42
Chad Plain 16
CHALUTCAM 131
Chari River 41
Chariot of the gods 2
Chirrapunji 26
Christaller, Walter 82
CIAC 114
CICAM 95, 113, 114, 177
CIMENCAM 113, 115, 177
Coastal Lowland Region 14, 157
Composite Cone 20, 193
Concentric Zone Model 79, 81, 193
Congo Basin 38, 40
Continental Platform 16
COOPAGRO 92
COPEMAR 131
Coriolis Force 26, 193
COTONEC 131
CRECAM 131
Credit of National Funds for Forestry and Pisicultural Development 138
Cretaceous 22
Cross River 38
Cuirasses 36
Custodian of Enemy Property 157

D
Daba 56
Death Rate 60
De Gaule, Charles 6
Degel 57
Delancey, Mark 186
Democrats Camerounais 7
Departements 75, 183
Derived Savanna 32
Development Bank of States of Central Africa 152
Dibamba River 40, 47
Dibongo 47
Dindi Kingdom 54
Distributaries 14, 38
Divisional Officer 183
Djerem River 38
Djottin 54
Douala Airport 144
Douala-Edea Reserve 124
Douala, Town 162
Douala, Tribe 45, 47
Doume 40
Dual Primacy 77
Duck-Bill Territory 4, 5
Dufoix, Georgina 70
Dykes 20

E
Ebele-Deido 47

Ebolowa 45
Edea, Town 164
Egwefi River 55
Ekona 18
Ekwu 54
Elf Serepca 106
Eluvial Zone 36
Emergent Nations 188
Emir 57
Enclosed Landscape 51
Endeley, E. M. L. 9, 10, 11
Equatorial Customs Union 152
Ethiopian Airlines 144
Ethnocentrism 193
Eton 50
Eucalyptus 121, 138
European Common Market 151
Evolutionaries 6
Evuzok 50
Ewale 47
Ewondo 50
Eyidi, Marcel 12

F
Fali 56, 66
Fang 50, 69
Faro Reserve 124
Federal Housing Administration 81
Fernando Po 2, 22
Ferrallitic Soils 36
Fertility Rate 60
Folepan 51
Fon 121, 185
FONADER 84
Foncha, John N. 10, 11
Fongo-Tongo 106
Fons 51, 127
Fonyonga II 54, 55
Forestry Ordinance 133
Fotsing, Raphael 9
Fotso II 169
French High Commissioner 6
French Union 5, 6, 177, 186
Fulani 50, 56, 58, 104
Fulbe (Foulbe) 44, 45, 55, 56, 57, 66
Fumeroles 17, 194
Fungom 54

G
Gainakos 58
Galega I 54
Garous Town 179
Garoua Airport 144
Gawolbe 54
Gbetnkom 53
Gendarmerie 9, 75
General Delegation for Tourism 119
Geysers 17, 194
Ghettoes 189, 190
Giant Cavendish 92
Gobir 57
Gotel Mts. 16
Goulet, Denis 192
Grand Cameroon 4
Grande Capitaine 123
Grassfield Bantu 50
Grassfields 33
Green Revolution 84
Greenwich Meridian 1
Gros Michel 92
Growth Rate 61
Guider 55
Guingue Mts 16
Guinea Savanna 32, 36
Guinea Type 26
Guiziga 55
Gulf Oil of Cameroon 106

H
Habib, Abba 10
Hagen, Everett 191
Half-orange 16
Hamites 58
Hanno, Carthaginian Navigator 2
Hardes 179
Hard-pan 36, 172
Harmattan 26, 194
Harris 81
Heterolis 132
HEVECAM 95, 96, 157, 159
Hewett 3
Higher School of Business 118
Hilton 126
Hiroshima 190
Holiday Inn 126
Hossere Vokre 175
Hoyt, Homer 81
Hydro-desulphurisation 160
Hydro-electric Power 108

I
Iberia 144
Ijunasung 54
Ikiliwindi 45
Illuvial Zone 36
Industrial Forest 136
Infant Mortality Rate 61
Infertility 60
Inselbergs 16
Institut de Recherches Zootechniques 105
INTELCAM 146
International Brain-drain 118
International Commission of Jurists 9
Integrated Priority Action Zone 155
Intertropical Convergence Zone 24
Intertropical Front 24, 26
Iroko 133
Itinelaa 54

J
Jantzen 3
Java 92
Jihad 57
Joseph, Richard 47
Jua, Augustine 11

K
Kainji Dam 108
Kalamaloue, Reserve 122
Kale, Paul M 9, 10, 12
Kalfou Reserve 124
Kamerun National Democratic Party 10, 11, 12
Kameroon National Congress 10
Kamerun People's Party 10, 12
Kamerun United Commoners' Party 12
Kamerun United Party 12
Kanen-Bornu 139
Kaolin 36, 194
Kaolinite 36
Kapsiki 22, 55, 56
Kienke River 40
Kimbaw 54
Kimbi Reserve 124
Kingue, Albert 9
Kirdi 56
Kofele-Kale, Ndiva 186
Kom 53, 54, 101, 127
Koromanjang 55
Kotokos 55, 66
Kounden 51, 104, 169
Kouotou 53
Koviferm 54
Kufom 54
Kuk 54
Kum 55
Kumba, town 163
Kung 54
Kupe Mt. 16, 19, 38
Kwakwa 3, 45

L
Lagdo Dam xi, 108, 178
Laikom 54
Laissez-passer 12
Lake Chad Basin Commssion 43, 186
Lamibe 73, 185
Lamidat 75
Lamido 57
Lancaster House 10
Lapilli 22
Lassin 54
Laterite 36
Law of Diminishing Returns 71
League of Nations 5
Levine, Victor T. 12
Life Expectancy 61
Limba 47
Limbe Town 162
Liquified Natural Gas 106
Lobe River 40
Logbada 106
Logone River 40, 41, 42
Lokele River 38
Lokoundje, River 14, 40
Lome Convention 186
Lyttelton 10

M
Macpherson, John 9, 10
Mafa 55, 56, 68
Mahogany 133
Maison du Cycle 113
Maka 45
Makombe River 40
Malthus, Thomas 71, 72
Mambila Mts. 16
Mambou 169
Mambu 55
Mandate Territories 5, 194
Manenguba Mts 16
Mangakely 166
Mangrove 30
Manguissa 50
Mankwi 53
Manyu River 38
Maquisards 50, 76
Margui-Kapsiki 66
Marienberg Mission 2
Marine Transgression 22
Maritime Cameroon Type 26
Maritime Fishing 129
Maroua Town 179
Marty, Cardinal 9
Massa 55, 56, 66
Mayi-Matip, Theodore 8, 12
Mayo Farda 40
Mayo Godi 40
Mayo Kebi 40, 41
Mayo Rey 40
Mbakaou 108, 174
Mbakaou Lake 42
Mbalangi 45
Mbam Massif 16
Mbam-Minkom 16
Mbam River 51, 53
Mbankomo 16, 37
Mbapit Mts. 16
Mbartoua 156
Mbaw Plain 165, 166
Mbebeli 53
Mbedi 47
Mbeikouo 53
Mbekum 55
Mbelewi 55
Mbere River 40
Mbiame 54
Mbida, Andre-Marie 5, 7, 8, 12
Mbida-Mbane 50
Mbile, Namaso N. 9, 10
Mbinnon 54
Mbongo 47
Mbo Plain Rice Development Authority 166
Mbouda 51
Mbum 50, 93
Mbwembwe 53
Mefou Lake 42
Mejang 54
Mejung 54
Meme River 38
Mendankwe 53
Mengap 53
Menemo 55
Meridien Hotel 126
Merrick, Joseph 2
Messamena 45
Metamorphic Rocks 22
MIDEBOM 154
MIDEPECAM 129
MIDERIM 111
MIDEVIV 154, 155
Mindif Mts. 16, 21
Minim-Martap 106, 174
Mitterand, Francois 70
Mixed Economy 180

Mme 54
Modelle 55
Moghamo 55
Moliwe Factory 115
Moloundou 45
Molybdenum 106
Momo River 55
Mondoleh Island 162
Mondoni Oil Mill 115
Monogamy 63
MONOPRIX 149
Moumie, Felix 8, 9
Monoum Lake 163
Monsoon 26
Mounjou 53
Mubako 55
Mukuru 55
Multiple-Nuclei Model 79, 81
Muna, Solomon Tandeng 11, 12
Mundang 56, 66
Mundum 53, 55
Mungaka 55
Mungo River 38, 47
Musgum 55, 56, 66
Muturu 103
Mvae 50
Mvele 50
Mvog-Nyenge 50

N
Nachtigal Falls 153
Nagasaki 190
Namme, L. N. 9
Namshi 103
Nandong 54
National Cooperative College, The 169
National Council of Nigeria and the Cameroons 10, 12
National Electricity Corporation 108
National Investment Corporation 116, 136
Nchang 54
Nchare 51, 53, 54
Nchare Yen 51
Ndama 103
Ndian, River 14, 38
Ndiba 156
Ndikinimiki Plateau 40
Ndobo 53
Ndokayo 172
Ndongmo, Albert 9
Ndotitichia 54
Necks 20, 22
Neo-Marxist 70
Neu-Kamerun 4, 5
Ngapna 53
Ngaoundal 106, 174
Ngaoundere 174
Ngie 55
Ngoko
Ngoko River 40
Ngonso 54
Ngoohouo 53
Ngosso 14
Ngouh 53
Ngouloure
Ngoungoure 53
Ngouopou 53
Ngovayang Chain 16
Ngwaw 55
Niger Basin 38, 40
Nigerian Airlines 144
Ninine, Jules 7
Njanga 129
Ndjamena 43
Njassep, Mathieu 9
Njimuyah 53
Njine, Michael 7
Njoumessi, Mathias 7
Njoya, Arouna 5, 7
Njoya, Ibrahim 53
Njoya, Njimuluh 53
Nkam River 38
Nkar 54
Nkogam 16
Nkolkosse 38
Nkongsamba, town 162
Nkor 54
Nkoupit 51
Nlonako 16
Nlonako Hill 40
NOBRA 111
Nomadism 101
Nonalignment 186, 187
Northern Lowland Region 14, 16
North West Cooperative Association 169
Noun River 42, 51
Novotel Hotel 126
Nsango 53
Nsaw 53, 54
Nso 54
Nsungli 53
Ntanka 54
Ntem River
Ntumazah, Ndeh 11
Ntumu 50
Nyobe, Ruben Um 6, 7, 8, 9
Nyong River 40
Nyongpasi 54
Nyos Lake 54, 121, 165

O
Office National Camerounais 120
Okala, Charles 5, 6, 12
Oku 54, 165
Omvang 50
One Kamerun 11
Onge River 19
Ongiekum 55
Open Landscape 51
Operation Green Sahel 32
Organisation Camerounaise de la Banane 92, 96, 157
Organisation of African Unity 186
Oshie 55
Ossa Lake 42
Othman Dan Fodio 57
Ouandie, Ernest 6, 9

P
Pahouin (Pahuin) 45, 47, 50
Paleo-Sudanese 68
Pallotine Fathers 2
PAMOL 95, 96, 97, 157, 159
Pangar and Djerem Reserve 124
Parent Rock 36
Paysans Independents 7
Peace Corps 192
PECH-CAM 131
People's Republic of China 187
Permanent Field Cultivation 101
Peske Bori 16, 21, 22
Peul 101, 103
Phonolite 19
Pitti
Pinyin 47
Planned Liberalism 180
PLASTICAM 114
Plebiscite xi, 11, 194
Plutonic Rocks 22
Pluvial Period 43
Pollution 78, 79
Polyandry 63
Polygamy 63
Polygyny 63
Polytechnique Higher School 118
Population Pyramids 62
Popologozom 21
Pouss 56
POWERCAM 108
Poyo 92
Prairie 19
Prefet 183
Pre, Roland 6, 7
Primate City 77
Priso, Soppo 5, 6, 7
Produce Marketing Organisation 84, 92, 149
Provinces 183
Pygmies 44, 45, 50

Q
Quasi-Federal Territory 10
Quinquina 96

R
Radical Nationalists 6
Rain Forest 30
Rankers 37
R & W King 149
Rassemblement du Peuple Camerounais 6
Reagan, Ronald 187
Referendum xi
REGIFERCAM 142
Regolith 34, 194
Reunification Road 146
Rhizophoria 30
Richards, Arthur 9
Rifum 51
Rio-del-Rey 3, 14, 38, 129, 131, 145, 157, 106, 129, 131, 145, 157
Rio-dos-Cameroes xi, 2, 40, 129
River Capture 41
River of Prawns xi, 2, 40, 129
Robusta Coffee 91, 92
Rodney, Walter 189
Ruga 58
Rumpi Mt. 38
Rumsiki 16, 20
Rumsiki Taurin 103
Rural Training Centre 166

S
Sabga 165
SACTA 96
SAFACAM 95, 96, 97, 115, 159
Sahel Savanna 32
Saker, Alfred 2, 162
Salanga 131
SANAGA 41, 95
Sanaga Cigars 155
Sanaga River 38
Sangha River 38, 40
Santa Isabel 16
Sao Tome 16, 22, 47
SAPCAM 111
Schmidt, Edouard 3
SCOA 149
SCORE 149
Scoria 18
Sector Model 79, 81
Seidou, El-haji 53
Self-Reliant Development 180, 181, 182
Semi-Bantu 45, 47, 50, 51, 55
SEMRY 111, 177, 179
Sex Distribution 62
Sex Ratio 62
Shehu 57
Sheraton Hotel 126
Shifting-Field Cultivation 84, 85, 101, 137, 193
Shuwa Arabs 55, 57, 58, 66
Smog 78, 195
SOCADA 149
SOCAME 113
SOCAPALM 157, 159
SOCAPAR 113
SOCARSEL 111
SOCATRAL 113
Societe Camerounaise des Tabacs 97
SOCOPELI 131
SODEBLE 87, 173
SODECAO 89, 154, 155
SODECOTON 96, 111, 179
SODEPA 169, 172
SODERIM 166
SOFIBEL 136, 154
SOFIMEC 106
Soil Profile 34
Sokoto 11
SONARA 115, 159, 160
SONEL 108
Song-Loulou xi, 108
SOPAC 131
SOPAME 95
Sorghum 87
SOSUCAM 87, 97, 111, 116, 153, 154
SOTUC 77
Southern Plateau Region 14, 16, 153, 157
St. Helena 24, 26
Subsidiary Cone 20
Sudanese 45, 50, 55, 56
Sudan Savanna 32
Sugar Loaf 16
Summital Plateau 19

T
Tabeu, Gabriel 9
Tainan V 166

Takala, Celestin 9
Tarzan 190
Taurin 101, 103
Tchabal Gangha 16
Tchabal Mbabo 16
Teak 133
Theory of Demographic Transition 72
Thormahlen 3
Thunen, Von 82
Tikar 53, 55, 68
Tikari 53
Tikar Tribes 51, 53
Tilapia 132
Timber Industries Ltd. 136
Tobacco Exploitation Service 155
Todaro, Michael P. 69, 192
Tourou Mt. 21
Trachyte 19, 20, 22
Trans-African Highway Coordinating Committee 186
Trans-African Line 139, 141
Transcam 142
Trans-Cameroon Line 142
Trans-Cameroon Railway 64, 174
Transhumance 64, 101
Tripanosomiasis 105, 153
Tropical Grassland 32
Trust Territories 5, 9, 195
Trusteeship Council 9

U
UCCAO 92, 167, 170
UDEAC 117, 151, 161, 186
Ullman 81
UNALOR 111, 136
UNESCO 186
UNICEF 186
Union Douaniere Equatoriale 152
Union Camerounaise 7, 12
Union des Populations du Cameroun 6, 7, 8, 9, 12
United Nations Organisation 5, 8, 9, 11, 63, 186
United Republic of Cameroon 12
UNVDA 111, 166
Urban Hierarchy 77
Urbanism 73
Urban Primacy 77
Urban Sprawl 79
USS Fairfax County 187
U Thant 9

V
VAPSAN 149
Victoria, Queen 3, 162
Vieter 2, 40
Vogt, Francois Xavier 2
Voss, Johannes 3

W
WADA 167
Wakwa 172
Walter 2
Wassande 89, 111, 173
WAZA 121, 122, 123
Weber, Max 191
Western Highlands 14
Widikum Tribe 55
William of Bimbia, King 162
Woermann 3
Wolfram 106
Wouri River 38, 47

Y
Yaeres 122, 175
Yambassi 68
Yaounde, town 155
Yaounde, University of xi, 155, 174
Yebekanga 50
Yekaba 50
Yekebolo 50
Yelinda 50
Yembama 50
Yengono 50
Yesum 50
Yetudi 50
Yeye Mt. 37
Yokadouma 45
Yu 54

Z
Zaman 50
ZAPI 64
Zebu 101, 103
Zintgraff 54, 55
Zone of Transition 76, 81